Edward Arber, John Lyly

Euphues

The anatomy of wit. Editio princeps, 1579. Euphues and his England. Editio

princeps, 1580. Collated with early subsequent editions.

Edward Arber, John Lyly

Euphues

The anatomy of wit. Editio princeps, 1579. Euphues and his England. Editio princeps, 1580. Collated with early subsequent editions.

ISBN/EAN: 9783337164683

Printed in Europe, USA, Canada, Australia, Japan

Cover: Foto ©berggeist007 / pixelio.de

More available books at **www.hansebooks.com**

Eng. Sem

English Reprints.

Anna R.

Vol. 3

JOHN LYLY, M.A.

EUPHUES. THE ANATOMY OF WIT.
Editio princeps. 1579.

EUPHUES AND HIS ENGLAND.
Editio princeps. 1580.

OLL^TED WITH EARLY SUBSEQUENT EDITIONS.

Eloquent and wittie *Iohn Lilly.*'
                    F. MERES. *Palladis Tamia.* 1598.

Edited by EDWARD ARBER, F.S.A.
:llow of King's College, London; Hon. Member of the Virginia Historical Society;
Examiner in English Language and Literature, Victoria University, Manchester;
Professor of English Language and Literature,
Sir Josiah Mason's College, Birmingham.

BIRMINGHAM
35 WHEELYS ROAD.
1 October, 1868.
No. 9.

CONTENTS.

CHRONICLE

of

fome of the principal events

3

in the

L I F E, W O R K S, and T I M E S

of

JOHN LYLY, M.A.,

Author, Wit, Poet, Dramatist.

---

* Probable or approximate dates.

1553. July 6. 𝕸ary succeeds to the crown.

*1553 or *1554. JOHN LYLY born. 'Touching whose [Mary's] life, I can say little because I was scarce borne.' *p.* 451.

1558. Nov. 17. 𝕰lizabeth begins to reign.

1560. Jan. 12. Sir Thomas Benger appointed Master of the Revels.— *Collier. Hist. Dram. Poetry,* i. 172.

1569. æt. 16. John Lylie or Lylly, a Kentish man born, became a student in Magd. coll. [which house was seldom or never without a Lilye (understand me not that it bears three lilyes for its arms) from the first foundation thereof to the latter end of queen Elizabeth. *Wood* i. 302] in the beginning of 1509, aged 16, or thereabouts, and was afterwards, as I conceive, either one of the demies or clerks of that house.—*A. à-Wood Ath. Oxon* i. 676. *Ed.* 1813.

John Lilly, born in the Weald of Kent in 1553 or 1554, became a student of Magdalen College, Oxford, in 1569, but

1571. Oct. 8. was not matriculated till 8 Oct. 1571, when he was entered
æt. 18. as *plebii filius.—Cooper. Ath. Cantab.* ii. 525. *Ed.* 1861.

[The Rev. Dr. Bloxam, formerly of Magdalen college, and who has made its history his especial study, informs me " Wood was probably right when he supposes Lylly to have entered college in 1569 for, as 1571 was the first year of matriculation and all the members of the college, old and young were matriculated together,—the matriculation would not fix the date of entrance. Lylly might have been a poor Scholar but there is no reason to suppose that he was either a Demy or Clerk."]

Wood reports, apparently in part copying from Blount, see p. 17, that Lyly was "always averse to the crabbed studies of logic and philosophy. For so it was that his genie being naturally bent to the pleasant paths of poetry, (as if Apollo had given to him a wreath of his own bays, without snatching or struggling,) did in a manner neglect academical studies, yet not so much but that he took the degrees in arts, that of master being compleated 1575. At which time, as he was esteemed at the university a noted wit so afterwards was in the court of Q. Elizabeth, where he was also reputed a rare poet, witty, comical, and facetious."—*Ath. Oxon* i. 276.

1573. Apr. 27. ' John Lilye of Magd. coll.' takes his B.A.—*Wood. Fasti*
æt. 20. *Oxon. Ed.* 1815.

1574. May 16. Lyly writes a Latin letter to Lord Burghley, which is now
æt. 21. in the British Museum, *Lansdowne MS.* 19, *Art.* 16. It is beautifully written on pencilled lines. On the back it is thus endorsed. "16. May 1574 John Lilie, a scholar of Oxford, an epistle For ye Queens letters to Magdalen College to admit him fellow." The letter is reprinted in Mr. Fairholt's *Dramatick works of John Lilly.* i.xii. *Ed.*

(left margin) Childhood. At Oxford.

Holding some appointment from Lord Burleigh.

1858. The application was unsuccessful in its immediate request: but Lord Burleigh seems afterwards to have extended his patronage to Lyly; *see p.* 441.

1575 June 1. 'John Lilye of Magd. coll.' takes his M.A.—*Wood.*
     æt. 22. *Fasti Oxon.*

1577. Mar. Sir T. Benger dies —*Collier. H. D. L. i.* 206.

*1578. æt 25. Lyly writes *Euphues. The Anatomy of Wit.* 'My first counterfaite,' *p.* 213, 'hatched in the hard winter with the Alcyon,' *p.* 215. 'Of the first I was deliuered, before my friendes thought me conceiued,' 'the one I sent to a nobleman to nurse.' *p.* 214.

Dec. 2. " Gabriel Cawood. Licensed vnto him the Anatomie of witt, compilled by Iohn Lyllie, under the hande of the bishopp of London. . . . . xij^{d}."
     Collier—Reg. of Stat. Co. Ed. 1848. The last clause intimates that the book was licensed by some one authorized by the Bp. of London.

Dec. 30. A Privy Seal was granted to Thomas Blagrave, Esqre appointing him *chief officer* of the Revels —*Collier, Hist. Dram. Lit.* i. 239.

1579 *Spring. *Euphues The Anatomy of Wit* is published.
     æt. 26. Lyly is incorporated M.A. of Cambridge. *Ath. Cantab.*

July 24. Edmund Tylney Esq. appointed *Master* of the Revels, which office he holds for 31 years, until his death in Oct. 1610.

July 24. " G. Cawood. Lycenced vnto him and —— the second part of euphues. . . . . vj^{d}."—*Collier, Reg. of Stat. Co.*

1580. Spring. Lyly probably writing *Euphues and his England* 'the
     æt. 27. other not daring to budde till the colde were past.' *p.* 215, *see also p.* 214. As to the dates in the story, *see p.* 210.

*1582. Thomas Watson, in this year published, *The ἑκατομπαθία or Passionate Centurie of Loue. Diuided into two parts: whereof, the first expresseth the Authors sufferance in Loue: the latter, his long farewell to Loue and all his tyrannie.* To this is prefixed the following letter.

     æt. 29. JOHN LYLY TO THE AUTHOUR HIS FRIEND.

My good friend, I haue read your new passions, and they haue renewed mine old pleasures, the which brought to me no lesse delight, then they haue done to your selfe commendations. And certes had not one of mine eies about serious affaires beene watchfull, both by being too too busie had beene wanton: such is the nature of persuading pleasure, that it melteth the marrow before it scorch the skin, and burneth before it warmeth: Not vnlike vnto the oyle of Ieat, which rotteth the bone and neuer ranckleth the flesh, or the Scarab flies, which enter into the roote and neuer touch the rinde.

And whereas you desire to haue my opinion, you may imagine my stomake is rather cloyed, then quesie, and therfore mine appetite of lesse force then mine affection, fearing rather a surfet of sweetenes, then desiring a satisfying. The repeating of Loue, wrought in me a remembrance of liking, but serching the very vaines of my hearte, I could fiude nothing but a broad scarre, where I left a deepe wounde; and loose stringes, where I tyed hard knots: and a table of steele, where I framed a plot of wax.

Whereby I noted that young swannes are grey, and the old white, young trees tender, and the old tough, young men amorous, and growing in yeeres, either wiser or warier. The Corall in the water is a soft weede, on the land a hard stone : a sworde frieth in the fire like a blacke ele, but layd in earth like white snowe: the heart in loue is altogether passionate, but free from desire, altogether carelesse.

But it is not my intent to inueigh against loue, which women account but a bare word, and that men reuerence as the best God ; onely this I would add without offence to Gentlewomen, that were not men more supersticious in their praises, then women are constant in their passions : Loue would either shortly be worne out of vse, or men out of loue, or women out of lightnes. I can condemne none but by coniecture, nor commend any but by lying, yet suspicion is as free as thought, and as farre as I see as necessary, as credulitie.

Touching your Mistres I must needes thinke well, seeing you haue written so well, but as false glasses shewe the fairest faces, so fine gloses amend the baddest fancies. Apelles painted the Phenix by hearesay not by sight, and Lysippus engraued Vulcan with a streight legge, whome nature framed with a poult foote, which prooueth men to be of greater affection then iudgement. But in that so aptly you haue varied vppon women, I will not vary from you, for confesse I must, and if I should not, yet mought I be compelled, that to Loue were the sweetest thing in the earth : If women were the faithfullest, and that women would be more constant if men were more wise. And seeing you haue vsed me so friendly, as to make me acquainted with your passions, I will shortly make you pryuie to mine, which I woulde be loth the printer shoulde see, for that my fancies being neuer so crooked he would put them in streight lines, vnfit for my humor, necessarie for his art, who setteth downe, blinde, in as many letters as seeing.   Farewell.

Lansdowne MS. 36. Art. 76. consists of the following
1582. July. letter, endorsed " July 1582 Iohn Lilly to my L."
æt 29. My duetie (right honorable) in most humble manner remembred.

It hath plesed my Lord vpon what colour I cannot tell, certaine I am vpon no cause, to be displesed with me, ye grief whereof is more then the losse can be. But seeing I am to liue in ye world, I must also be iudged by the world, for that an honest seruaunt must be such as Cæsar wolo haue his wif, not only free from synne, but from suspicion. And for that I wish nothing more then to commit all my waies to your wisdome, and the deuises of others to your iudgment, I here yeld both my self and my soule, the one to be tried by your honnor, the other by the iustic of god ; and if I doubt not but my dealings being sifted, the world shall find whit meale, where others thought to show cours branne. It may be manie things wilbe obiected, but yf any thing can be proued I doubt, I know your L. will soone smel deuises from simplicity, trueth from trecherie, factions from iust seruis. And god is my witnes, before whome I speak, and before whome for my speach I sha l aunswer, yat all my thoughtes concerning my L haue byne ever reuerent, and almost relligious. How I haue dealt god knoweth and my Lady can coniecture, so faithfullie, as I am as vnspotted for dishonestie, as a suckling from theft. This conscius of myne maketh me presume to stand to all trialls, ether of accomptes, or counsell, in the one I neuer vsed falshood, nor in the other dissembling. My most humble suit therfore vnto your L. is, yat my accusations be not smothered and I choaked in ye smoak, but that they maie be tried in ye fire, and I will stand to the heat. And my only comfort is, yat ye yat is wis shall iudg trueth, whos nakednes shall manifest her noblenes. But I will not troble your honorable eares, with so meinie idle words only this vpon my knees I ask. yat your L. will vousalf to talk with me, and in all things will I shew my self so honest, yat my disgrac shall bring to your L. as great meruell, as it hath done to me grief, and so thoroughly will I satisfie euerie obiection, yat your L. shall think me faithfull, though infortunat. That your honnor rest persuaded of myne honest mynd, and my Lady of my true seruis, that all things may be tried to ye vttermost, is my desire, and the only reward I craue for my iust, (I iust I dare tearme it) scruis. And thus in all humility submitting my caus to your wisdome and my consins to ye trieall. I commit your L. to the Almightie. Your L. most dutifullie to commaund. Ihon Lyly.

for yat I am for some few daies going into the countrie yf your L. be not at leasure to admit me to your speach, at my returne I will giue my most dutifull attendaunce. at which time, it may be my honesty may ioyne with your L. wisdome, and both preuent, that nether would allow. In the meane season what color soeuer be alledged, if I be not honest to my L. and so meane to he during his pleasure, I desire but your L. secret opinion, for as [I know] my Lord to be most honorable, so I beseech god in time he be not abused. Loth I am to be a prophett, and to be a wiche I loath. Most dutifull to command. Ihon Lyly. To ye right honorable, ye L. Burleigh,
L. High Tresorer of England.

Before 1589, Lily wrote nine dramatic pieces—seven in prose, one in rhyme, and one in blank verse.—*Collier. Hist. Dram. Lit.* iii. 176. Of these two were published soon after they were acted ; the others in or after 1591 : all in 4to. In each of these plays there were two or three songs which do not appear in 4tos : but were first published by the book-seller Edward Blount in 1632. in his reprint "*Sixe Court Comedies*. Often Presented and Acted before Queene Elizabeth, by the Children of her Maiesties Chappell, and the children of Paules. Written By the onely Rare Poet of that time, The Wittie, Comicall, Facetiously-Quicke and vnparalleled Iohn Lilly, Master of Arts." These songs show Lyly's powers in a different sphere, and are otherwise intrinsicly beautiful. An insertion of a few of them may therefore assist our estimate of his genius. The occasions of the several plays are given as in the titles of the 4tos.

*1584. Jan. 1. æt. 31. (1) *Campaspe.* Played before the Queenes Maiestie on new yeares day at night, by her Maiesties Children, and the Children of Paules. [Prose.] London, 1584. [Reprinted 1591.] This play has two prologues. The first when performed at the Court : the second when at the *Blackfriars* theatre. It was written in a hurry : "We feare . . . that our labours slylye glaunced on, will breede some content, but examined to the proofe, small commendation. The haste in performing shall be our excuse."—*Prol. at the Blackfriers.* In it is the famous Song by *Appelles.*

CVpid and my Campaspe playd,
At Cardes for kisses, *Cupid* payd ;
He stakes his Quiuer, Bow, and Arrows,
His Mothers doues, and teeme of sparows,
Looses them too, then, down he throwes
The corrall of his lippe, The rose
Growing on's cheek, (but none knows how)
With These, the cristall of his Brow,
And then the dimple of his chinne,
All These did my Campaspe winne.

At last hee set her, both his eyes
Shee won, and *Cupid* Blind did rise.
O Loue ! has shee done this to Thee ?
What shall (Alas !) become of Mee !

*1584. Shrove (2) *Sapho. and Phao* Played beefore the Queenes Maiestie
Tuesday. on Shrouetewsday by her Maiesties Children and the
Boyes of Paules [Prose] London 1584 [Reprinted 1591]
In this play is the following Song.

Sap. O Cruell Loue ! on thee I lay,
My curse which shall strike blinde the Day,
Neuer may sleepe with veluet hand
Charme thine eyes with Sacred wand,
Thy Iaylours shalbe Hopes and Feares,
Thy Prison-mates, Grones, Sighes, and Teares ;
Thy Play to weare out weary times,
Phantasticke Passions, Vowes, and Rimes,
Thy Bread bee frownes, thy Drinke bee Gall,
Such as when you *Phao* call,
The Bed thy lyest on by [be ?] Despaire,
Thy sleepe, fond dreames, thy dreames long Care,
Hope (like thy foole) at thy Beds head,
Mockes thee, till Madnesse strike thee Dead,
As *Phao*, thou dost mee with thy proud Eyes,
In thee poore *Sapho* liues, for thee shee dies.

1584. Lyly owes 23s. 10d for his battels : as appears by the
following entry in the Day book of the bursars of Magdalen
college Oxford. 1584. ' Mr. Iohn Lillie communarius debet
pro communis et batellis 23s 10d.'—*Ath. Cantab. idem.*

Prior to the year 1591, but how much earlier cannot be
ascertained, the performances by the children of Paul's, in
their singing school, were suppressed. . . . The conclusion,
from all the existing evidence, seems to be, that the inter-
diction was imposed about 1589 or 1590, and withdrawn
about 1600.—*Collier. H. D. L.* i. 279. 282.

' Since the Plaies in Paules were dissolued, there are
certaine Commedies come to my handes by chaunce, which
were presented before her Maiestie at seuerall times by the
children of Paules. This [*Endimion*] is the first, and if in
any place it shall dysplease, I will take more paines to
perfect the next.'—*Printer to the Reader.*

(3) *Endimion, The Man in the Moone* Playd before
the Queenes Maiestie at Greenwich on Candlemas day at
night, by the Chyldren of Paules : [Prose.] London 1591.
In this, we select the third Song, by *Fairies.*

Omnes. Pinch him, pinch him, blacke and blue,
Sawcie mortalls must not view
What the Queene of Stars is doing,
Nor pry into our Fairy woing.
1 *Fairy.* Pinch him blue.
2. *Fairy.* And pinch him blacke.
3. *Fairy.* Let him not lacke
Sharpe nailes to pinch him blue and red,
Till sleepe has rock'd his addle head.
4 *Fairy.* For the trespasse hee hath done,
Spots ore all his flesh shall runne.
Kisse *Endimion,* Kisse his eyes,
Then to our Midnight Heidegyes.

(4) *Gallathea.* As it was playde before the Queenes
Maiestie at Greene-wiche on Newyeeres day at Night (∴)
By the Chyldren of Paules. [Prose] London 1592.
In Act IV. *Cupid, Telusa, Eurota, Larissa,* enter
singing.

Te. O Yes, O Yes, if any Maid,
    Whom lering *Cupid* has betraid
    To frownes of spite, to eyes of scorne,
    And would in madnes now see torne
All 3. The Boy in Pieces, let her come
    Hither, and lay on him her doome.

Eur. O Yes, O Yes, has any lost,
    A Heart which many a sigh hath cost,
    Is any cozened of a teare,
    Which (as a Pearle) disdaine does weare?
All 3. Here stands the Thiefe, let her but come
    Hither, and lay on him her doome.

Lar. Is any one vndone by fire,
    And Turn'd to ashes through desire?
    Did euer any Lady weepe,
    Being cheated of her golden sleepe?
All 3. Stolne by sicke thoughts! the pirats found
    And in her teares, hee shalbe drownd.
    Reade his Inditement, let him heare,
    What hees to trust to: Boy giue eare.

(5) *Midas.* Plaied before the Qveenes Maiestie vpon tvvelfe day at night, By the Children of Paules. [Prose] London 1592. In Act IV. *Apollo* and *Pan* contend for sovereignty in music, before *Midas* and some Nymphs. *Apollo* sings

    A Song of *Daphne* to the Lute.
Apol. My *Daphne's* Haire is twisted Gold,
    Bright starres a-piece her Eyes doe hold,
    My *Daphne's* Brow inthrones the Graces,
    My *Daphne's* Beauty staines all Faces,
    On *Daphne's* Cheeke grow Rose and Cherry,
    On *Daphne's* Lip a sweeter Berry,
    Daphne's snowy Hand but touch'd does melt,
    And then no heauenlier Warmth is felt,
    My *Daphne's* voice tunes all the Spheres,
    My *Daphne's* Musick charmes all Eares.
    Fond am I thus to sing her prayse,
    These glories now are turn'd to Bayes.
Pan pipes and then sings
    Pan's Syrinx was a Girle indeed,
    Though now shee's turn'd into a Reed,
    From that deare Reed *Pan's* Pipe does come,
    A Pipe that strikes *Apollo* dumbe;
    Nor Flute, nor Lute, nor Gitterne can,
    So chant it, as the Pipe of *Pan*;
    Crosse-gartred Swaines, and Dairie girles,
    With faces smug, and round as Pearles,
    When *Pans* shrill Pipe begins to play,
    With dancing weare out Night and Day:
    The Bag-pipes Drone his Hum layes by,
    When *Pan* sounds vp his Minstrelsie,
    His Minstrelsie! O Base! This Quill
    Which at my mouth with winde I fill,
    Puts me in minde though Her I misse,
    That still my *Syrinx* lips I kisse.
The nymphs decide for *Apollo*, *Midas* for *Pan*. *Apollo* incensed gives *Midas* asses' ears.

(6) *Mother Bombie.* As it was sundrie times plaied by the Children of Powles. [Prose.] London. 1594 [Reprinted 1598.] In which *Memphio* and *Stellio* sing this song:—
Memp O *Cupid!* Monarch ouer Kings,
    Wherefore hast thou feete and wings?

About the court, writing plays, &c.

It is to shew how swift thou ait,
When thou wound'st a tender heart,
Thy wings being clip'd, and feete held still,
Thy Bowe so many could not kill.
Stel. It is all one in *Venus* wanton schoole,
  Who highest sits, the wiseman or the foole :
    Fooles in loues colledge
    Haue farre more knowledge
    To Reade a woman ouer.
    Than a neate prating louer.
      Nay, tis confest,
      That fooles please women best.
We have no accounts from the office of the Revels since
1589.—*Collier, H. D. L. i.*, 301.

A book was anonymously published in the Martin Mar-
prelate controversy, of which the short title is "*Pappe
with an hatchet*, Alias *A figge for my God sonne*. Or
cracke me this nut. Or *A Countrie cuffe*," &c.

Gabriel Harvey, in the second part of his *Pierce's Supererogation* [the book was published in 1593 ; but this part is
dated At Trinitie Hall. 5. November 1589] thus charges
Lyly with its authorship. [*p.* 69.]

Pap-hatchet (for the name of thy good nature is pittyfully
growen out of request) thy olde acquaintance in the Sauoy,
when young Euphues hatched the egges, that his elder
freendes laide,(surely Euphues was someway a pretty fellow:
would God, Lilly had alwaies bene Euphues, and neuer Pap-
hatchet ;) that old acquaintance, now somewhat straungely
saluted with a new remembrance, is neither lullabied with
thy sweete Papp, nor scarre-crowed with thy sower hatchet.

In *Harl. MS.* 1877 *fol.* 71, is a transcript of the following
undated petitions to Queen Elizabeth.

A PETICION OF JOHN LILLY TO THE QUEENES MAIESTIR.
*Tempora si numeres quæ nos numeramus
Non venit ante suam, nostra quærela diem.*
Most gratious and drad soueraigne, I dare not pester
your highnes with many words and want witt to wrapp vpp
much matter in fewe. This age Epitomies the pater-noster
thrust into the compasse of a penny, the world into the
modell of a Tennis ball, All science malted into sentence
I would I were so compendious as to expresse my hopes,
my fortunes, my ouerthirts [? thwarts] in two sillables, as
marchants do riches in fewe Ciphers, But I feare to comitt
the error I discomend, tediousnes, like one that vowed to
search out what tyme was, spent all his, and knewe yt not.
I was entertayned your Maiesties seruant by your owne
gratious fauour, strengthened with condicions that I should
ayme all my courses at the Reuells (I dare not saye with a
promise but a hopefull Item to the reuercion) for which
these 10 years I haue attended with an vnwearyed patience,
And nowe I knowe not what Crabb tooke me for an Oyster
that in the midest of your sunshine of your most gratious
aspect hath thrust a stone between the shells to eate me
aliue that onely liue on dead hopes. If your sacred Maiestie
thinke me vnworthy and that after x yeares tempest, I
must att the Court suffer shypwrack of my tyme, my wittes,
my hopes, vouchsafe in your neuer-erring iudgement, some
Plank, or rafter to wafte me into a Country wherein my
sadd and settled deuocion I may in euery corner of a
thatcht Cottage write prayers in stead of Plaies, prayer for
your longe and prosprous life, and a repetaunce that I
haue played the foole so longe, and yett like.
  *Quod petimus pœna est nec etiam miser esse recuso,
  Sed precor vt possem, mitius esse miser.*

About the court.

JOHN LILLIES SECOND PETICION TO THE QUEENE.

*1593.

æt 39.

Most gratious and dread soueraigne, tyme cannot worke my peticions, nor my peticions the tyme. After many years seruice yt pleased your Maiestie to except against Tents and Toyles, I wish that for Teants I might putt in Tenements, so should I be eased of some toyles. Some lande some good fines, or forfeitures that should fall by the iust fall of these most false traitors, that seeing nothing will come by the Revells, I may pray vppon the Rebells. Thirteene years your hignes seruant but yet nothing, Twenty freinds that though they saye they wilbe sure I find them sure to be slowe. A thowsand hopes but all nothing, a hundred promises but yet nothing. Thus casting vpp the Inventary of my freinds, hopes, promises, and tymes, the summa totalis amounteth to iust nothing. My last will is shorter than myne invencion : but three legacies, patience to my Creditors, Melancholie without measure to my friends, and beggerie without shame to my family.

Si placet hoc merui quod ô tua fulmina cessent
     Virgo parens Princeps.

In all humilitie I entreate that I may dedicate to your sacred Maiestie *Lillie de tristibus* wherein shalbe seene patience labours and misfortunes.

      Quorum si singula nostrum.

Frangere non poterant, poterant temen omnia mentem.

The last and the least, that if I bee borne to haue nothing, I may haue a protection to pay nothinge, which suite is like his that haueing folloued the Court tenn years for recompence of his seruis committed a Robberie and tooke it out in a pardon.

Occupation not known.

Mr. Collier, *Biblio. Cata. i.* 503, Ed. 1865, gives the following particulars as to Lyly's family, who, he states, seems to have lived in the parish of St. Bartholomew the Less, London.

1596. Sept. 10. ' John, the sonne of John Lillye. gent., was baptised.' —*Regr. St. Bartholomew.*

1597. Aug. 20. This son was buried at St. Botolph, Bishopsgate.

1597. Henry Lock, or Lok, publishes *Ecclesiastes, otherwise* æt. 43. *called The Preacher.* Among the prefatory poetry are—

AD SERENISSIMAM REGINAM ELIZABETHAM.

Regia Virgineæ soboles dicata parenti,
Virgo animo, patriæ mater, Regina quidquid optas?
Chara domi, metuenda foris, Regina quid optas?
Pulchra, pia es, princeps, fœlix, Regina quid optas?
Cœlum est? Certò at serò sit Regina quod optas.
               Ioh. Lily.

AD LOCKUM EIUSDEM.

Ingenio et genio locuples, dic Locke quid addam?
Addo, quod ingenium quondam preciosius auro.

(7) *The Woman in the Moone.* As it was presented before her Highnesse. By Iohn Lyllie, maister of Arts. [Blank verse.] London, 1597.

(8) *The Maydes Metamorphosis.* As it hath bene sundrie times Acted by the Children of Powles. [Chiefly in rhyme.] London, 1600.

(9) *Love's Metamorphosis.* A Wittie and Courtly Pastorall, written by Mr. *Iohn Lyllie.* First playd by the Children of Paules, and now by the Children of the Chappell. [Prose.] London, 1601.

From *Register of St. Bartholomew,* quoted by Mr. Collier.

1600. July 3. ' John, sonne of John Lillye, gent., was baptized.'
1603. May 21. ' Frances, daughter of John Lyllye, gent., was baptized.'
1606. Nov 30. æt. 52. ' John Lyllie, gent., was buried.'

EUPHUES.

INTRODUCTION.

He prefent work is a reprint of a great bibliographical rarity. *Euphues*, once so famous, has almoft difappeared from among Englifh books. Even now the number of its various editions cannot be determined with abfolute certainty. No one library has a fet even of its afcertained iffues; the copies of which are fcattered through the public and private collections of the kingdom.

The laft edition was printed in 1636—two hundred and thirty-two years ago. During this period, this work has been fubjected to increafing obloquy; and for the laft hundred years, in fo far as it has been referred to at all, it has, for the moft part, been treated as an abfurdity, a byword, a literary fcare-crow. Yet in the greateft age of Englifh literature, Lyly held a high place. *Euphues* was his firft work. It, at once, made him famous : fo famous indeed, that it is furprifing that fimple curiofity did not provoke an earlier reprint.

A brief account of the prefent iffue may be advifable. Mr. Henry Morley, then a Profeffor of the Englifh Language and Literature at King's College, now of Univerfity College, London, in preparing his article on *Euphuism*, which appeared in the *Quarterly Review* for April, 1861, commiffioned the well-known bookmerchants in the Strand, to obtain for him a copy of *Euphues*. In due time one was fupplied : the parts of which—unhappily wanting the firft five leaves of the firft part, and the laft leaf of the fecond—proved to be of the years 1579 and 1580 : dates earlier than thofe generally known, but not than thofe which have long fince been in the Malone collection, in the Bodleian Library at Oxford.

It was not till the prefent month, September, 1868, that an infpection of thefe Bodleian copies eftablifhed— what was in part known to Malone—that there were two editions of *each* part, in their firft years of pub- lication : and that of thefe Profeffor Morley's copies happened to be the earlier: in fact, the only known copies of the *Editiones principes* of the entire book.

Unfortunately, this unexpected information came too late to be made ufe of in the prefent edition. Theoretically, what is required is Profeffor Morley's texts, collated with the Bodleian copy, (No. 713): and then again, with the next earlieft editions printed, fay 1580 (?) and 1581 (?) refpectively. Six editions in the firft two years.

What the reader now has, is Profeffor Morley's texts collated with the earlieft editions previoufly acceffible to me, viz., 1581 and 1582 refpectively, *i.e.*, years fubfequent to the original iffue, in each cafe.

This collation, however, proves that Lyly's correc- tions were almoft entirely verbal and grammatical, and that the original text was never fubftantially altered by him : alfo that his only augmentation was his addrefs to the ‘Gentlemen Schollers of Oxford,’ which he affixed to the fecond edition of *Euphues. The Anatomy of Wit,*—the Bodleian copy of 1579.†

In the prefent work, the fources of each part of the text have been clearly indicated. The prefatory portion of the firft part,—having been taken from a later edition,—has been affixed to it : it being uncertain to what extent, if any, the two firft prefaces were fubfequently varied : the third being indubitably an addition to the original iffue. Variations or additions of words, and of important letters in words, from the firft editions, are inferted between [ ]. Words in thofe editions, fubfequently omitted are afterifked *.

The refult of the whole is that a perufal of the prefent work will probably convince the reader, that he has not only *Euphues*—the miffing leaves excepted—

† See p. 30.

as it was firſt iſſued from the preſs, but alſo as John Lyly afterwards reviſed it.

A book may be of great bibliographical rarity, yet of no hiſtoric intereſt or intrinſic value. *Euphues* is of all three.

When a book, heavily abuſed, is thus recovered from oblivion, and found to be not ſo bad as it has been repreſented, the tendency may be to over-eſtimate it. It may be uſeful, therefore, to gather together the principal opinions expreſſed upon Lyly and *Euphues*, in his own age and ſince : not ſo much to try the book by the critics, as the critics by the book ; giving the quotations pretty fully, to exhibit the occaſion, tone, and general purport of the criticiſm as well as the preciſe reference. The earlier opinions are but evidence of the influence of *Euphues*, and the reputation of its author : the later will incidentally give its poſition in the Elizabethan literature, as realized by ſome of our modern Engliſh ſcholars.

Euphues appeared in 1579 and 1580, and by 1586 each part had probably gone through five editions.

In 1586, WILLIAM WEBBE, Graduate, publiſhed *A Diſcourſe of Engliſh Poetrie*—of which only two copies are known, one of which is in the Bodleian †—in which he adduces *Euphues* as a proof of the capa- bilities of Engliſh language for Heroic verſe ; ſinc more than demonſtrated by Milton.

Nowe will I ſpeake ſomewhat, of that princelie part of Poetrie, wherein are diſplaied the noble actes and valiant exploits of puiſſaunt Captaines, expert ſouldiers, wiſe men, with the famous reportes of auncient times, ſuch as are the Heroycall workes of *Homer* in Greeke, and the heauenly verſe of *Virgils Æneidos* in Latine : which workes, comprehending as it were the ſumme and grounde of all Poetrie, are verilie and incomparably the beſt of all other. To theſe, though wee haue no Engliſh worke aunſwer- able, in reſpect of the glorious ornaments of gallant handling : yet our auncient Chroniclers and reporters of our Countrey affayres, come moſt neere them : and no doubt, if ſuch regarde of our Engliſh ſpeeche, and curious handling of our verſe, had

† W. C. Hazlitt. *Handbook.* Ed. 1867.

beene long fince thought vppon, and from time to time beene
pollifhed and bettered by men of learning, iudgement, and
authority, it would ere this, haue matched them in all refpects.
A manifeft example thereof, may bee the great good grace and
fweet vayne, which Eloquence hath attained in our fpeeche,
becaufe it hath had the helpe of fuch rare and finguler wits, as
from time to time myght ftill adde fome amendment to the
fame.   Among whom I thinke there is none that will gainfay,
but Mafter *Iohn Lilly* hath deferued mofte high commendations,
as he hath ftept one fteppe further therein than any either before
or fince he firft began the wyttie difcourfe of his *Euphues*.   Whofe
workes, furely in refpecte of his finguler eloquence and braue
compofition of apt words and fentences, let the learned examine
and make tryall thereof thorough all the parts of Rethoricke, in
fitte phrafes, in pithy fentences, in gallant tropes, in flowing
fpeeche, in plaine fence, and furely in my iudgement, I think
he wyll yeelde him that verdict, which *Quintillian* giueth of
bothe the beft Orators *Demofthenes* and *Tully*, that from the
one, nothing may be taken away, to the other, nothing may be
added.   But a more neerer example to prooue my former affertion
true, (I meane ye meetneffe of our fpeeche to receiue the beft
forme of Poetry). . . .   *E, i. b. Ed.* 1586.

In 1588, JOHN ELIOT—who publifhed *Ortho-epia-
Gallica, Eliots Fruits for the French* in 1593—prefixed
the following to Greene's *Perimedes. The Blacke Smith.*

Au R. Greene Gentilhomme.
Sonnet.

E Vphues qui a bien connu fils-aifné d'Eloquence,
 Son propre frere puifné te pourroit reconnoiftre
  Par tes beaux efcrits, GREENE, tu fais apparoiftre
Que de la docte Sœur tu as pris ta naiffance.
Marot et de-Mornay pour le langage Francois :
 Pour l'Efpaignol Gueuare, Boccace pour le Tofcan :
  Et le gentil Sleidan refait l'Allemand :
GREENE et Lylli tous deux raffineurs de l'Anglois.
GREENE a fon Marefchal monftrant fon arte diuine,
 Moulé d'vne belle Idée : fa plume efforée
  Vole vifte et haute en parolle empennée ;
Son ftile d'vn beau difcours portant la vraie mine.
Courage, donc ie-dis, mon amy GREENE, courage,
 Mefprife des chiens, corbeaux et chathuans la rage :
  Et (glorieux) endure leur malignante furie.
Zoyle arriere, arriere Momus chien enragé,
 Furieux maftin hurlant au croiffant argenté,
  A GREENE iamais nuyre fauroit ta calomnie.   I. Eliote.

On 9 Dec. 1588, was licenfed to John Wolfe, one of
Robert Greene's many works, entitled *Alcida Greenes*

Metamorphofis, but of this edition no copy is known. A fecond edition was publifhed in 1617, of which there is a copy in the Bodleian. Among the prefatory poems, is the following :—

In laudem Roberti Greni, *Cantab. in Artibus Magiflri.*

OLim præclaros fcripfit *Chaucerus* ad Anglos,
Aurea metra fuis patrio fermone refundens :
Poft hunc *Gowerus*, poft hunc fua carmina *Lydgate*,
Poftque alios alij fua metra dedere Britannis.
Multis poft annis, coniungens carmina profis,
Floruit *Afcamus, Chekus, Gafcoynus*, et alter
Tullius Anglorum nunc viuens *Lillius*, illum
Confequitur *Grenus*, præclarus vterque Poëta.

ROBERT GREENE and THOMAS LODGE took up the fubject of *Euphues*, where Lyly left off.

In 1589, (? firft edition, 1587) Greene publifhed *Mena-phon. Camillas alarum to flumbering Euphues, in his melancholie Cell at Silexedria* ; prefixed to which are fix ftanzas by HENRY UPCHEAR, gentleman. *In laudem Au-thoris. Diflichon amoris :* of which the third runs thus :—

Of all the flowers a *Lillie* once I lou'd,
Whofe labouring beautie brancht it felfe abroade ;
But now old age his glorie hath remoud,
And Greener obiectes are my eyes abroade.

In 1587, Greene alfo publifhed *Euphues his cenfure to Philautus*, &c.

In 1590, Lodge publifhed his *Rofalynde Euphues Golden Legacie found after his death in his Cell at Silexedra.* This work is the foundation of Shake-fpeare's *As you Like it.*

In paffing by Gabriel Harvey's counter-abufe of Lyly, in *Pierces Supererogation or A New Prayfe of The Old Affe* [*i. e.*, T. Nafh] 1593, to Lyly's *Pappe with an Hatchet* of 1589 : we jot the following fample of the amenities of literature then current.

Nafh, the Ape of Greene, Greene the Ape of Euphues, Euphues, the Ape of Enuie. *p.* 141.

In 1596, [Epiftle dated Nov. 5], Lodge, in a work entitled *Wits Miferie, and the VVorlds Madneffe : Dif-covering the Deuils Incarnate of this Age :* thus writes :—

Diuine wits, for many things as fufficient as all antiquity (I fpeake it not on flight furmife, but confiderate iudgemert). . . .

Lilly, the famous for facility in difcourfe : *Spencer*, beft read in ancient Poetry : *Daniel*, choife in word, and inuention : *Draiton* diligent and formall : *Th. Nafh*, true Englifh Aretine. *p.* 57.

In 1598, FRANCIS MERES, M.A. of both Univerfities in his *Palladis Tamia. Wits Treasury.* Being the Second Part of *Wits Commonwealth*, thus fpeaks of Lyly, twenty years after the compofition of *Euphues.*

The beft for Comedy amongft vs bee, *Edward* Earle of Oxforde, Doctor *Gager* of Oxforde, Maifter *Rowley* once a rare Scholler of learned Pembrooke Hall in Cambridge, Maifter *Edwarde* one of her Maiefties Chappell, eloquent and wittie *Iohn Lilly.* *Lodge, Gafcoyne, Greene, Shakefpeare, Thomas Nafh, Thomas Heywood, Anthony Mundye* our beft plotter, *Chapman, Porter, Wilfon, Hathway,* and *Henry Chettle. fol.* 284.

In 1599, was firft acted BEN JONSON's comedy *Every Man out of His Humour*, in which he is fuppofed to have ridiculed Euphuifm, in the character of Faftidious Brifk, who is thus defcribed in the preface to the piece.

A neat fpruce, affecting courtier, one that wears clothes wells and in fafhion : practifed by his glafs how to falute ; fpeak, good remnants, notwithftanding the bafe viol and tobacco : fwears terfly, and with variety ; cares not what lady's favour he belies, or great man's familiarity : a good property to perfume the boot of a coach.   He will borrow another man's horfe to praife, and backs him as his own.   Or, for a need, on foot can poft himfelf into credit with his merchant, only with the gingle of his fpur, and the jerk of his wand.

In Act V., Scene X., Jonfon makes Fallace (Deliro's wife, and idol) thus fpeak to this courtier :—

O Mafter Brifk (as 'tis in Euphues) ' Hard is the choice, when one is compell'd either by filence to die with grief or by eaking to living with fhame.*

On ' 30 November 1606 John Lyllie gent was buried' at St. Bartholomew the Lefs, London.

In 1623, was publifhed the firft folio edition of Shakefpeare's plays.   In BEN JONSON's well-known pre-fatory verfes, Lyly occupies rather a prominent pofition.

> My *Shakefpeare*, rife ; I will not lodge thee by
>   *Chaucer*, or *Spenfer*, or bid *Beaumont* lye
> A little further, to make thee a roome :
>   Thou art a Moniment, without a tombe,
> And art aliue ftill, while thy Booke doth liue,
>   And we haue wits to read, and praife to giue.

* p. 354.

That I not mixe thee fo, my braine excufes;
I mean with great, but difportion'd *Mufes:*
For, if I thought my iudgement were of yeeres,
I fhould commit thee furely with thy peeres,
And tell, how farre thou didſt our *Liſy* out-fhine,
Or fporting *Kid*, or *Marlowes* mighty line.

In 1627, MICHAEL DRAYTON publiſhed a folio
'olume of poems, the firſt of which is entitled *The
Battaile of Agincourt.* At the end of this volume,
mong *Elegies upon fundry occafions*, is a poem 'To
ny moſt dearely loued friend Henery Reynold of
Poets and Poefie.' This piece of rather fevere criticifm
has the following :—

Gafcoine and *Churchyard* after them againe
In the beginning of *Eliza's* raine,
Accoumpted were great Meterers many a day,
But not infpired with braue fier, had they
Liu'd but a little longer, they had feene,
Their workes before them to haue buried beene.

Graue morrall *Spencer* after thefe came on
Then whom I am perfwaded there was none
Since the blind *Bard* his *Iliads* vp did make,
Fitter a tafke like that to vndertake,
To fet downe boldly, brauely to inuent,
In all high knowledge, furely excellent.

The noble *Sidney*, with this laſt arofe,
That *Heroe* for numbers, and for Profe.
That throughly pac'd our language as to fhow,
The plenteous *Englifh* hand in hand might goe
With *Greeke* and *Latine*, and did firſt reduce
Our tongue from *Lillies* writing then in vfe ;
Talking of Stones, Stars, Plants, of fifhes, Flyes,
Playing with words, and idle Similies,
As th' *Englifh*, Apes and very Zanies be
Of euery thing, that they doe heare and fee,
So imitating his ridiculous tricks,
They fpake and writ, all like meere lunatiques. *p.* 205.

In 1632,—in ſtrong contraſt to Drayton—EDWARD
BLOUNT, the bookfeller, reprinted fix of Lyly's plays,
under the title of *Six Court Comedies*, to which he
prefixed the following ' Epiſtle Dedicatorie ' :—

To the right honovrable Richard Lvmley, Vifcount Lvmley
of Waterford. *My noble Lord :* It can be no difhonor, to liften
to this Poets Mufike, whofe Tunes alighted in the Eares of a
great and euer-famous Queene : his Inuention, was fo curioufly
ſtrung, that *Elizaes* Court held his notes in Admiration. Light

B

Ayres are now in fafhion; And thefe being not fad, fit the
feafon, though perchance not fute fo well with your more ferious
Contemplations.   The fpring is at hand, and therefore I
prefent you a Lilly, growing in a Groue of Lawrels.   For this
Poet, fat at the *Sunnes* Table : *Apollo* gaue him a wreath of
his owne *Bayes* ; without fnatching.   The *Lyre* he played on,
had no borrowed ftrings.   I am (my LORD) no executor, yet
I prefume to diftribute the Goods of the Dead : Their value beeing
no way anfwerable to thofe Debts of dutie and affection, in
which I ftand obliged to your Lordfhip.   The greateft treafure
our Poet left behind him, are thefe fix ingots of refined inuention :
richer than Gold.   Were they Diamonds they are now yours.
Accept them (Noble Lord) in part, and Mee
Your Lordfhips euer Obliged and Deuoted   Ed. Blount.

He adds the following addrefs ' To the Reader ' :—

Reader, I haue (for the loue I beare to Pofteritie) dig'd vp the
Graue of a Rare and Excellent Poet, whom *Queene Elizabeth* then
heard, Graced, and Rewarded.   Thefe Papers of his, lay like
dead Lawrels in a Churchyard ; But I haue gathered the fcattered
branches vp, and by a Charme (gotten from *Apollo*) made them
greene againe, and fet them vp as Epitaphes to his Memory.   A
finne it were to fuffer thefe Rare Monuments of wit, to lye
couered in Duft, and a fhame, fuch conceipted Comedies, fhould
be Acted by none but wormes.   *Obliuion* fhall not fo trample
on a fonne of the *Mufes;* And fuch a fonne, as they called their
Darling.   Our Nation are in his debt, for a new Englifh which
hee taught them.   *Euphues* and his England began firft, that
language : All our Ladies were then his Schollers ; And that
Beautie in Court, which could not Parley, *Euphueifme*, was as
little regarded ; as fhe which now there, fpeaks not French.

Thefe his playes Crown'd him with applaufe, and the Spectators
with pleafure.   Thou canft not repent the Reading of them ouer :
when Old *Iohn Lilly*, is merry with thee in thy Chamber, Thou
fhalt fay, Few (or None) of our Poets now are fuch witty Companions : And thanke mee, that brings him to thy Acquaintance.   *Thine.* ED. BLOVNT.

It may be doubted whether thefe effufions have not
hitherto done more harm than good to the memory
of Lyly.   For Blount is Lyly gone mad ; and fubfequent critics have fometimes quoted him, inftead of
reading *Euphues.*

Though another edition of *Euphues* appeared in
1636 ; with the exception of bare catalogues of his
plays, almoft a century of oblivion now refts upon Lyly
and his works. We pafs at a jump into the laft century.

WILLIAM OLDYS, in his MS. notes to a copy in the

Britifh Mufeum, of Gerald Langbaine's *Account of the English Dramatick Poets*, Oxford, 1691, has the following criticifm of Lyly :—

Lillye was a man of great reading, good memory, ready faculty of application and uncommon eloquence ; but he ran into a vaft excefs of allufion : in fentence and conformity of ftyle he feldom fpeaks directly to the purpofe ; but is continually carried away by one odd allufion or fimile or other (out of natural hiftory,— that yet is fabulous and not true in nature) and that ftill overborne by more, thick upon the back of one another, and thro' an eternal affectation of fententioufnefs keeps to fuch a formal meafure of his periods as foon grows tirefome, and fo by confining himfelf to fhape, his fenfe fo frequently into one artificial cadence, however ingenious or harmonious, abridges that variety which the ftyle fhould be admired for. *p.* 328.

In 1756, PETER WHALLEY—late Fellow of St. John's College, Oxford—brought out an edition of Ben Jonfon's *Works.* Upon Fallace's fpeech above quoted, he notes:

Euphues is the title of a romance, wrote by one Lilly, that was in the higheft vogue at this time. The court ladies had all the phrafes by heart. The language is extremely affected ; and like the fpecimen here quoted, confifts chiefly of antithefis in the thought and expreffion. *i.* 286.

In the *Literary Magazine* for May 1758, in a concluding paper on *The Hiftory of our own language*, is the following notice of our author :—

We muft not leave the times preceding the reftoration of *Charles* the Second, without mentioning one *Lilly*, who was author of fome pieces which he called plays, one of which is printed in Mr. *Dodfley*'s collection. His ftile is a kind of prodigy for neatnefs, clearnefs and precifion. But thofe were no recommendations to the times in which he liv'd. The learned of thofe days thought they indicated levity and flightnefs. He is, it is true, full of antithefes, and he carries the neatnefs of his language fometimes to a ridiculous affectation; yet a judicious head may receive great improvement by reading his works, which are now fcarcely ever mentioned. *p.* 197.

In 1777, JOHN BERKENHOUT, M.D., in his *Biographia Literaria*, is fimply ignorant and violent, when thus fpeaking of *Euphues.*

This romance, which Blount, the editor of the fix plays, fays introduced a new language, efpecially among the ladies, is in fact a moft contemptible piece of affectation and nonfenfe : neverthelefs it feems very certain, that it was in high eftimation by the women of fafhion of thofe times, who, we are told by Whalley the editor of Ben Jonfon's works, had all the phrafes by heart. As to Lilly's dramatic pieces, I have not feen any of them; but from the ftyle of this romance, I have no doubt but they are wretched performances. *i. p.* 377, *note* (*a*).

In 1816, WILLIAM GIFFORD, the firft Editor of the *Quarterly Review*, publifhed an edition of Ben Jonfon's works, in which he thus amplifies Whalley's note on Fallace's quotation :—

This was written by John Lilly, the author of feveral plays, which were once in high favour. Its title was "Euphues; the Anatomie of Wit, verie pleafant for all gentlemen to read, and moft neceffarie to remember, &c." 1580. Two years afterwards came out, "Euphues and his England, containing his Voyage and Adventures, &c." Thefe notable productions were full of pedantic and affected phrafeology, (as Whalley truly fays,) and of high-ftrained antithefes of thought and expreffion. Unfortunately they were all well received at court, where they did incalculable mifchief, by vitiating the tafte, corrupting the language, and introducing a fpurious and unnatural mode of converfation and action, which all the ridicule in this and the following drama [Ben Jonfon's *Cynthia's Revels*, acted in 1600] could not put out of countenance. *ii.* 205.

In 1817, NATHAN DRAKE, M.D., in his *Shakefpeare and his Times*, takes Berkenhout to tafk for his violence.

In 1581, John Lilly, a dramatic poet, publifhed a Romance in two parts, of which the firft is entitled, *Euphues*, The Anatomy of Wit; and the fecond, *Euphues and his England.* This production is a tiffue of antithefis and alliteration, and therefore juftly entitled to the appellation of *affected;* but we cannot with Berkenhout confider it as a moft *contemptible piece of nonfenfe.* The moral is uniformly good; the vices and follies of the day are attacked with much force and keennefs; there is in it much difplay of the manners of the times, and though, as a compofition, it is very meretricious, and fometimes abfurd in point of ornament, yet the conftruction of its fentences is frequently turned with peculiar neatnefs and fpirit, though with much monotony of cadence. *i.* 441.

In 1820, Sir WALTER SCOTT publifhed *The Monaftery.* Writing years afterwards—on 1 Jan. 1831— his Introduction to a new edition of *The Abbot:* he candidly announced that he confidered *The Monaftery* 'as fomething very like a failure,' referring to that romance as a whole.

In *The Monaftery*, Sir W. Scott has endeavoured to depict what he thought a Euphuift was, in the character of Sir Piercie Shafton: in which he has but

mifreprefented, if indeed he ever underftood, either
fpoken Euphuifm, as in Elizabeth's court, or written
Euphuifm fuch as Lyly might poffibly have written.
After the following note, in his own perfon

Notwithftanding all exaggeration, Lylly was really a man of wit
and imagination, though both were deformed by the moft unnatural
affectation that ever difgraced a printed page. *ii.* 44. *Ed.* 1820.

he introduces Sir Piercie Shafton, talking this balderdafh,
which he intends for Euphuifm.

Ah that I had with me my Anatomy of Wit—that all-to-be-
unparalleled volume—that quinteffence of human wit—that trea-
fury of quaint invention—that exquifitely-pleafant-to-read, and
inevitably-neceffary-to-be-remembered manual of all that is
worthy to be known—which indoctrines the rude in civility, the
dull in intellectuality, the heavy in jocofity, the blunt in gentility,
the vulgar in nobility, and all of them in that unutterable perfection of human utterance, that eloquence which no other eloquence
is fufficient to praife, that art which, while we call it by its own
name of Euphuifm, we beftow on it its richeft panegyric. *ii.* 49.

In 1831, in an Introduction to *The Monaftery;*
Sir W. Scott endeavours at length to palliate his failure,
as beft he can; which is chiefly by drawing attention
to the Euphuifm of France, a century later.

The extravagance of Euphuifm, or a fymbolical jargon of the
fame clafs, predominates in the romances of Calprenade and
Scuderi, which were read for the amufement of the fair sex of
France during the long reign of Louis XIV., and were fuppofed
to contain the only legitimate language of love and gallantry.
In this reign they encountered the fatire of Molière and Boileau.
A fimilar diforder, fpreading into private fociety, formed the
ground of the affected dialogue of the *Precieufes*, as they were
ftyled, who formed the coterie of the Hôtel de Rambouillet, and
afforded Molière matter for his admirable comedy, *Les Precieufes
Ridicules*. In England, the humour does not feem to have long
furvived the acceffion of James I.

The author had the vanity to think that a character, whofe
peculiarities fhould turn on extravagances which were once uni-
verfally fafhionable, might be read in a fictitious ftory with a
good chance of affording amufement to the exifting generation,
who, fond as they are of looking back on the actions and manners
of their anceftors, might be alfo fuppofed to be fenfible of their
abfurdities. He muft fairly acknowledge that he was difap-
pointed, and that the Euphuift, far from being accounted a well-
drawn and humorous character of the period, was condemned as
unnatural and abfurd. *i.* xxi. *Ed.* 1831.

The character of Sir Piercie Shafton, however, by

fo accepted a writer, defpite its failure, recalled public
attention to *Euphues*.

In 1831, Mr. JOHN PAYNE COLLIER, in his *Hiſtory
of Dramatic Poetry*, thus expreffes his then eftimate of
our author :—

John Lyly was an ingenious fcholar, with fome fancy; but if
poetry be the heightened expreffiou of natural fentiments and
impreffions, he has little title to the rank of a poet. His
thoughts and his language are ufually equally artificial, the
refults of labour and ftudy, and in fcarcely a fingle inftance
does he feem to have yielded to the impulfes of genuine feeling.
. . . . Lyly became fo fafhionable, that better pens, as in
the cafe of Robert Greene and Thomas Lodge, followed his
example, and became his imitators. The chief characteriftic of
his ftyle, befides its fmoothnefs, is the employment of a fpecies or
fabulous or unnatural natural philofophy, in which the exiftence
of certain animals, vegetables, and minerals with peculiar properties is prefumed, in order to afford fimiles and illuftrations. *iii.* 173.

In 1839, Mr. HENRY HALLAM firft publifhed the
fecond volume of his *Introduction to the Literature of
Europe*, in which he gives the following meagre account
of Englifh polite literature in the Elizabethan age.
Mr. Hallam feems to have accepted *Euphues* as the
firft attempt in England at elegant writing. His defcription of *Euphues* is in the old groove, and will not
ftand the teft of a perufal of the prefent work.

In the fcanty and obfcure productions of the Englifh prefs
under Edward and Mary, or in the early years of Elizabeth, we
fhould fearch, I conceive, in vain for any elegance or eloquence
in writing. Yet there is an increafing expertnefs and fluency,
and the language infenfibly rejecting obfolete forms, the manner
of our writers is lefs uncouth, and their fenfe more pointed and
perfpicuous than before. Wilfon's Art of Rhetorique is at leaft
a proof that fome knew the merits of a good ftyle, if they did
not yet bring their rules to bear on their own language. In
Wilfon's own manner there is nothing remarkable. The firft
book which can be worth naming at all, is Afcham's School-
mafter, publifhed in 1570, and probably written fome years
before. Afcham is plain and ftrong in his ftyle, but without
grace or warmth ; his fentences have no harmony of ftructure.
He ftands, however, as far as I have feen, above all other
writers in the firft half of the queen's reign. The beft of thefe,
like Reginald Scot, exprefs their meaning well, but with no
attempt at a rhythmical ftructure or figurative language ; they
are not bad writers, becaufe their folid fenfe is aptly conveyed

to the mind ; but they are not good, becaufe they have little feleftion of words, and give no pleafure by means of ftyle. Puttenham is perhaps the firft who wrote a well-meafured profe ; in his Art of Englifh Poefie, publifhed in 1586, he is elaborate, ftudious of elevated and chofen expreffion, and rather diffufe, in the manner of the Italians of the fixteenth century, who affefted that fulnefs of ftyle, and whom he probably meant to imitate. But in thefe later years of the queen, when almoft every one was eager to be diftinguifhed for fharp wit or ready learning, the want of good models of writing in our own language gave rife to fome perverfion of the public Tafte. Thoughts and words began to be valued, not as they were juft and natural, but as they were removed from common apprehenfion, and moft ex-clufively the original property of thofe who employed them. This in poetry fhowed itfelf in affefted conceits, and in profe led to the pedantry of recondite mythological allufion, and of a Latinifed phrafeology.

The moft remarkable fpecimen of this clafs is the Euphues of Lilly, a book of little value, but which deferves notice on account of the influence it is recorded to have had upon the court of Elizabeth; an influence alfo over the public tafte, which is manifefted in the literature of the age. It is divided into two parts, having feparate titles; the firft "Euphues, the Anatomy of Wit ;" the fecond, "Euphues and his England." This is a very dull ftory of a young Athenian, whom the author places at Naples in the firft part and brings to England in the fecond; it is full of dry commonplaces. The ftyle which obtained celebrity is antithetical, and fententious to affeftation ; the perpetual effort with no adequate fuccefs rendering the book equally difagreeable and ridiculous, though it might not be diffi-cult to find paffages rather more happy and ingenious than the reft. The following fpecimen is taken at random, and though fuffi-ciently charafteriftic, is perhaps rather unfavourable to Lilly, as a little more affefted and empty than ufual. [Paffages on pp. 377-8 from 'The fharpeft north-eaft wind' to 'wax green,' quoted; alfo on p. 447, 'The Lords and gentlemen' to 'revenge them.'] Lilly pays great compliments to the ladies for beauty and modefty, and overloads Elizabeth with panegyric. [Paffage at p. 457, 'Touching the beauty' to 'in the water.']

It generally happens that a ftyle devoid of fimplicity, when firft adopted, becomes the objeft of admiration for its imagined ingenuity and difficulty ; and that of Euphues was well adapted to a pedantic generation who valued nothing higher than far-fetched allufions and fententious precepts. All the ladies of the time, we are told, were Lilly's fcholars ; " fhe who fpoke not Euphuifm being as little regarded at court as if fhe could not fpeak French." "His invention," fays one of his editors, who feems well worthy of him, "was fo curioufly ftrung that

Elizabeth's court held his notes in admiration." Shakſpeare has ridiculed his ſtyle in Love's Labour Loſt, and Jonſon in Every Man out of his Humour ; but, as will be ſeen on comparing the extraɔts I have given above, with the language of Holofernes and Faſtidious Briſk, a little in the tone of caricature, which Sir Walter Scott has heightened in one of his novels, till it bears no great reſemblance to the real Euphues. I am not ſure that Shakſpeare has never caught the Euphuiſtic ſtyle, when he did not intend to make it ridiculous, eſpecially in ſome ſpeeches of Hamlet. *pp.* 408-411.

The tide of opinion now turns from the ebb to the flow.

In 1855, the Rev. CHARLES KINGSLEY publiſhed *Weſt-ward Ho !,* probably the beſt hiſtorical romance of the preſent generation. He thus opens his account ' how the noble brotherhood of the Roſe was founded ':—

If this chapter ſhall ſeem to any Quixotic and fantaſtical, let them recolleɔt that the generation who ſpoke and aɔted thus in matters of love and honour were, neverthelefs, praɔtiſed and valiant ſoldiers, and prudent and crafty politicians ; that he who wrote the Arcadia was at the ſame time, in ſpite of his youth, one of the ſubtleſt diplomatiſts of Europe ; that the poet of the Faery Queene was alſo the author of The State of Ireland ; and if they ſhall quote againſt me with a ſneer Lilly's Euphues itſelf, I ſhall only anſwer by aſking—Have they ever read it ? For if they have done ſo, I pity them if they have not found it, in ſpite of occaſional tedioufneſs and pedantry, as brave, righteous, and pious a book as man need look into ; and wiſh for no better proof of the nobleneſs and virtue of the Elizabethan age, than the fact that " Euphues " and the " Arcadia," were the two popular romances of the day. It may have ſuited the purpoſes of Sir Walter Scott, in his cleverly drawn Sir Piercie Shaſton, to ridi-cule the Euphuiſts, and that *affectatum comitatem* of the travelled Engliſh of which Lanquet complains : but over and above the anachroniſm of the whole charaɔter (for, to give but one instance, the Euphuiſt knight talks of Sidney's quarrel with Lord Oxford at leaſt ten years before it happened), we do deny that Lilly's book could, if read by any man of common ſenſe, produce ſuch a coxcomb, whoſe ſpiritual anceſtors would rather have been Gabriel Harvey and Lord Oxford,—if indeed the former has not ma-ligned the latter, and ill-tempered Tom Naſh maligned the ma-ligner in his turn.

But, indeed, there is a double anachroniſm in Sir Piercie ; for he does not even belong to the days of Sidney, but to thoſe worſe times which began in the latter years of Elizabeth, and after breaking her mighty heart, had full licence to bear their crop of fools' heads in the profligate plays of James. Of them, perhaps, hereafter. And in the meanwhile, let thoſe who have not read

"Euphues" believe that, if they could train a son after the pattern of his Ephœbus, to the great faving of their own money and his virtue, all fathers, even in thefe money-making days, would rife up and call them blessed. Let us rather open our eyes, and fee in these old Elizabeth gallants our own anceftors, fhowing forth with the luxuriant wildnefs of youth, all the virtues which ftill go to the making of a true Englifhman. Let us not only fee in their commercial and military daring, in their political aftute-nefs, in their deep reverence for law, and in their folemn fenfe of the great calling of the Englifh nation, the antetypes, or rather the examples of our own ; but let us confefs that their chivalry is only another garb of that beautiful tendernefs and mercy which is now, as it was then, the twin fifter of Englifh valour ; and even in their often extravagant fondnefs for Continental manners and literature, let us recognise that old Anglo-Norman teachablenefs and wide-heartednefs, which has enabled us to profit by the wifdom and the civilization of all ages, and of all lands, without prejudice to our own diftinctive national character. *pp.* 275-277.

In the autumn and winter of 1860-61, Mr. GEORGE PERKINS MARSH—at prefent the United States Minifter to Italy—delivered a feries of lectures at the Lowell Inftitute, in Bofton, U.S. ; which he publifhed in London, in 1862, under the title of *The Origin and Hiftory of the Englifh Language, and of the Early Literature it embodies.* He gives this account of Lyly :—

Stanihurft flourifhed in that brief period of philological and literary affectation which for a time threatened the language, the poetry, and even the profe of England with a degradation as complete as that of the fpeech and the literature of the laft age of imperial Rome. This quality of ftyle appears in its moft offenfive form in the naufeous rhymes of Skelton, in the moft elegant in Lillie, in its moft quaint and ludicrous in Stanihurft. Spenfer and Shakefpeare were the *Dei ex machina* who checked the ravages of this epidemic ; but it ftill fhowed virulent fymptoms in Sylvefter, and the ftyle of glorious Fuller and of gorgeous Browne is tinted with a glow which is all the more attractive becaufe it is recognifed as the flufh of convalefcence from what had been a dangerous malady. . . . . *p.* 539.

I have fpoken of the literary and philological affectation of Stanihurft's time, as having affumed its moft elegant form in the works of Lillie, the Euphuift. Though the quality of ftyle called Euphuifm has more or lefs prevailed in all later periods of Englifh literature, the name which defignates it had become almoft obfolete and forgotten, until Scott revived it in his character of Sir Piercie Shafton. The word is taken from Euphues,† the name of the hero of a tale by John Lillie, the

† The Greek εὐφυής means well-grown, symmetrical; also clever, witty, and this is the sense in which Lillie applies it to his hero.

first part of which is entitled Euphues, the anatomie of Wit ; the fecond, Euphues and his England. It confifts of the niftory and correfpondence of a young Athenian, who, after fpending fome time in Italy, vifits England, in the year 1579 ; and as this was the period when the author flourifhed, it was, of courfe, a ftory of the time of its appearance. The plot is a mere thread for an endlefs multitude of what were efteemed fine fayings to be ftrung upon, or, as Lillie himfelf expreffes it, ' fine phrafes, fmooth quips, merry taunts, jefting without meane and mirth without meafure.' The formal charaᶜteriftics of Euphuifm are alliteration and verbal antithefis. Its rhetorical and intelleᶜtual traits will be better underftood by an example, than by a critical analyfis. An extraᶜt from the dedication of the fecond edition to the author's 'Very good friends the gentlemen Scholers of Oxford' may ferve as a fpecimen. It is as follows. [*see pp.* 207-8.]

The fuccefs of Euphues was very great. The work was long a *vade-mecum* with the fafhionable world, and confidered a model of elegance in writing and the higheft of authorities in all matters of courtly and polifhed fpeech. It contains, with all its affeᶜtations, a great multitude of acute obfervations, and juft and even profound thoughts ; and it was thefe ftriking qualities, not lefs than the tinfel of its ftyle, which commended it to the praᶜtical good fenfe of contemporary England. *pp.* 544-6.

In April 1861, appeared the article in the *Quarterly Review* on *Euphuism.* In writing which, Profeffor Henry Morley feems to have been under the impreffion that *Euphues* was dead and buried for ever. Yet more than any other, he has contributed to its refurreᶜtion : not only by the loan of his texts, but by being my fofter-father in Englifh literature. In his *Englifh Writers,* Profeffor Morley is giving the beft hiftory of our national language and literature ; in which, and in his power to fruᶜtify others' minds, with his willingnefs to promote, in every way, others' labours in the fame field ; he is doing the worthy work of a worthy Englifhman.

In his article he thus writes of *Euphues :*—

The work paffed through ten editions in fifty-fix years, and then was not again reprinted. Of thefe editions, the firft four were iffued during twenty-three years of Elizabeth's reign, the next four appeared in the reign of James, and the laft two in the reign of Charles I. ; the lateft edition being that of the year 1636, eleven years after that king's acceffion. Its readers were the men who were difcuffing Hampden's ftand againft fhip-money. During all this time, and for fome years beyond it, worfhip of conceits was in this country a literary paganifm, that gave ftrength to the ftrong as well as weaknefs to the weak, lafting from Surrey's days until the time when Dryden was in mid career. It was of this *culte* that the Euphuift undoubtedly afpired to be the high prieft, but it was not of his eftablifhing. Still lefs, of course, are we entitled to accept the common doctrine that it had its origin in Donne's fashionable poetry, and in the pedantry of James I.

Such is a brief hiftory of the opinion upon Lyly and his works. Let the reader now difmifs it all from his judgment; and turn to *Euphues* itfelf. What is it?

It is a very clever book, upon Friendfhip, Love, Education, and Religion. A ftory and difcourfes of love of Lyly's peculiar workmanfhip, are followed by a treatife on Education, that Afcham might have written: which is fucceeded by a fummary expofition of the Chriftian faith, that reminds one of Latimer. Then follow letters of counfel, how with Chriftian philofophy, to bear bereavement, exile and the like. So the firft part comes to an end. The fecond is unlike to it. 'Twinnes they are not, but yet Brothers.'† At a time when Englifhmen were feeking adventure upon every fea, Lyly tells us the ftory of Caffander and Callimachus, of which it is his 'whole drift, either neuer to trauaile, or fo to trauaile, as though ye purffe be weakened, ye minde may be ftrengthened.'§ Then comes the converfation with Fidus, and his account of Iffida—the moft charming character in the whole book. Then follows Philautus' fuit to Camilla, who is apparently intended as a type of the ladies of Elizabeth's court, and the ftory virtually clofes with Lady Flavia's fupper party and its attendant difcourfes on love. Finally, by way of appendix, is inferted Euphues' *Glafs for Europe;* wherein in more earneft than jeft, Lyly holds up to honour his country, its court, and his Queen.

The book throughout, a book for ladies. '*Euphues* had rather lye fhut in a Ladyes cafket, then open in a Schollers ftudie.'‡ 'This I haue diligently obferued that there fhall be nothing found, that may offend the chaft minde with vnfeemely tearmes, or vncleanly talke.'||

Space forbids a further purfuit here of the fubject.

† p. 215. § p. 245. ‡ p. 220. || p. 221.

All editions down to 1636 are in 4to. * Editions not

EUPHUES. THE ANATOMY OF WIT.

1. **1579.** *Editio princeps.* Has no title page. Text, as in present work, without [ ]. Colophon, as on page 198. *Professor H. Morley.*

2. **1579.** ¶ EUPHVES THE ANATOMY OF WIT. Very pleasant for all Gentle- | *men to reade, and most neces-* | sary to remember.| *wherein are conteined the delights* | that Wit followeth in his youth, by the | pleasantnesse of loue, and the hap- | pinesse he reapeth in | age, by | the perfectnesse of | Wisedome |
¶ By Iohn Lylly Master | of Art. | Corrected and augmented |
¶ Imprinted at London for | *Gabriell Cawood, dwel-* | ling in Paules Church- | yard. Colophon, the same as in first edition, see p. 198. *Bodleian.*

3. ***1580.** *see p. 30.*

4. **1581.** Title reprinted at p. 201. Important variations of text, within [ ]. Colophon at p. 198. *B. Museum.*

5. **1585.** EVPHVES. THE ANATOMY OF WIT. &c. By Iohn Lyly, Maister of Art. Corrected and augmented. AT LONDON Printed for Gabriel Cavvood, dvvelling in Paules Church-yard. Colophon. AT LONDON printed by Thomas East for Gabriel Cawood, dwelling in Paules Churchyard. 1585. *H. Pyne, Esq.*

6. **[1597.]** EVPHVES THE ANATOMY OF VVIT &c. By Iohn Lylie, Maister of Art. *Corrected and augmented. AT LON-DON.* Printed by I. Roberts for Gabriell Cawood, dwelling in Paules Churchyard. No colophon. *B. Museum.*

7. **1607.** EVPHVES. THE ANATOMIE OF VVIT. &c. By *Iohn Lylie*, Maister of Art. *Corrected and augmented. AT LON-DON.* Printed for *William Leake*, dwelling in Paules Church-yard, at the Signe of the Holy Ghost 1607. No colophon. *B. Museum.*

8. **1613.** EVPHVES. THE ANATOMY OF WIT &c. By IOHN LILIE, Master of Art. *Corrected and augmented. AT LON-DON,* Printed for *William Leake*, dwelling in Paules Church-yard, at the Signe of the Holy-ghost 1613. No colophon. *B. Museum, Bodleian.*

9. **1617.** EVPHVES. THE ANATOMY OF WYT. &c. By IOHN LILIE, Master of Art. *Corrected and augmented.* Printed at London by *G. Eld*, for *W. B.* and are to be sold by *Arthur Iohnson.* 1617. No colophon. *B. Museum, Bodleian.*

10. **[1623.]** EVPHVES. THE ANATOMY OF WIT. &c. By IOHN LYLIE, Master of Art. Corrected and augmented. Printed at London by IOHN BEALE for IOHN PARKER. No colophon. *B. Museum, Bodleian.*

11. ***1626.** } *see p. 30.*
    ***1630.**

12. **1636.** EVPHVES THE ANATOMIE OF WIT. By IOHN LYLLIE, Master of Art. *Corrected and Augmented.* LONDON, Printed by Iohn Haviland 1636. No colophon. *B. Museum, Bodleian.*

1718. 8vo. The false friend and inconstant Mistress; An instructive Novel to which is added, Loves diversion, &c. London. 1718. *B. Museum.*

13. **1868. Oct. 1.** *English Reprints:* see title at p. 1.

.'. The list is but tentative.

GRAPHY.

seen. The black figures (1.) denote corresponding editions.

EUPHUES AND HIS ENGLAND.

1. 1580. *Editio princeps.* Title as at p. 211. [Wants last leaf.] Text, as in present work without [ ]. *Professor H. Morley.*

2. 1580. Title as No. 1. Colophon as at *p.* 478. *Bodleian.*

3. *1581. *See p.* 30.
4. 1582. EVPHVES AND HIS ENGLAND. &c. ¶ By Iohn Lyly, Maister of Arte. Commend it or amend it. ¶ *Imprinted at London for* Gabriel Cawood, dwelling in Paules Church-yard. 1582. [Imperfect copy, *see p.* 209.] Important variations of text, within [ ]. *H. Pyne, Esq.*

5. 1586. EVPHVES AND HIS ENGLAND. &c. By Iohn Lyly, Maister of Arte. Commend it, or amend it. Printed at London for Gabriel Cawood, dwelling in Paules Churchyard. 1586. No colophon. *H. Pyne, Esq.*

6. 1597. EVPHVES AND HIS ENGLAND. &c. By Iohn Lyly, Maister of Art. *Commend it or amend it.* At London, Printed by I. R. for Gabriell Cavvood, and are to be sold at his shop in Paules Churchyarde. 1597. No colophon. *B. Museum.*

7. 1606. EVPHVES AND HIS ENGLAND. &c. ¶ By *Iohn Lily,* Master of Art. *Commend it or amend it.* AT LONDON, Printed for *William Leake* dwelling in Pauls church-yard, at the signe of the Holy-ghost. 1606 No colophon. *B. Museum.*

8. 1613. EVPHVES AND HIS ENGLAND &c. ¶ *By Iohn Lily,* Master of Art. *Commend it or amend it. AT LONDON* Printed for *William Leake,* dwelling in Paules Church-yard, at the Signe of the Holy-Ghost. 1613. No colophon. *Bodleian.*

9. 1617. EVPHVES AND HIS ENGLAND. &c. By IOHN LILIE, Master of Art. *Commend it, or amend it.* Printed at London by *G. Eld,* for *W. B.* and are to be sold by *Arthur Iohnson.* 1617. No colophon. *Bodleian.*

10. 1623. EVPHVES AND HIS ENGLAND. &c. By IOHN LYLIE Master of Art. *Commend it, amend it.* Printed at London by IOHN BEALE, for IOHN PARKER. 1623. No colophon. *B. Museum, Bodleian.*

11. 1631. EVPHVES AND HIS ENGLAND. &c. By IOHN LILIE, Master of Arts. *Commend it, or amend it.* Printed at London by *I. H.* and are to be sold by *Iames Boler* 1631. No colophon. *B. Museum.*

12. 1636. EVPHVES AND HIS ENGLAND. &c. By IOHN LILIE, Master of Arts. *Commend it, or amend it.* Printed at LONDON by *Iohn Haviland.* 1636. No colophon. *B. Museum, Bodleian.*

13. 1868. Oct. 1. *English Reprints:* see title at page 1.

BIBLIOGRAPHY.

Note on the Earliest Editions of 'Euphues.' 1579 and 1580.

An inspection of the Bodleian copies, in September, 1868, convinced me that there were two issues of each part in the first years of their publication ; which was known to Malone, so far as the first part was concerned : the following rough memorandum in his handwriting being on a loose piece of paper, now inserted in *Euphues*, Number 713 of his collection :—

"Lilly's Euphues, or Anatomy of Wit, &c.
1579, two editions.
1580, both parts. 3rd Ed. of *Euphues*, and 1st of *Euphues and his England*.
1581—1588.

1595.	1623.
1605, both parts.	1626.
1606.	(1630-31.)
1617.	1636, both parts.

Ten editions, at least, besides that of the first part in '79, probably more."

The evidences that the Bodleian copies are *second* editions of their respective years, are briefly these :—

EUPHUES. THE ANATOMY OF WIT, 1579.

(1) The title-page of the Malone copy has on it 'Corrected and augmented.' *See p.* 28.

(2) It has also the Address to the 'Gentlemen Schollers of Oxford' *affixed* to it ; and this address is of a piece with the rest of the first part. Lyly having given offence, takes the earliest opportunity of trying to remove it.

(3) The type on the reverse of folio 90 is somewhat differently set up.

EUPHUES AND HIS ENGLAND. 1580.

(1) By the following variations :—

	Prof. Morley's Copy.	Malone's Copy, No. 713.	Difference.
	(wanting last leaf.)	(perfect.)	
Lines to a full page	35	36	1 line.

The making up of the type is consequently different.
Last *page* of text would be the *even p.* of 141st fol., is *odd p.* of 131st fol.: the last, or *even p.* being blank.

Exact folios of *text* would be	141 folios,	are 130½ folios,	10½ fols.

The difference of one line a *page*, = 282 lines. would reduce the Professor Morley's text by four folios.

.'. A minuter collation—impossible then, the first editions being in the hands of the printers—would probably but confirm this result.

Euphues. The Anatomy of Wit.

TEXT. *Editio princeps*, 1579.

Profeffor Morley's copy.

Completed (Title-page, prefaces, &c.) from
the Grenville copy, 1581.

COLLATION. Edition, 1581.

The Grenville copy, in the Britifh Mufeum.

EUPHUES. THE ANATOMY OF WIT.

THE PRINCIPAL CHARACTERS PRESENT IN THE ACTION.

EUPHUES, *a young gentleman of* ATHENS.

PHILAUTUS, *a young gentleman of* NAPLES.

EUBULUS, *an old gentleman of* NAPLES.

DON FERARDO, *one of the chief governors of* NAPLES.

LUCILLA, *daughter of* DON FERARDO.

LIVIA, *a lady of* NAPLES, *in the houfe of* DON FERARDO, *afterwards at the Emperor's court.*

SCENE.
NAPLES *and* ATHENS.

TIME.
Not defined.

EVPHVES.

Here dwelt in *Athens* a young gen-
tleman of great patrimony, and of fo
comelye a perfonage, that it was
doubted whether he were more bound
to Nature for the liniaments of his
perfon, or to Fortune for the increafe
of his poffeffions.   But Nature impatient of comparifons, and as it were
difdaining a companion or copartner in hir working,
added to this comelyneffe of his bodye fuch a fharpe
capacity of minde, that not onely fhe proued Fortune
counterfaite, but was halfe of that opinion that fhe hir
felfe was onely currant.   This young gallaunt of more
witte then wealth, and yet of more wealth then wifedome, feeing himfelfe inferiour to none in pleafant conceits, though himfelfe fuperiour to all his [in] honeft
conditions, infomuch that he thought himfelfe fo apt to
all thinges that he gaue himfelfe almoft to nothing but
practifing of thofe thinges commonly which are indicent
[incident] to thefe fharpe wittes, fine phrafes, fmooth
quippes, merry tauntes, [vfing] ieftinge without meane,
and abufing mirth without meafure.   As therefore the
fweeteft Rofe hath his prickell, the fineft veluet his
bracke, the faireft flower his branne, fo the fharpeft
wit hath his wanton will, and the holieft head his

c

wicked way. And true it is that fome men write and
moft men beleeue, that in al perfect fhapes, a blemmifh bringeth rather a lyking euery way to the eyes,
then a loathing any way to the minde. *Venus* had hir
Mole in hir cheeke which made hir more amiable :
Helen hir Scarre in hir chinne, which *Paris* called *Cos
Amoris*, the whetftone of loue, *Ariftippus* his Wart,
Lycurgus his Wen : So likewife in the difpofition of the
minde, either vertue is ouerfhadowed with fome vice, or
vice ouercaft with fome vertue. *Alexander* valyant in
warre, yet giuen to wine. *Tullie* eloquent in his glofes,
yet vaineglorious. *Salomon* wife, yet to[o] too wanton.
Dauid holy, but yet an homicide. None more wittie
then *Euphues*, yet at the firft none more wicked. The
frefheft colours fooneft fade, the teeneft Rafor fooneft
tourneth his edge, the fineft cloth is fooneft eaten with
[the] Moathes, and the Cambricke fooner ftayned then
the courfe Canuas : which appeared wellinthis *Euphues*,
whofe wit beeing like waxe, apt to receiue any impreffion, and bearing the head in his owne hande, either to
vfe the rayne or the fpurre, difdayning counfaile, leauing
his country, loathinge his olde acquaintance, thought
either by wit to obteyne fome conqueft, or by fhame
to abyde fome conflict, who preferring fancy before
friends, and [t]his prefent humor, before honourtocome,
laid reafon in water being to[o] falt for his taft, and followed vnbrideled affection, moft pleafant for his tooth.
When parents haue more care how to leaue their children wealthy then wife, and are more defirous to haue
them mainteine the name, then the nature of a gentleman : when they put gold into the hands of youth,
where they fhould put a rod vnder their gyrdle, when
in fteed of awe they make them paft grace, and leaue
them rich executors of goods, and poore executors of
godlynes, then is it no meruaile, yat the fon being left
rich by his fathers Will, become retchles by his owne
will. But it hath bene an olde fayde fawe, and not of
leffe truth then antiquitie, that wit is the better if it be
the deerer bought : as in the fequele of this hiftory

fhall moft manifeftly appeare. It happened this young Impe to ariue at *Naples* (a place of more pleafure then profit, and yet of more profit then pietie), the very wallʳ and windowes whereoff, fhewed it rather to be the Tabernacle of *Venus*, then the Temple of *Vefta*. Ther was all things neceffary and in redynes, that might either allure the mind to luft or entice ye heart to folly : a court more meete for an *Atheyft*, then for one of *Athens :* for *Ouid*, then for *Ariftotle :* for a grace-leffe louer, then for a godly liuer : more fitter for *Paris* then *Hettor*, and meeter for *Flora* then *Diana.* Heere my youth (whether for wearineffe he could not, or for wantonnes would not go any farther) determined to make his abode, whereby it is euidently feene that the fleeteft fifh fwalloweth the delicateft bait : that the higheft foaring Hauke traineth to ye lure : and that ye wittieft braine, is inuegled with the fodeine view of alluring vanities. Heere he wanted no com-panyons, which courted him continually with fundryꞓ kindes of deuifes, whereby they might either foake his purffe to reape commoditie, or footh his perfon, to winne credite : for he had gueftes and companions of all forts.

Ther frequented to his lodging, as well the Spider to fucke poyfon of his fine wit, as the Bee to gather Hunny : as well the Drone as the Doue : the Foxe as the Lambe : as wel *Damocles* to betray him, as *Damon* to be true to him. Yet he behaued himfelfe fo warily, that hee fingled his game wifelye. Hee coulde eafily difcerne *Appollos* Muficke, from *Pan* his Pype, and *Venus* beautie from *Iunos* brauerye, and the faith of *Lælius*, from the flattery of *Ariftippus*, hee welcommed all, but trufted none, hee was mery but yet fo wary, that neither the flatterer coulde take aduauntage to entrap him in his talke, nor ye wifeft any affurance of his friendfhip : who being demaunded of one what countryman he was, he anfwered, what countryman am I not? if I be in *Crete*, I can lye, if in *Greece* J can fhift, if in *Italy* I can court it : if thou afke whofe

sonne I am also, I aske thee whose sonne I am not. I
can carous with *Alexander*, abstaine with *Romulus*,
eate with the *Epicure*, fast with the *Stoyck*, sleepe with
Endimion, watch with *Chrisippus*, vsing these speaches
and other like. An olde Gentleman in *Naples* seeing
his pregnant wit, his eloquent tongue somwhat taunt-
ing yet with delight: his mirth without measure, yet
not without wit: his sayings vaineglorious, yet pithie:
began to bewaile his Nurture, and to muse at his
Nature, beeing incensed against ye one as most pernitious, and enflamed with the other as most precious:
for he well knew that so rare a wit would in time,
either breed an intollerable trouble, or bring an in-
comperable treasure to the common weale: at the one
he greatly pitied, at the other he reioysed.

Hauing therefore gotten opportunitie to communicate with him his minde, with watrye eyes, as one
lamenting his wantonnesse and smiling face, as one
louing his wittinesse, encountered him on this
manner.

Young gentleman, although my acquaintaunce bee
small to entreat you, and my authoritie lesse to commaund you, yet my good will in giuing you good
counsaile should induce you to beleeue mee, and my
hoarye haires (ambassadors of experience) enforce you
to follow me, for by how much the more I am a
straunger to you, by so much the more you are beholding to me, hauing therefore opportunitie to vtter
my minde, I meane to be importunate with you to
follow my meaning. As thy byrth doth shewe the
expresse and liuely Image of gentle bloud, so thy
bringing vp seemeth to mee to bee a great blotte to
the lynage of so noble a brute, so that I am enforced
to thinke that either thou diddest want one to giue
thee good instructions, or that thy parents made thee
a wanton with too much cockering: eyther they were
too foolish in vsing no discipline, or thou too froward
in reiecting their doctrine: either they willing to haue
thee idle, or thou wilful to be il employed. Did they

not remember that which no man ought to forgette, that the tender youth of a childe is like the tempering of new Waxe, apt to receiue any forme? Hee that will carye a Bull with *Milo*, muſt vſe to carye him a Calfe alſo, hee that coueteth to haue a ſtraight Tree, muſt not bow him beeing a twigge. The Potter faſhioneth his clay when it is ſoft, and the Sparrow is taught to come when he is young: As therefore the yron, beeing hot receiueth any forme with the ſtroake of the hammer, and keepeth it beeing colde for euer, ſo the tender witte of a childe, if with diligence it be inſtructed in youth, will with induſtrie vſe thoſe qualyties in his* age.

They might alſo haue taken example of the wiſe huſbandmen, who in their fatteſt and moſt fertil ground ſow Hempe before Wheat, a graine that dryeth vp the ſuperfluous moyſture, and maketh the ſoyle more apt for corne : Or of good Gardeiners who in their curious knots mixe Hiſoppe with Time, as ayders the one to the growth of the other, the one beeing drye, the other moyſt : Or of cunning Painters, who for the whiteſt worke caſt the blackeſt ground, to make ye picture more amiable. If therefore thy Father had bene as wiſe an huſbandman as he was a fortunate huſbande, or thy Mother as good a huſwife as ſhe was a happy wife, if they had bene both as good Gardeiners to keepe their knotte, as they were grafters to bring forth ſuch fruit, or as cunning Painters, as they wer happie parents, no doubt they had ſowed Hempe before Wheat, that is diſcipline before affection, they had ſet Hiſoppe with Time, that is manners with witte, the one to ayde the other, and to make thy dexteritie more, they had caſt a blacke grounde for their white worke, that is, they hadde mixed threates with faire lookes. But things paſt, are paſt calling againe : it is too late to ſhutte the ſtable doore when the ſteede is ſtolne. The *Troyans* repented too late when their towne was ſpoyled : Yet the remembraunce of thy former follyes, might breede in thee a remorce

of confcience, and bee a remedie againſt farther concupifcence. But now to thy prefent time. The
Lacedemonians were wont to fhewe their children
dronken men and other wicked men, that by feing their
filth, they might fhunne the lyke fault, and auoyd the
lyke [fuch] vices when they were at the lyke ſtate. The
Perſians to make their youth ahhorre gluttony would
paint an *Epicure* fleeping with meate in his mouth,
and moſt horribly ouerladen with wine, that by the
view of fuch monſtrous fights, they might efchew the
meanes of the lyke exceſſe. The *Parthians*, to caufe
their youth to loathe the alluring traines of womens
wiles and deceiptful entifements, hadde moſt curiouſly
carued in their houfes, a young man blynde, befides
whome was adioyned a woman fo exquifite, that in
fome mens iudgement *Pigmalions* Image was not halfe
fo excellent, hauing one hande in his pocket as noting
hir theft, and holding a knife in the other hande to
cut his throate. If the fight of fuch vgly fhapes caufed
a loathing of ye like fins, then my good *Euphues* confider their plight, and beware of thine owne perill.
Thou art heere in *Naples* a young foiourner, I an olde
fenior: thou a ſtraunger, I a Citizen: thou fecure
doubting no mifhappe, I forrowfull dreading thy miffortune. Heere mayſt thou fee that which I figh to
fee: dronken fottes wallowing in euery houfe [corner]?
in euery chamber, yea, in euery channel. Heere mayſt
thou beholde that which I cannot without blufhing
beholde, nor without blubbering vtter: thofe whofe
bellyes be their Gods, who offer their goodes as Sacrifice to their guttes: Who fleepe with meate in their
mouthes, with finne in their heartes, and with fhame
in their houfes. Heere, yea, heere *Euphues*, mayſt
thou fee, not the carued viſarde of a lewde woman,
but the incarnate vyſage of a lafciuious wantonne:
not the fhaddowe of loue, but the fubſtaunce of luſt.
My hearte melteth in droppes of bloud to fee a[n] harlotte
with the one hande robbe fo many cofers, and with the
other to rippe fo many corfes. Thou arte heere amid-

deſt the pykes betweene *Scylla* and *Carybdis*, ready if thou ſhunne *Syrtes*, to ſinke into *Semphlagades*. Let the *Lacedemonian*, the *Perſian*, the *Parthian*, yea the *Neapolitan*, cauſe thee rather to deteſt ſuch villany, at the ſight and viewe of their vanitie. Is it not farre better to abhorre ſinnes by the remembraunce of others faults, then by repentaunce of thine owne follyes? Is not hee accompted moſt wiſe, whome other mennes harmes doe make moſt warie? But thou wilt happely ſaye, that although there bee many things in *Naples* to be iuſtly condempned, yet there are ſome things of neceſſitie to bee commended: and as thy will doth leane vnto the one, ſo thy witte woulde alſo embrace the other. Alas *Euphues* by how much the more I loue [ſee] the high clymbing of thy capacitie, by ſo much the more I feare thy fall. The fine Chriſtall is ſooner craſed then the hard Marble: the greeneſt Beech, burneth faſter then the dryeſt Oke: the faireſt ſilke is ſooneſt ſoyled: and the ſweeteſt Wine, tourneth to the ſharpeſt Vineger. The Peſtilence doth moſt riſeſt infeɛt the cleareſt compleɛtion, and the Caterpiller cleaueth vnto the ripeſt fruite: the moſt delycate witte is allured with ſmall enticement vnto vice, and moſt ſubieɛt to yeelde vnto vanitie. If therefore thou doe but hearken to the *Syrenes*, thou wilt be enamoured: if thou haunt their houſes and places, thou ſhalt be enchaunted. One droppe of poyſon infeɛteth the whole tunne of Wine: one leafe of *Colloquintida*, marreth and ſpoyleth the whole pot of porredge: one yron Mole, defaceth the whole peece of Lawne. Deſcend into thine owne conſcience, and conſider with thy ſelfe, the great difference betweene ſtaring and ſtarke blynde, witte and wiſedome, loue and luſt: be merry, but with modeſtie: be ſober, but not too ſullen: be valyaunt, but not too venterous. Let thy attyre bee comely, but not coſtly: thy dyet wholeſome, but not exceſſiue: vſe paſtime as the word importeth to paſſe the time in honeſt recreation. Miſtruſt no man without cauſe, nether be thou credulus without proofe : be not lyght to follow euery mans opinion, nor obſtinate to ſtande in thine owne conceipt. Serue GOD, loue God, feare God, and God will ſo bleſſe thee, as eyther [thy] heart canne wiſh, or thy friends deſire: And ſo I ende my counſayle, beſeeching thee to beginne to follow it. This olde gentleman hauing finiſhed his diſcourſe, *Euphues* began to ſhape him an aunſwere in this ſort.

Father and friend (your age ſheweth the one, your honeſtie the other) I am neither ſo ſuſpitious to miſtruſt your good wil, nor ſo ſottiſh to miſlike your good counſayle, as I am therfore to thanke you for the firſt, ſo it ſtandes me vpon to thinke better on the latter: I meane not to cauil with you, as one louing ſophiſtrie: neither to controwle you, as one hauing ſuperioritie, the one woulde bring my talke into the ſuſpition of fraude, the other conuince me of folly.

Whereas you argue I know not vpon what pro-babilyties, but ſure I am vpon no proofe, that my bringing vp ſhould be a blemmiſh to my birth. I aunſwere and ſweare to that, you were not therin a little ouerſhot, either you gaue too much credite to the report of others, or too much libertie to your owne iudgement: You conuince my parents of peeuiſhnes in making me a wanton, and me of lewdneſſe in reiecting correction. But ſo many men ſo many mindes, that may ſeeme in your eye odious, which in an others eye may be gracious. *Ariſtippus* a Philoſo-pher yet who more courtly? *Diogenes* a Philoſopher, yet who more carterly? Who more popular then *Plato*, retayning alwayes good company? Who more enuious then *Tymon*, denouncing all humaine ſocietie? Who ſo ſeuere as the *Stoickes*, which lyke ſtocks were moued with no melodie? Who ſo ſecure as the *Epicures*, which wallowed in all kind of lycentiouſneſſe? Though all men bee made of one mettall, yet they bee not caſt all in one moulde, there is framed of the ſelfe ſame clay as wel the tile to keepe out water, as the potte to conteine licour, the Sunne doth harden the durte, and melte the waxe, fire maketh the golde to fhine, and the ftrawe to fmother, Perfumes doth refrefh the Doue, and kill the Betill, and the nature of the man dif-pofeth that confent of the manners. Now whereas you feeme to loue my nature, and loath my nurture, you bewraye your owne weakeneffe, in thinking that nature may any wayes be altered by education, and as you haue enfamples to confirme your pretence, fo I haue moft euident and infallible arguments to ferue for my purpofe. It is natural for the vine to fpread, the more you feeke by Art to alter it, the more in the ende you fhal augment it. It is proper for the Palme tree to mount, the heauier you loade it, the higher it fprowteth. Though yron be made foft with fire, it returneth to his hardnes: though the Fawlcon be re-claimed to the fift, fhe retyreth to hir haggardneffe : the whelpe of a Maftife wyll neuer be taught to retriue the Partridge: education can haue no fhewe, where the excellencye of Nature doth beare fway. The filly Moufe will by no manner of meanes be tamed : the fubtill Foxe may well be beaten, but neuer broken from ftealing his pray : if you pownd Spices they fmell the fweeter : feafon the woode neuer fo well the wine will taft of the cafke : plante and tranflate the crabbe tree, where, and whenfoeuer it pleafe you, and it wyll neuer beare fweete Apple, vnleffe you graft it by Arte, which nothing toucheth nature.

Infinite and innumerable were the examples I coulde alledge and declare to confirme the force of Nature, and confute thefe your vaine and falfe forgeryes, were not the repeticion of them needeleffe, hauing fhewed fufficient, or booteleffe feeinge thofe alleaged will not perfwade you. And can you bee fo vnnatural, whome dame Nature hath nouryfhed and brought vp fo many years, to repine as it were againft Nature.

The fimilytude you rehearfe of the waxe, argueth your waxinge and meltinge braine, and your example of the hotte and harde yron, fheweth in you but

colde and weake difpofition. Doe you not knowe
that which all men doe affirme and know, that blacke
will take no other coulour? That the ftone *Abefton*
beeing once made hot will neuer be made colde?
That fyre cannot be forced downewarde? That
Nature will haue courfe after kinde? That euery
thing will difpofe it felfe according to Nature? Can
the *Aethiope* chaunge or alter his fkinne? or the
Leopard his hiew? Is it poffible to gather grapes of
thornes, or figges of thiftles, or to caufe any thing
to ftriue againft Nature? But why goe I about to
praife Nature, the which as yet was neuer any Impe
fo wicked and barbarous, any Turke fo vyle and
brutifhe, any beaft fo dull and fenceleffe, that coulde,
or woulde, or durft difprayfe or contemne? Doth
not *Cicero* conclude and allowe, that if we followe and
obey Nature, we fhall neuer erre? Doth not *Ariftotle*
alledge and confirme, that Nature frameth or maketh
nothing in any point rude, vaine, or vnperfect?

Nature was had in fuch eftimation and admiration
among the Heathen people, that fhe was reputed for
the onely Goddeffe in heauen: If Nature then haue
largely and bountifully endewed me with hir gyftes,
why deeme you me fo vntoward and graceleffe? If
fhe haue dealt hardely with me, why extoll you fo
much my byrth? If nature beare no fway, why vfe
you this adulation? If nature worke the effect, what
booteth any education? If nature be of ftrength or
force, what auaileth difcipline or nurture? If of none,
what helpeth nature? But let thefe fayings paffe as
knowen euidently, and graunted to be true, which
none can or may deny vnleffe he be falfe, or that he be
an enimye to humanitie.

As touching my refidence and abiding heere in
Naples, my youthlye affections, my fportes and
pleafures, my paftymes, my common dalyaunce, my
delyghtes, my reforte and companye, which dayly vfe
to vyfile me, although to you they breede more
forrow and care, then folace and comfort, bicaufe of

your crabbed age : yet to me they bring more comfort
and ioye, then care and griefe : more blyffe then bale,
more happineffe then heauineffe : bicaufe of my youthfull gentleneffe. Either you wold haue all men olde
as you are, or els you haue quite forgotten that you
your felfe wer young, or euer knewe young dayes :
eyther in your youth you were a very vicious and ·
vngodly man, or now being aged very fuperfticious
and deuout aboue meafure.

Put you no difference betweene the young flourifhing Bay tree, and the olde withered Beach ? No
kinde of diftinction betweene the waxinge and the
wayninge of the Moone ? And betweene the rifinge
and the fetting of the Sunne ? Doe you meafure the
hot affaults of youth, by the colde fkirmifhes of age ?
whofe yeares are fubiect to more infirmities then our
youth. We merry, you melancholy : we zealous in
affection, you iealous in all your doings : you teftie
without caufe, we haftie for no quarrell : you carefull,
wee careleffe, we bolde, you fearefull : we in all
poynts contrary vnto you, and yee in all poynts vnlyke
vnto vs. Seeing therefore we be repugnaunt eache to
the other in Nature, would you haue vs alyke in qualyties ? Would you haue one potion miniftred to the
burning Feuer, and to the colde Palfey ? One playfter
to an olde iffue and a frefh Wound ? one falue for all
fores ? one fauce for all meates ? No no *Eubulus*,
but I wil yeeld to more, then either I am bounde to
graunt, either thou able to proue. Suppofe that which
I neuer will beleeue, that *Naples* is a cankered ftore-
houfe of all ftrife, a common ftewes for all ftrumpettes,
the finke of fhame, and the verye Nurfe of all finne :
fhall it therefore follow of neceffitie, that all that are
wo[o]ed of loue fhould be wedded to luft : will you conclude, as it were *ex confequenti*, that whofoeuer arriueth
heere fhall be enticed to follye, and beeing enticed of
force fhal be entangled ? No no, it is the difpofition of
the thought, that altereth the nature of the thing. The
Sunne fhineth vpon the dounghil, and is not corrupted :

the Diamond lyeth in the fire, and is not confumed :
the Chriftall toucheth the Toade and is not poyfoned :
the birde *Trochilus* lyueth by the mouth of the Crocodile
and is not fpoyled : a perfeȼt wit is neuer bewitched
with leaudeneffe, neither entifed with lafciuioufneffe.

Is it not common that the Holme Tree fpringeth
amidft the Beech ? That the Iuie fpreadeth vpon the
hard ftones ? That the foft fetherbed breaketh the
hard blade ? If Experience haue not taught you this,
you haue liued long and learned little : or if your
moift brain haue forgot it, you haue learned much, and
profited nothing. But it may be, that you meafure my
affeȼtions by your owne fancies, and knowing your
felfe either too fimple to raife the fiege by pollicie, oɪ
too weake to refift the affault by proweffe, you deemc
me of as lyttle wit as your felf, or of leffe force : either
of fmall capacitie, or of no courage. In my iudgement
Eubulus, you fhal affoone catch a Hare with a taber,
as you fhal perfwade youth with your aged and ouer-
worn eloquence, to fuch feueritie of life, which as yet
ther was neuer *Stoicke* in preceptes fo ftriȼt, neither
any in lyfe fo precife, but woulde rather allowe it in
wordes, then follow it in workes, rather talke of it then
try it. Neither were you fuch a Saint in your youth,
that abandoning all pleafures, all paftimes and delyghts, you would choofe rather to facrifice the firft
fruits of your lyfe to vayne holineffe then to youthly
affeȼtions. But as to the ftomack quatted with dain-
ties, al delicates feeme queafie, and as he that furfetteth
with wine, vfeth afterward to allay with water : fo thefe
old huddles hauing ouercharged their gorges with fancie,
accompt al honeft recreation meere folly, and hauing
taken a furfet of delight, feeme now to fauour it with
defpight. Seing therefore it is labour loft for me to
perfwade you, and winde vainly wafted for you to
exhort me, heere I found you, and heere I leaue you,
hauing neither bought nor fold with you, but chaunged
ware for ware : if you haue taken litle pleafure in my
reply, fure I am that by your counfel I haue reaped

leſſe profite. They that vſe to ſteale Hoiny burne Hemlocke to ſmoake the Bees from their hiues, and it may bee, that to get ſome aduauntage of me, you haue vſed theſe ſmoakie arguments, thinking thereby to ſmother me with the conceipt of ſtrong imagination. But as the *Camelion* though he haue moſt guttes draweth leaſt breath, or as the Elder tree though hee bee fulleſt of pith, is fartheſt from ſtrength : ſo though your reſons ſeeme inwardly to your ſelfe ſomewhat ſubſtantiall, and your perſwaſions pithie in your owne conceipte, yet beeing well wayed without, they be ſhadows without ſubſtaunce, and weake without force. The Birde *Taurus* hath a great voyce, but a ſmal body : the thunder a great clap, yet but a lyttle ſtone : the emptie veſſell giueth a greater ſound then the full barrell. I meane not to apply it, but looke into your ſelf and you ſhall certeinely finde it, and thus I leaue you ſeeking it, but were it not that my company ſtay my comming I would ſurely helpe you to looke it, but I am called hence by my acquaintaunce.

Euphues hauing thus ended his talke, departed leauing this olde gentleman in a great quandarie : who perceiuing that he was more enclined to wantonnes then to wiſdome, with a deepe ſigh the teares trickling downe his cheekes, ſayd : Seeing thou wilt not buye counſel at the firſt hande good cheape, thou ſhalt buye repentaunce at the ſecond hande, at ſuch an vnreaſonable rate, that thou wilt curſſe thy hard penyworth, and ban thy harde heart. Ah *Euphues* little doſt thou know that if thy wealth waſt, thy wit will giue but ſmall warmth, and if thy wit encline to wilfulnes, that thy wealth will doe thee no great good. If the one had bene employed to thrift, the other to learning, it had bene harde to coniecture, whether thou ſhouldeſt haue ben more fortunate by riches, or happie by wiſdome, whether more eſteemed in ye common weale for welth to maintaine warre, or for counſell to conclude peace. But alas why doe I pitie that in thee which thou ſeemeſt to praiſe in thy ſelf. And ſo ſaying, he immediatly

went to his owne houfe, heauily bewayling the young
mans vnhappineffe.

Heere ye may behold Gentlemen, how leaudly wit
ftandeth in his owne light, how he deemeth no penny
good filuer but his owne, prefering the bloffome before
the fruite, the budde before the floure, the greene blade
before the ripe eare of Corne, his owne wit before all
mens wifedomes. Neither is that geafon, feeing for
the moft part it is proper to all thofe of fharpe capacitie
to efteeme of themfelues as moft proper : if one be hard
in conceiuing, they pronounce him a dowlte, if giuen
to ftudie, they proclaime him a dunce : if merry, a
iefter : if fad, a Saint : if full of words, a fot : if without fpeach, a Cipher. If one argue with them boldly,
then he is impudent : if coldly, an innocent : If
there be reafoning of diuinitie, they cry, *Quæ fupra
nos, nihil ad nos* : If of humanitie, *Sententias loquitur
carnifex.*

Heereoff commeth fuch great familyaritie between
the ripeft wittes, when they fhall fee the difpofition the
one of the other, the *Sympathia* of affeċtions, and as it
were but a paire of fheeres to goe betweene their
natures, one flattereth an other in his owne folly, and
layeth cufhions vnder the elbow of his fellow when
he feeth him take a nappe with fancie, and as their
wit wrefteth them to vice, fo it forgeth them fome feat
excufe to cloake their vanitie.

Too much ftudie doth intoxicate their braines, for
(fay they) although yron the more it is vfed the brighter
it is, yet filuer with much wearing doth waft to nothing :
though the Cammocke the more it is bowed the better
it ferueth, yet the bow the more it is bent and occupied,
the weaker it waxeth : though the Camomill the more
it is troden and preffed downe, the more it fpreadeth,
yet the Violet the oftner it is handeled and touched,
the fooner it withereth and decayeth. Befides this, a
fine witte, a fharpe fence, a quicke vnderftanding, is
able to attaine to more in a moment or a very little
fpace, then a dull and blockifh head in a month. The

fithe cutteth farre better and fmoother then the fawe,
the waxe yeeldeth better and fooner to the feale, then
the fteele to the ftampe : the fmoothe and playne Beech
is eafier to be carued then the knottie Boxe.

For neither is there any thing but that hath his contraries. Such is the Nature of thefe nouifes, that thinke
to haue learning without labour, and treafure without ·
trauaile : either not vnderftanding or els not remembring, that the fineft edge is made with the blunt
whetftone : and the faireft Iewel fafhioned with the
hard hammer. I goe not about (Gentlemen) to inueigh
againft wit, for then I wer witleffe, but frankly to confeffe mine owne little wit. I haue euer thought fo
fuperfticioufly of wit, that I feare I haue committed
Idolatrie againft wifedome, and if Nature had dealt fo
beneficially with mee to haue giuen mee anye wit, I
fhoulde haue bene readier in the defence of it to haue
made an Apologie, then any way to tourne to Apoftacie.
But this I note, that for the moft parte they ftand fo on
their pantuffles, that they be fecure of periis, obftinate
in their own opinions, impatient of labour, apt to conceiue wrong, credulous to beleeue the worft, redy to
fhake off their olde acquaintaunce without caufe, and
to condemne them without coulour : All which humors
are by fo much the more eafier to be purged, by how
much the leffe they haue feftred the finewes. But
returne [turne] we again to *Euphues.*

Euphues having foiourned by the fpace of two
monethes in *Naples,* whether he were moued by the
courtefie of a young gentleman named *Phila[u]tus,* or
inforced by deftany : whether his pregna[n]t wit, or his
pleafant conceits wrought the greater lyking in [of] the
minde of *Euphues,* I know not for certeintie : But
Euphues fhewed fuch entyre loue towards him, that
he feemed to make fmall accompt of any others,
determining to enter into fuch an inuiolable league of
friendfhip with him, as neither time by peecemeale
fhould impaire, neither fancie vtterly defolue, nor any
fufpition infringe. I haue read (faith he) and well I

beleeue it, that a friend is in profperitie a pleafure, a folace in aduerfitie, in griefe a comfort, in ioy a merry companion, at al times an other I, in all places the expreffe Image of myne owne perfon : infomuch that I cannot tell wether the immortall Gods haue beftowed any gift vpon mortall men, either more noble [able] or more neceffary then friendfhip. Is there any thing in the world to be reputed (I will not fay compared) to friendfhip ? Can any treafure in this tranfitory pilgrimage be of more valew then a friend ? in whofe bofome thou maift fleepe fecure without feare, whom thou maift make partner of al thy fecrets without fufpition of fraude, and partaker of all thy miffortune without miftruft of fleeting, who will accompt thy bale his bane, thy mifhap his mifery, the pricking of thy finger the percing of his heart. But whether am I caryed ? Haue I not alfo learned yat one fhould eate a bufhel of falt with him whom he meaneth to make his friend ? that tryal maketh truft ? that ther is falfhood in felowfhip ? and what then ? Doth not the fimpathy of manners make the coniunction of mindes ? Is it not a by word lyke will to lyke ? Not fo common as commendable it is, to fee young Gentlemen choofe them fuch friendes, with whom they may feeme being abfent to be prefent, being a funder to be conuerfant, being dead to be aliue. I will therefore haue *Philautus* for my pheere, and by fo much the more I make my felfe fure to haue *Philautus*, by how much the more I view in him the liuely image of *Euphues.*

Although there be none fo ignoraunt that doth not know, neither any fo impudent that will not confeffe, friendfhip to be the iewell of humaine ioye : yet whofoeuer fhal fee this amitie grounded vpon a little affection, will foone coniecture that it fhall be diffolued vpon a light occafion : as in the fequele of *Euphues* and *Philautus* you fhall [foon] fee, whofe hot loue waxed foone colde : For as the beft Wine doth make the fharpeft vineger, fo the deepeft loue turneth to the deadlyeft hate. Who deferued the moft blame, in mine opinion, it* is doubtful and fo difficult, that I dare not prefume to giue verdit. For loue being the caufe for which fo many mifchiefes haue ben attempted, I am not yet perfwaded, whether of them was moft to be blamed, but certeinely neither of them was blameleffe. I appeale to your iudgement Gentlemen, not that I thinke any of you of the lyke difpofition, able to decide the queftion, but being of deeper difcretion then I am, are more fit to debate ye quarrell. Though the dif-courfe of their friendfhip and falling out be fomwhat long, yet being fomwhat ftrange, I hope the delight-fulneffe of the one wil attenuate the tedioufneffe of the other.

Euphues had continual acceffe to the place of *Philautus*, and no little familiaritie with him, and finding him at conuenient leafure, in thefe fhort termes vnfolded his minde vnto [to] him.

Gentleman and friend, the tryall I haue had of thy manners cutteth off diuers termes, which to an other I wold haue vfed in the lyke matter. And fithens a long difcourfe argueth folly, and delicate words incurre the fufpition of flattery, I am determined to vfe neither of them, knowing either of them to breede offence. Wayinge with my felfe the force of friendfhippe by the effeéts, I ftudyed euer fince my firft comming to *Naples* to enter league with fuch a one as might direét my fteps being a ftranger, and refemble my manners being a fcholler, the which two qualities as I find in you able to fatiffie my defire, fo I hope I fhal finde a heart in you willinge to accomplifh my requeft. Which if I may obteine, affure your felfe, that *Damon* to his *Pythias*, *Pilades* to his *Oreftes*, *Tytus* to his *Gyfippus*, *Thefius* to his *Pirothus*, *Scipio* to his *Lælius*, was neuer founde more faithfull, then *Euphues* will bee to *Philautus*.

Philautus by how much the leffe he looked for this difcourfe, by fo much the more he lyked it, for he fawe all qualities both of body and minde, in *Euphues*, vnto whom he replyed as followeth.

D

Friend *Euphues* (for fo your talke warranteth me to
term you) I dare neither vfe a long proceffe, neither a
louing fpeach, leaft vnwittingly I fhold caufe you to
conuince me of thofe things which you haue already
condemned. And verily I am bold to prefume vpon
your curtefie, fince you your felf haue vfed fo little
curiofitie : perfwading my felfe that my fhort anfwere
wil worke as great an effect in you, as your few words
did in me. And feeing we refemble (as you fay) each
other in qualities, it cannot be yat the* one fhould differ
from the other in curtefie, feing the fincere affection
of the minde cannot be expreffed by the mouth, and
that no art can vnfold the entire loue of ye heart, I am
earneftly to befeech you not to meafure the firmeneffe
of my faith, by ye fewnes of my wordes, but rather
thinke that the ouerflowing waues of good wil, leaue
no paffage for many words. Triall fhall proue truft,
heere is my hand, my hart, my lands and my life at
thy commaundement. Thou maift wel perceiue that
I did beleeue thee, that fo foone I did loue thee : and
I hope thou wilt the rather loue me, In that I did
beleeue thee. Either *Euphues* and *Phila[u]tus* ftoode in
neede of frindfhippe, or were ordeined to be friendes :
vpon fo fhort warning, to make fo foone [fine] a conclufion might feeme in mine opinion if it continued
myraculous, if fhaken off, ridiculous.

But after many embracings and proteftations one to
an other, they walked to dinner, wher they wanted
neither meat, neither Muficke, neither any other
paftime : and hauing banqueted, to digeft their fweete
confections, they daunced all that after noone, they
vfed not onely one boorde but one bed, one booke (if
fo be it they thought not one too many.) Their
friendfhip augmented euery day, infomuch that the one
could not refraine the company of the other one
minute, all things went in common betweene them,
which all men accompted commendable.

Phila[u]tus being a towne borne childe, both for his
owne countenaunce, and the great countenaunce which

his father had while he liued, crept into credit with
Don Ferardo one of the chiefe gouernours of the citie,
who although he had a courtly crew of gentlewomen
foiourning in his pallaice, yet his daughter, heire to his
whole reuenewes ftayned ye beautie of them al, whofe
modeft bafhfulnes caufed the other to looke wanne for
enuie, whofe Lilly cheekes dyed with a Vermilion red,
made the reft to blufh for fhame. For as the fineft
Ruby ftaineth ye coulour of the reft that be in place,
or as the Sunne dimmeth the Moone, that fhe cannot
be difcerned, fo this gallant girle more faire then fortunate, and yet more fortunate then faithful, eclipfed
the beautie of them all, and chaunged their colours.
Vnto hir had *Philautus* acceffe, who wan hir by right
of loue, and fhould haue worne hir by right of law, had
not *Euphues* by ftraunge deftenie broken the bondes
of mariage, and forbidden the banes of Matrimony.

It happened that *Don Ferardo* had occafion to goe
to *Venice* about certeine [of] his owne affaires, leauing his
daughter the onely fteward of his houfehold, who fpared
not to feaft *Philautus* hir friend, with al kinds of
delights and delycates, referuing only hir honeftie as
the chiefe ftay of hir honour. Hir father being gone
fhe fent for hir friend to fupper, who came not as hee
was accuftomed folitarilye alone, but accompanyed
with his friend *Euphues*. The Gentlewoman whether
it were for niceneffe, or for nigardneffe of courtefie,
gaue him fuch a colde welcome, that he repented that
he was come.

Euphues though he knewe himfelfe worthy euerye
way to haue a good countenaunce, yet coulde he not
perceiue hir willing any way to lende him a friendly
looke. Yet leaft he fhould feeme to want geftures, or
to be dafhed out of conceipt with hir coy countenaunce,
he addreffed him to a Gentlewoman called *Liuia*, vnto
whome he vttered this fpeach. Faire Ladye, if it be
the guife of *Italy* to welcome ftraungers with ftrangnes,
I muft needes fay the cuftome is ftrange and the
countrey barbarous, if the manner of Ladies to falute

Gentlemen with coyneffe, then I am enforced to think the women without [voyde of] courtefie to vfe fuch welcome, and the men paft fhame that will come. But heereafter I will either bring a ftoole on mine arme for an vnbidden gueft, or a vifard on my face, for a fhameleffe goffippe. *Liuia* replyed.

Sir, our country is ciuile, and our gentlewomen are curteous, but in *Naples* it is compted a ieft, at euery word to fay, In faith you are welcome. As fhe was yet talking, fupper was fet on the bord, then *Philautus* fpake thus vnto *Lucilla*. Yet Gentlewoman, I was the bolder to bring my fhadow with me, (meaning *Euphues*) knowing that he fhould be the better welcome for my fake: vnto whom the Gentlewoman replyed. Sir, as I neuer when I faw you, thought that you came without your fhadow, fo now I cannot a lyttle meruaile to fee you fo ouerfhot in bringing a new fhadow with you. *Euphues*, though he perceiued hir coy nippe, feemed not to care for it, but taking hir by the hand faid.

Faire Lady, feeing the fhade doth [fo] often fhield your beautie from the parching Sunne, I hope you will the better efteeme of the fhadow, and by fo much the leffe it ought to be offenfiue, by how much the leffe it is able to offende you, and by fo much the more you ought to lyke it, by how much the more you vfe to lye in it.

Well Gentleman, aunfwered *Lucilla*, in arguing of the fhadow, we forgoe the fubftaunce: pleafeth it you therefore to fit downe to fupper. And fo they all fate downe, but *Euphues* fed of one difh, which [was] euer ftoode* before him, the beautie of *Lucilla*.

Heere *Euphues* at the firft fight was fo kindled with defire, that almoft he was like to burn to coales. Supper beeing ended, the order was in *Naples*, that the Gentlewomen would defire to heare fome difcourfe, either concerning loue, or learning: And although *Philautus* was requefted, yet he pofted it ouer to *Euphues*, whome he knewe moft fit for that purpofe: *Euphues* beeing thus tyed to the ftake by their importunate intreatie, began as followeth.

He that worſt may is alway enforced to holde the candell, the weakeſt muſt ſtill to the wall, where none will, the Diuell himſelfe muſt beare the croſſe. But were it not Gentlewomen, that your luſt ſtandes for law, I would borrow ſo much leaue as to reſigne mine office to one of you, whoſe experience in loue hath made you learned, and whoſe learninge hath made you ſo louely : for me to intreat of the one being a nouiſe, or to diſcourſe of the other being a trewant, I may well make you weary, but neuer the wiſer, and giue you occaſion rather to laugh at my raſhneſſe, then to lyke my reaſons : Yet I care the leſſe to excuſe my boldneſſe to you, who were the cauſe of my blindneſſe. And ſince I am at mine owne choyce, either to talke of loue or of learning, I had rather for this time bee deemed an vnthrift in reiecting profite, then a *Stoicke* in renouncing pleaſure.

It hath bene a queſtion often diſputed, but neuer determined, whether the qualities of the minde, or the compoſition of the man, cauſe women moſt to lyke, or whether beautie or wit moue men moſt to loue. Certes by how much the more the minde is to be preferred before the body, by ſo much the more the graces of the one are to be preferred before ye gifts of the other, which if it be ſo, that the contemplation of the inward qualitie ought to bee reſpected, more then the view of the outward beautie, then doubtleſſe women either do or ſhould loue thoſe beſt whoſe vertue is beſt, not meaſuring the deformed man, with the reformed minde.

The foule Toade hath a faire ſtone in his head, the fine golde is found in the filthy earth : the ſweet kernell lyeth in the hard ſhell : vertue is harboured in the heart of him that moſt men eſteeme miſhapen. Contrariwiſe, if we reſpect more the outward ſhape, then the inward habit, good God, into how many miſchiefes do wee fall ? into what blindneſſe are we ledde ? Doe we not commonly ſee that in painted pottes is hidden the deadlyeſt poyſon ? that in the greeneſt graſſe is ye greateſt Serpent? in the cleereſt water the vglyeſt Toade? Doth not experience teach vs, that in the moſt curious Sepulcher are encloſed rotten bones? That the Cypreſſe tree beareth a faire leafe, but no fruite? That the Eſtridge carieth faire feathers, but ranke fleſh? How frantick are thoſe louers which are caried away with the gaye gliſtering of the fine face? the beautie whereoff is parched with the ſummers blaze, and chipped with the winters blaſt : which is of ſo ſhort continuance, that it fadeth before one perceiue it flouriſh : of ſo ſmal profit, that it poyſoneth thoſe that poſſeſſe it : of ſo litle value with the wife, that they accompt it a delicate baite with a deadly hooke : a ſweet *Panther* with a deuouring paunch, a ſower poyſon in a ſiluer potte. Heere I could enter into diſcourſe of ſuch fine dames as being in loue with their owne lookes, make ſuch courſe accompt of their paſſionate louers : for commonly if they be adorned with beautie, they be ſtraight laced, and made ſo high in the inſteppe, that they diſdaine them moſt that moſt deſire them. It is a worlde to ſee the doating of their louers, and their dealing with them, the reueling of whoſe ſubtil traines would cauſe me to ſhed teares, and you Gentlewomen to ſhut your modeſt eares. Pardon me Gentlewomen if I vnfolde euery wile and ſhew euery wrinkle of womens diſpoſition. Two things do they cauſe their ſeruants to vow vnto them, ſecrecie, and ſouereintie : the one to conceale their entiſing ſleights, by the other to aſſure themſelues of their only ſeruice. Againe, but hoe there : if I ſhoulde haue waded anye further, and ſownded the depth of their deceipt, I ſhould either haue procured your diſpleaſure, or incurred the ſuſpicion of fraud : either armed you to praċtiſe the like ſubtiltie, or accuſed my ſelfe of periury. But I meane not to offend your chaſt mindes, with the rehearſal of their vnchaſt manners : whoſe eares I perceiue to glow, and hearts to be grieued at that which I haue alredy vttered : not that amongſt you there be any ſuch, but that in your ſexe ther ſhould be any ſuch. Let not

Gentlewomen therefore make to[o] much of their painted
fheath, let them not be fo curious in their owne conceit,
or fo currifh to their loyal louers. When the black
Crowes foote fhall appeare in their eye, or the blacke
Oxe treade on their foote, when their beautie fhall be
lyke the blafted Rofe, their wealth wafted, their bodies
worne, their faces wrinkled, their fingers crooked, who
wil like of them in their age, who loued none in their
youth? If you will be cherifhed when you be olde, be
courteous while you be young: if you looke for comfort in your hoarie haires, be not coye when you haue
your golden lockes: if you would be imbraced in ye
wayning of your brauerie, be not fqueymifh in the
waxing of your beautie: if you defire to be kept lyke
the Rofes when they haue loft their coulour, fmel fweete
as the Rofe doth in the budde: if you woulde bee
tafted for olde Wine, bee in the mouth a pleafaunt
Grape: fo fhall you be cherifhed for your courtefie,
comforted for your honeftie, embraced for your amitie,
fo fhall you [ye] be preferued with the fweete Rofe, and
dronke with the pleafant wine. Thus farre I am
bolde gentlewomen, to counfel thofe that be coy, that
they weaue not the web of their owne woe, nor fpinne
the threede of their own thraldome, by their own
ouerthwartnes. And feeing we are euen in the bowells
of loue, it fhal not be amiffe, to examine whether man
or woman be fooneft allured, whether be moft conftant
the male or the female. And in this poynte I meane
not to be mine owne caruer, leaft I fhould feeme either
to picke a thanke with men, or a quarel with women.
If therefore it might ftand with your pleafure (Miftres
Lucilla) to giue your cenfure, I would take the contrarie: for fure I am though your iudgement be found,
yet affection will fhadow it.

Lucilla feeing his pretence, thought to take aduauntage of his large profer, vnto whom fhe faide. Gentleman in my opinion, women are to be wonne with euery
wind, in whofe fexe ther is neither force to withftand
the affaults of loue, neither conftancy to remaine faith·

full. And bicaufe your difcourfe hath hetherto bred delight, I am loth to hinder you in the fequele of your deuifes. *Euphues*, perceiuing himfelfe to be taken napping, aunfwered as followeth.

¶ Miftres *Lucilla*, if you fpeake as you thinke, thefe gentlewomen prefent haue little caufe to thanke you, if you caufe me to commend women, my tale will be accompted a meere trifle, and your wordes the plaine truth : Yet knowing promife to be debt, I will paye it with performance. And I woulde the Gentlemen heere prefent were as ready to credit my proofe, as the gentlewomen are willing to heare their own prayfes, or I as able to ouercome, as Miftres *Lucilla* would be content to be ouerthrowne, howe fo euer the matter fhall fall out, I am of the furer fide : for if my reafons be weake, then is our fexe ftrong: if forcible, then [is] your iudgement feeble : if I finde truth on my fide, I hope I fhall for my wages win the good will of women : if I want proofe, then gentlewomen of neceffitie you muft yeeld to men. But to the matter.

Touching the yeelding to loue, albeit their heartes feeme tender, yet they harden them lyke the ftone of *Sicilia*, the which the more it is beaten the harder it is : for being framed as it were of the perfection of men, they be free from all fuch cogitations as may any way prouoke them to vncleanenffe, infomuch as they abhorre the light loue of youth, which is grounded vppon luft, and diffolued, vpon euery light occafion. When they fee the folly of men turne to fury, their delyght to doting, their affection to frencie, when they fee them as it were pine in pleafure, and to wax pale through their own peeuifhnes, their futes, their feruice, their letters, their labours, their loues, their liues, feeme to them fo odyous, that they harden their hearts againft fuch concupyfence, to the ende they might conuert them from rafhneffe to reafon : from fuch lewde difpofition, to honeft difcretion. Heereoff it commeth that men accufe woemen of cruelty, bicaufe they them-felues want ciuility : they accompt them full of wyles,

in not yeelding to their wickednes: faithleſſe for
refiſting their filthynes. But I had almoſt forgot my
felfe, you ſhal pardon me Miſtres *Lucilla* for this time, if
this[thus]abruptlye, I finiſh my difcourfe: it is neither for
want of good wil, or lack of proofe, but yat I feele in
my felf fuch alteration, yat I can fcarcely vtter one
worde. Ah *Euphues, Euphues.* The gentlewomen
were ſtrooke into fuch a quandary with this fodeine
chaunge, that they all chaunged coulour. But *Euphues*
taking *Philautus* by the hande, and giuing the gentlewomen thankes for their patience and his repaſt, bad
them al farewell, and went immediatly to his chamber.
But *Lucilla* who nowe began to frye in the flames of
loue, all the companye being departed to their lodgings,
entered into thefe termes and contrarieties.

Ah wretched wench *Lucilla*, how art thou perplexed?
what a doubtfull fight doſt thou feele betwixt [betweene]
faith and fancy? hope and feare? confcience and concupifcence? O my *Euphues*, lyttle doſt thou knowe the
fodeyn forrowe that I fuſteine for thy fweete fake :
Whofe wyt hath bewitched me, whofe rare qualyties
haue depryued me of myne olde qualytie, moſt curteous
behauiour without curiofitie, whofe comely feature,
wythout fault, whofe filed fpeach without fraud, hath
wrapped me in this miſfortune. And canſt thou *Lucilla*
be fo light of loue in forfaking *Philautus* to flye to
Euphues? canſt thou prefer a ſtraunger before thy
countryman? a ſtarter before thy companion? Why,
Euphues doth perhappes [perhappes doeth] defire my
loue, but *Philautus* hath deferued it. Why, *Euphues*
feature is worthy as good as I, but *Philautus* his faith
is worthy a better. I, but the latter loue is moſt feruent, I, but ye firſt ought to be moſt faythfull. I,
but *Euphues* hath greater perfe6tion, I, but *Philautus*
hath deeper affe6tion.

Ah fonde wench, doeſt thou thincke *Euphues* will
deeme thee conſtant to him, when thou haſt ben
vnconſtant to his friend? Weeneſt thou that he will
haue no miſtruſt of thy faithfulnes, when he hath had

tryall of thy fickleneffe? Wil he haue no doubt of
thine honour, when thou thy felfe calleft thine honeftie
in queftion? Yes, yes, *Lucilla*, well doth he knowe
that the glaffe once crafed, will with the leaft clappe
be cracked, that the cloth which ftayneth with milke,
will foone loofe his coulour with Vineger: that the
Eagles wing will waft the feather as well of the *Phœnix*,
as of the Pheafaunt: that fhe that hath beene faithleffe
to one, will neuer be fa[i]thfull to any. But can *Euphues*
conuince me of fleeting, feeing for his fake I break my
fidelitie? Can he condemne me of difloyaltie, when
he is the only caufe of my difliking? May he iuftly
condemne me of trechery, who hath this teftimony as
tryal of my good wil? Doth not he remember that
the broken bone once fet together, is ftronger than
euer it was? That the greateft blot is taken off with
the Pommice? That though the Spider poyfon the
flye, fhee cannot infect the Bee? That although I haue
bene light to *Philautus*, I may be louely to *Euphues*?
It is not my defire, but his defertes that moueth my minde
to this choyfe: neither the want of the lyke good will
in *Philautus*, but the lacke of the lyke good qualyties that remoueth my fancie from the one to the
other.

For as the Bee that gathereth Honnye out of the
weede, when fhee efpieth the fayre floure flyeth to the
fweeteft: or as the kinde fpaniell though he hunt after
Birds, yet forfakes them to retriue the Partridge: or as
we commonly feede on beefe hungerly at the firft, yet
feeing the Quaile more daintie, chaunge our dyet:
So I, although I loued *Philautus* for his good properties, yet feeing *Euphues* to excell him, I ought by
Nature to lyke him better. By fo much the more
therefore my chaunge is to be excufed, by how much
the more my choyce is excellent: and by fo much the
leffe I am to be condemned by how much the more
Euphues is to be commended. Is not the Diamond of
more valew then the Rubie bicaufe he is of more
vertue? Is not the Emeraulde preferred before the

Saphire for his wonderfull propertie? Is not *Euphues*
more prayfe worthy then *Philautus* being more wittie.
But fye *Lucilla*, why doft thou flatter thy felfe in thine
owne folly? Canft thou faine *Euphues* thy friend,
whom by thine owne words thou haft made thy foe?
Diddeft not thou accufe women of inconftancie? Diddeft
not thou accompt them [thy felfe] eafie to be won?
Diddeft not thou condemne them of weakenes, what
founder argument can he haue againft thee then thine
own aunfwere? What better proofe then thine owne
fpeach? What greater tryall then thine owne talke?
If thou haft belyed women, he will iudge thee vnkinde :
if thou haue reuealed the troth, he muft needes thinke
thee vnconftant : if he perceiue thee to be wonne with
a Nut, he wil imagine that thou wilt be loft with an
Apple, if he finde thee wanton before thou be wo[o]ed,
he wil geffe thou wilt be wauering when thou art
wedded.

But fuppofe that *Euphues* loue thee, that *Philautus*
leaue thee, wil thy Father thinkeft thou giue thee
libertie to lyue after thine owne luft? Wil he efteeme
him worthy to enherite his poffeffions, whome he
accompteth vnworthy to enioy thy perfon? Is it lyke
that hee will match thee in mariage with a ftraunger,
with a *Grecian*, with a meane man? I, but what
knoweth my father whether he be wealthy, whether
his reuenews be able to counteruaile my fathers landes,
whether his birth be noble yea, or no? Can any one
make doubt of his gentle bloud, that feeth his gentle
conditions? Can his honour be called into queftion,
whofe honeftie is fo great? Is he to be thought thrift-
leffe, who in all qualyties of the minde is peereleffe?
No no, the tree is known by his fruit, the gold by his
touch, the fonne by the fire. And as the foft waxe
receiueth whatfoeuer print be in the feale, and fheweth
no other impreffion, fo the tender babe being fealed
with his fathers gifts, reprefenteth his Image moft
liuely. But were I once certeine of *Euphues* [his] good
will, I would not fo fuperfticioufly accompt of my

fathers ill will. Time hath weaned me from my
mothers teat, and age ridde me from my fathers
correction, when children are in their fwathe cloutes,
then are they fubiect to the whip, and ought to
be carefull of the rigour of their parents. As for
me feeing I am not fedde with their pap, I
am not to be ledde by their perfwafions. Let my
father vfe what fpeaches he lyft, I will follow mine
owne luft. Luft *Lucilla*, what fayft thou? No no,
mine owne loue I fhould haue fayd, for I am as farre
from luft, as I am from reafon, and as neere to loue
as I am to folly. Then fticke to thy determination,
and fhew thy felfe, what loue can doe, what loue dares
doe, what loue hath done. Albeit I can no way
quench the coales of defire with forgetfulneffe, yet
will I rake them vp in the afhes of modeftie : Seeing
I dare not difcouer my loue for maidenly fhamefaftneffe,
I will diffemble it till time I haue opportunitie. And
I hope fo to behaue my felfe, as *Euphues* fhall thinke
me his owne, and *Philautus* perfwade himfelf I am
none but his. But I would to God *Euphues* would
repaire hether that the fight of him might mitigate
fome parte of my martirdome.

She hauing thus difcourfed with hir felfe, hir owne
miferies, caft hir felfe on the bedde and there lette hir
lye, and retourne we to *Euphues*, who was fo caught
in the ginne of folly, that he neither could comfort
himfelfe, nor durft afke counfaile of his friend, fufpecting that which in deede was true, that *Philautus* was
corriual with him and cooke-mate with *Lucilla*.
Amiddeft therefore thefe his extremities, betweene
hope and feare, he vttered thefe or the lyke
fpeaches.

What is he *Euphues*, that knowing thy witte, and
feeing thy folly, but will rather punifh thy leaudneffe,
then pittie thy heauineffe? Was ther euer any fo
fickle fo foone to be allured? any euer [euer anie] fo
faithleffe to deceiue his friend? euer any fo foolifh to
bathe himfelfe in his owne miffortune? Too true it is,

that as the ﾃ｣ea Crab ﾃ｣wimmeth alwayes againﾃ｡ the
ﾃ｡reame, ﾃ｣o wit alwayes ﾃ｡riueth againﾃ｡ wiﾃ｣edome :
And as the Bee is oftentimes hurt with hir owne
Honny, ﾃ｣o is witte not ﾃ｣eldome plagued with his owne
conceipt.

O ye Gods, haue ye ordeyned for euery malady a
medicine, for euery ﾃ｣ore a ﾃ｣alue, for euery paine a
pla[y]ﾃ｡er, leauing onely loue remedileﾃ｣ﾃ｣e? Did ye
deeme no man ﾃ｣o mad to be entangled with deﾃ｣ire,
or thought ye them worthie to be tormented that were
ﾃ｣o miﾃ｣ledde? haue ye dealt more fauourably with
brute beaﾃ｡es, then with reaﾃ｣onable creatures.

The filthy Sow when ﾃ｣he is ﾃ｣icke, eateth the Sea-
Crab, and is immediatly recured : the Torteyﾃ｣e hauing
taﾃ｡ted the Viper, ﾃ｣ucketh *Origanum* and is quickly
reuiued : the Beare ready to pine licketh vp the Ants,
and is recouered : the Dog hauing ﾃ｣urfetted to procure
his vomitte, eateth graﾃ｣ﾃ｣e and findeth remedy : the
Hart beeing perced with the dart, runneth out of hand
to the hearb *Diﾃ｡anum*, and is healed. And can men
by no hearbe, by no art, by no way, procure a remedie
for the impatient diﾃ｣eaﾃ｣e of loue? Ah well I perceiue
that Loue is not vnlyke the Figge tree, whoﾃ｣e fruite is
ﾃ｣weete, whoﾃ｣e roote is more bitter then the clawe of a
Bitter : or lyke the Apple in *Perﾃ｣ia*, whoﾃ｣e bloﾃ｣ﾃ｣ome
fauoreth lyke Honny, whoﾃ｣e budde is more ﾃ｣ower then
Gall.

But O impietie. O broad blaﾃ｣phemie againﾃ｡ the
heauens. Wilt thou be ﾃ｣o impudent *Euphues*, to accuﾃ｣e
the Gods of iniquitie? No fonde foole, no. Neither
is it forbidden vs by the Gods to loue, by whoﾃ｣e diuine
prouidence we are permitted to liue : neither do wee
want remedies to recure our maladies, but reaﾃ｣on to
vﾃ｣e the meanes. But why goe I about to hinder the
courﾃ｣e of loue, with the diﾃ｣courﾃ｣e of law? haﾃ｡ thou not
read *Euphues*, that he that loppeth the Vine, cauﾃ｣eth it
to ﾃ｣pread faire : that he that ﾃ｡oppeth the ﾃ｡reame, forceth
[cauﾃ｣eth] it to ﾃ｣well higher? that he that caﾃ｡eth water
on [in] the fire in [at] the Smithes forge, maketh it to flame

fiercer? Euen fo he that feeketh by counfaile to
moderate his ouerlafhing affections, encreafeth his
own miffortune. Ah my *Lucilla*, would thou wer
either leffe faire, or I more fortunate : either I wifer,
or thou milder : either I would I were out of this mad
moode, either I would we wer both of one minde.
But how fhould fhe be perfwaded of my loyaltie, that
yet had neuer one fimple proofe of my loue? will fhe
not rather imagine me to be entangled with hir beautie,
then with hir vertue. That my fancie being fo lewdly
chaunged [chayned] at ye firft, will be as lyghtly
chaunged at the laft : that nothing violent, can bee
permanent. Yes, yes, fhee muft needes coniecture fo,
although it bee nothing fo : for by howe much the
more my affection commeth on the fodeine, by fo
much the leffe will fhe thinke it certeine. The ratling
thunderbolt hath but his clap, the lightning but his
flafh, and as they both come in a moment, fo doe they
both ende in a minuite.

I, but *Euphues*, hath fhe not hard alfo that the dry
touchewoode is kindled with lyme? that the greateft
Mufhrompe groweth in one night? that the fire quickly
burneth the flaxe? that loue eafily entereth into the
fharpe wit without refiftance, and is harboured there
without repentaunce.

If therefore the Gods haue endewed hir with as
much bountie as beautie, if fhe haue no leffe witte
then fhe hath comelineffe : certes fhee wyll neyther
conceiue finifterly of my fodeine fute, neither be coye
to receiue me into hir feruice, neither fufpect me of
lyghtneffe in yeelding fo lyghtly, neither reiect me
difdainefully, for louing fo haftely? Shall I not then
hazarde my life to obteine my loue? and deceiue
Philautus to receiue *Lucilla*? Yes *Euphues*, where
loue beareth fway, friendfhip can haue no fhewe : As
Philautus brought me for his fhadowe the laft fupper,
fo will I vfe him for my fhadow till I haue gained his
Saint. And canft thou wretch be falfe to him that is
faithful to thee? Shall his curtefie bee caufe of thy

crueltie? Wilt thou violate the league of fayth, to enherite the lande of folly? Shall affe&tion be ·of more force then friendfhip, loue then lawe, luft then loyaltie? Knoweft thou not that he that lofeth his honeftie, hath nothing els to loofe.

Tufh the case is lyght, where reafon taketh place, to loue and to lyue well, is not graunted to *Iupiter*. Who fo is blynded with the caule of beautie, difcerneth no colour of honefty. Did not *Giges* cut *Candaules* a coat by hys owne meafure? Did not *Paris*, though he were a welcome gueft to *Menelaus*, ferue his hoaft a flippery pranke? If *Philautus* had loued *Lucilla*, hee would neuer haue fuffered *Euphues* to haue feene hir. Is it not the pray that enticeth the theefe to rifle? Is it not the pleafaunt bayte that caufeth the fleeteft fifh to byte? Is it not a by worde amongft vs, that gold maketh an honeft man an ill man? Did *Philautus* accompt *Euphues* too [fo] fimple to decypher beautie, or [fo] fuperftitious not to defire it? Did he deeme him a faint in reie&ting fancy, or a fot in not difcerning? Thought he him a *Stoycke*, that he woulde not be moued, or a ftocke that he could not?

Well, wel, feeing the wound that bleedeth inwardly is moft daungerous, that the fyre kept clofe burneth moft furious, that ye Ouen dammed vp, baketh fooneft, that fores hauing no vent fefter fecretly, it is hyghe tyme to vnfolde my fecret loue to my fecret friend. Let *Philautus* behaue himfelf neuer fo craftely, he fhal know that it muft be a wyly Moufe that fhall breede in the Cats eare : and bicaufe I refemble him in wit, I meane a little to diffemble with him in wyles. But O my *Lucilla*, if thy heart be made of that ftone which may be mollified onely with bloud, would I had fipped of that ryuer in *Caria*, which turneth thofe that drinke of it to ftones. If thyne eares be anoynted with the oyle of *Syria* that bereaueth hearing, would mine eyes had bene rubbed with the firop of the Cedar tree, which taketh away fight.

If *Lucilla* be fo proude to difdayne poore *Euphues*,

woulde *Euphues* were fo happye to denye *Luciila*, or
if *Lucilla* be fo mortyfied to lyue without loue, woulde
Euphues were fo fortunate to lyue in hate.   I but my
colde welcome foretelleth my colde fuit, I but hir
priuie glaunces fignifie fome good Fortune.   Fye
fonde foole *Euphues*, why goeſt thou about to alleadge
thofe thinges to cutte off thy hope which fhe perhaps
woulde neuer haue founde, or to comfort my felfe
with thofe reafons which fhee neuer meaneth to pro-
pofe :   Tufh it were no loue if it were certeyne, and a
fmall conqueſt it is to ouerthrowe thofe that neuer
refiſteth.

In battayles there ought to be a doubtfull fight, and
a defperat ende, in pleadinge a diffyculte enteraunce,
and a defufed determination, in loue a lyfe wythout
hope, and a death without feare.   Fyre commeth out
of the hardeſt flynte wyth the ſteele.   Oyle out of the
dryeſt Ieate by the fyre, loue out of the ſtonieſt hearte
by fayth, by truſt, by tyme.   Hadde *Tarquinus* vfed
his loue with coulours of countenuaunce, *Lucretia*
woulde eyther wyth fome pitie haue aunfwered hys
defyre, or with fome perfwafion haue ſtayed hir death.
It was the heate of hys luſt, that made hyr haſt to ende
hir lyfe, wherefore loue in neyther refpecte is to bee
condempned, but hee of rafhneffe to attempte a Ladye
furiouflye, and fhee of rygor to punifhe hys follye in
hir owne flefhe, a fact (in myne opinion) more worthy
the name of crueltie then chaſtitie, and fitter for a
Monſter in the defartes, then a Matrone of *Rome*.
Penelope no leffe conſtaunt then fhee, yet more wyfe,
woulde bee wearie to vnweaue that in the nyght, fhee
fpunne in the daye, if *Vlyſſes* hadde not come home
the fooner.   There is no woeman, *Euphues*, but fhee
will yeelde in time, bee not therefore difmaied either
with high lookes or frowarde words.

Euphues hauing thus talked with himfelfe, *Philautus*
entered the chamber, and finding him fo worne and
waſted with continuall mourning, neither ioying in hys

meate, nor reioycing in his friend, with watry eyes vttered this fpeach.

FRiend and fellow, as I am not ignoraunt of thy prefent weakenes, fo I am not priuie of the caufe : and although I fufpect many things, yet can I affure my felf of no one thing. Therfore my good *Euphues*, for thefe doubts and dumpes of mine, either remoue the caufe, or reueale it. Thou haft hetherto founde me a cheerefull companion in thy myrth, and nowe fhalt thou finde me as carefull with thee in thy moane. If altogether thou maift not be cured, yet maift thou bee comforted. If ther be any thing yat either by my friends may be procured, or by my life atteined, that may either heale thee in part, or helpe thee in all, I proteft to thee by the name of a friend, that it fhall rather be gotten with the loffe of my body, then loft by getting a kingdome. Thou haft tried me, therefore truft me : thou haft trufted me in many things, therfore try me in this one thing. I neuer yet failed, and now I wil not fainte. Be bolde to fpeake and blufh not : thy fore is not fo angry but I can falue it, the wound not fo deepe but I can fearch it, thy griefe not fo great [fore] but I can eafe it. If it be ripe it fhalbe lawnced, if it be broken it fhalbe tainted, be it neuer fo defperat it fhalbe cured. Rife therefore *Euphues*, and take heart at graffe, younger thou fhalt neuer be : plucke vp thy ftomacke, if loue it felfe haue ftoung thee, it fhal not ftifle thee. Though thou be enamoured of fome Lady, thou fhalt not be enchaunted. They that begin to pine of a confumcion, without delay preferue themfelues with culliffes : he that feeleth his ftomack enflamed with heat, cooleth it eftfoones with conferues : delayes breede daungers, nothing fo perillous as procraftination. *Euphues* hearing this comfort and friendly counfaile, diffembled his forrowing heart with a fmiling face, aunfwering him forthwith as followeth.

True it is *Philautus* that hee which toucheth the

E

Nettle tenderly, is fooneſt ſtoung : that the Flye which playeth with the fire, is ſinged in the flame, that he that dalyeth with women is drawne to his woe. And as the Adamant draweth the heauie yron, the Harpe the fleete Dolphin, ſo beautie allureth the chaſt minde to loue, and the wiſeſt witte to luſt : The example whereoff I woulde it were no leſſe profitable, then the experience to me is lyke to be perillous. The Vine watered with Wine, is ſoone withered : the bloſſome in the fatteſt ground, is quickly blaſted : the Goat the fatter ſhee is, the leſſe fertile ſhe is : yea man, the more wittie he is, the leſſe happy he is. So it is *Philautus* (for why ſhould I conceale it from thee, of whome I am to take counſayle) that ſince my laſt and firſt being with thee at the houſe of *Ferardo*, I haue felt ſuch a furious battayle in mine owne body, as if it be not ſpeedely repreſſed by pollicie, it wil cary my minde (the graund captaine in this fight) into end-leſſe captiuitie. Ah *Liuia, Liuia*, thy courtly grace with out coyneſſe, thy blazing beautie without blemiſh, thy curteous demeanor without curioſitie, thy ſweet ſpeech ſauoured with witte, thy comely mirth tempered with modeſtie ? thy chaſt lookes, yet louely : thy ſharp taunts, yet pleaſaunt : haue giuen me ſuch a checke, that ſure I am at the next viewe of thy vertues, I ſhall take thee mate : And taking it not of a pawne but of a Prince, the loſſe is to be accompted the leſſe. And though they be commonly in a great cholar that receiue the mate, yet would I willingly take euery minute tenne mates to enioy *Liuia* for my louing mate. Doubtleſſe if euer ſhe hir ſelfe haue bene ſcorched with the flames of deſire, ſhe wil be redy to quench the coales with curteſie in an other : if euer ſhe haue bene attached of loue, ſhe will reſcue him that is drenched in deſire : if euer ſhe haue ben taken with the feuer of fancie, ſhe will help his ague, who by a *quotidian* fit is conuerted into phrenſie : neither can ther be vnder ſo delycate a hue lodged deceipt, neither in ſo beautifull a mould, a malicious minde : True it

is that the difpofition of the minde, foloweth the compofition of the body ; how then can fhe be in minde
any way imperfeċt, who in body is perfeċt euery way,
I know my fucces will be good, but I know not how
to haue acces to my goddes : neither do I want
courage to difcouer my loue to my friend, but fome
colour to cloake my comming to the houfe of *Ferardo* :
for if they be in *Naples* as iealous as they bee in the
other parts of *Italy*, then it behoueth me to walke
circumfpeċtly, and to forge fome caufe for mine often
comming. If therefore *Philautus*, thou canft fet but
this fether to mine arrow, thou fhalt fee me fhoote fo
neere, that thou wilt accompt me for a cunning Archer.
And verily if I had not loued thee well, I would haue
fwallowed mine own forrow in filence, knowing yat in
loue nothing is fo daungerous as to perticipate the
meanes thereoff to an other, and that two may keepe
counfaile if one be away, I am therefore enforced perforce, to challenge that curtefie at thy hands, which
earft thou didft promife with thy heart, the performaunce whereoff fhall binde me to *Philautus*, and
prooue thee faithfull to *Euphues*. Now if thy cunning
be anfwerable to thy good will, praċtife fome pleafant
conceipt vpon thy poore patient : one dram of *Ouids*
art, fome of *Tibullis* drugs, one of *Propertius* pilles,
which may caufe me either to purge my new difeafe,
or recouer my hoped defire. But I feare me wher fo
ftraunge a ficknefse is to be recured of fo vnfkilfull
a Phifition, that either thou wilt be to bold to praċtife,
or my body too weake to purge. But feeing a
defperate difeafe is to be committed to a defperate
Doċtor, I wil follow thy counfel, and become thy cure,
defiring thee to be as wife in miniftring thy Phifick,
as I haue bene willing to putte my lyfe into thy
handes.

Philautus thinking al to be gold that gliftered, and
all to be Gofpell that *Euphues* vttered, anfwered his
forged gloafe with this friendly cloafe.

In that thou haft made me priuie to thy purpofe, I

will not conceale my practise : in yat thou crauest my
aide, assure thy selfe I will be the finger next thy
thombe : insomuch as thou shalt neuer repent thee of
ye one or the other, for perswade thy selfe that thou
shalt finde *Philautus* during life ready to comfort thee
in thy missortunes, and succour thee in thy necessitie.
Concerning *Liuia*, though she be faire, yet is she not
so amiable as my *Lucilla*, whose seruaunt I haue bene
the terme of three yeres : but least comparisons should
seeme odious, chiefely where both the parties be without comparison, I will omitte that, and seing that we
had both rather be talking with them, then tatling of
them, we will immediately goe to them.   And truly
Euphues, I am not a lyttle glad, that I shall haue thee
not only a comfort in my life, but also a companion in
my loue : As thou hast ben wise in thy choice, so I
hope thou shalt be fortunate in thy chaunce.   *Liuia* is
a wench of more wit then beautie, *Lucilla* of more
beautie then wit, both of more honestie then honour,
and yet both of such honour, as in all *Naples* there is
not orteous birth to be compared with any of them
both*. how much therefore haue wee to reioyce in
our choice.   Touching our accesse, be thou secure, I
will flappe *Ferardo* in the mouth with some conceipt,
and fil his olde head so full of new fables, that thou
shalt rather be earnestly entreated to repaire to his
house, then euill entreated to leaue it.   As olde men
are very suspicious to mistrust euery thing, so are they
verye credulous to beleeue any thing : the blynde
man doth eate manye a Flye, yea but sayd *Euphues*,
take heede my *Philautus*, that thou thy self swallow
not a Gudgen, which word *Philautus* did not mark,
vntil he had almost digested it.   But said *Euphues*, let
vs go deuoutly to ye shrine of our Saints, there to offer
our deuotion, for my books teach me, that such a
wound must be healed wher it was first hurt, and for
this disease we will vse a common remedie, but yet
comfortable.   The eye that blinded thee, shall make
thee see, the Scorpion that stung thee shall heale

thee, a fharpe fore hath a fhort cure, let vs goe : to the which *Euphues* confented willyngly, fmiling to himfelfe to fee how he had brought *Philautus*, into a fooles Paradife.

Heere you may fee Gentlemen, the falfehood in fellowfhip, the fraude in friendfhippe, the paynted fheath with the leaden dagger, the faire wordes that make fooles faine : but I will not trouble you with fuperfluous addition, vnto whom I feare mee I haue bene tedious with the bare difcourfe of this rude hiftorie.

Philautus and *Euphues* repaired to the houfe of *Ferardo*, where they founde Miftres *Lucilla* and *Liuia*, accompanied with other Gentlewomen, neyther bee-ing idle, nor well imployed, but playing at cardes. But when *Lucilla* beheld *Euphues*, fhe coulde fcarcely conteine hir felfe from embracing him, had not womanly fhamefaftnes and *Philautus* his prefence, ftayed hir wifedome.

Euphues on the other fide was fallen into fuch a traunce, that he had not ye power either to fuccor himfelfe, or falute the gentlewomen. At the laft *Lucilla*, began as one that beft might be bolde, on this manner.

Gentlemen, although your long abfence gaue mee occafion to think that you diflyked your late entertein-ment, yet your comming at the laft hath cut off my former fufpition : And by fo much the more you are welcome, by how much the more you were wifhed for. But you Gentleman (taking *Euphues* by the hande) were the rather wifhed for, for that your difcourfe being left vnperfect, caufed vs all to longe (as woemen are wont for thinges that lyke them) to haue an ende thereoff. Unto whome *Philautus* replyed as followeth.

Miftres *Lucilla*, though your curtefie made vs nothing to doubt of our welcome, yet modeftye caufed vs to pinch curtefie, who fhould firft come : as for my friende, I thinke hee was neuer wyfhed for heere fo earneftly of any as of himfelfe, whether it myght be to renewe his talke, or to recant his fayings, I cannot tell. *Euphues* takynge the tale out of *Philautus* mouth, aunfwered: Miftres *Lucilla*, to recant verities were herefie, and renewe the prayfes of woemen flattery : the onely caufe I wyfhed my felfe heere, was to giue thankes for fo good entertainment the which I could no wayes deferue, and to breede a greater acquaintaunce if it might be to make amendes. *Lucilla* inflamed with his prefence, faid, nay *Euphues* you fhall not efcape fo, for if my curtefie, as you fay, were ye caufe of your comming, let it alfo be ye occafion of ye ending your former difcourfe, otherwife I fhall thinke your proofe naked, and you fhall finde my rewarde nothinge. *Euphues* nowe as willing to obey as fhee to commaunde, addreffed himfelfe to a farther conclufion, who feeing all the gentlewomen readie to giue him the hearing, proceeded as followeth.

I haue not yet forgotten yat my laft talke with thefe gentlewomen, tended to their prayfes, and therefore the ende muft tye vp the iuft proofe, otherwife I fhold fet downe *Venus* fhadow without the liuely fubftance.

As there is no one thing which can be reckened either concerning loue or loyaltie wherin women do not excell men, yet in feruencye aboue all others, they fo farre exceede, that men are lyker to meruaile at them, then to imitate them, and readier to laugh at their vertues then emulate them. For as they be harde to be wonne without tryall of greate faith, fo are they hard to be loft without great caufe of fickleneffe. It is long before the colde water feeth, yet being once hot, it is long before it be cooled, it is long before falt come to his faltneffe, but beeing once feafoned, it neuer loofeth his fauour.

I for mine owne part am brought into a Paradife by the onely imagination of woemens vertues, and were I perfwaded that all the Diuelles in hell were woemen, I woulde neuer liue deuoutlye to enherite

heauen, or yat they were al Saintes in heauen, I
woulde liue more ftricktly for feare of hell. What
coulde *Adam* haue done in his Paradife before his fall
without a woeman, or howe woulde [coulde] he haue ryfe
agayne after his fall wyth[out] a woeman? Artificers are
wont in their laft workes to excell themfelues, yea,
God when he had made all thinges, at the laft, made
man as moft perfeċt, thinking nothing could be framed
more excellent, yet after him hee created a woman,
the expreffe Image of Eternitie, the lyuely picture of
Nature, the onely fteele glaffe for man to beholde
hys infirmities, by comparinge them wyth woemens
perfeċtions.   Are   they   not   more   gentle,   more
wittie, more beautifull then men? Are not men fo be-
wytched with their qualyties that they become madde
for loue, and woemen fo wife that they [doo] deteft
luft.

I am entred into fo large a fielde, that I fhall fooner
want time then proofe, and fo cloye you wyth varietie
of prayfes [phrafes], that I feare mee I am lyke to
infeċt women with pride, whiche yet they haue not,
and men with fpyte whyche yet I woulde not. For as
the horfe if he knew his owne ftrength were no wayes
to be brideled, or the Vnicorne his owne vertue, were
neuer to bee caught, fo woemen if they knewe what
excellency were in them, I feare mee men fhould
neuer winne them to their wills, or weane them from
their minde.

Lucilla beganne to fmyle, faying, in faith *Euphues*,
I woulde haue you ftaye there, for as the Sunne when
he is at the higheft beginneth to goe downe, fo when
the prayfes of women are at the beft, if you leaue not,
they wyll beginne to fayle, but *Euphues* (beinge rapt
with the fight of his Saint) aunfwered, no no *Lucilla*.
But whileft he was yet fpeakinge, *Ferardo* entered,
whome they all duetifully welcommed home, who
rounding *Philautus* in the eare, defired hym to accom-
panye hym immediatlye without farther paufinge, pro-
tefting it fhoulde bee as well for his preferment as for

his owne profite. *Philautus* confentinge, *Ferardo*
fayde vnto hys daughter.

Lucilla, the vrgent aff[a]yres I haue in hande, wyll
fcarce fuffer mee to tarrye with you one houre, yet my
returne I hope will bee fo fhort, that my abfence fhall
not breede thy forrowe : in the meane feafon I commit
all things into thy cuftody, wifhing thee to vfe thy
accuftomable curtefie.   And feeing I muft take
Philautus with mee, I will bee fo bolde to craue you
Gentleman (his friende) to fupply his roome, defiring
you to take this haftye warning for a hartye welcome,
and fo to fpend this time of mine abfence in honeft
myrth.   And thus I leaue you.

Philautus knewe well the caufe of thys fodeyne
departure, which was to redeeme certeine landes that
were morgaged in his Fathers time, to the vfe of
Ferardo, who on that condition had before time promifed him his daughter in mariage.   But returne we
to *Euphues*.

Euphues was furprifed with fuch increadible ioye
at this ftraunge euent, that he had almoft founded, for
feeing his coriuall to be departed, and *Ferardo* to giue
him fo friendly entertaynment, doubted not in time to
get the good wil of *Lucilla* : Whom finding in place
conuenient without company, with a bold courage
and comely gefture, he began to affay hir in this
fort.

Gentlewoman, my acquaintaunce beeing fo little, I
am afrayd my credite wyll be leffe, for that they commonly are fooneft beleeued, that are beft beloued, and
they lyked beft whom we haue knowen longeft, neuer-
theleffe the noble minde fufpecteth no guyle without
caufe, neither condemneth any wight* without proofe :
hauing therefore notife of your heroycall heart, I am
the better perfwaded of my good hap.   So it is
Lucilla, that comming to *Naples* but to fetch fire, as
the by[e] word is, not to make my place of abode, I
haue founde fuch flames that I can neither quench
them with ye water of free will, neither coole them

with wifdome. For as the Hoppe, the poale beeing
neuer fo hye, groweth to the ende, or as the drye
Beech kindled at the roote, neuer leaueth vntill it
come to the toppe : or as one droppe of poyfon
difperfeth it felfe into euery vaine, fo affeĉtion hauing
caught holde of my heart, and the fparkles of loue
kindled my Lyuer, wyll fodeynelye, though fecretly,
flame vp into my heade, and fpreade it felfe into
euerye finewe. It is your beautie (pardon my abrupte
boldneffe) Lady, that hath taken euery parte of me
prifoner, and brought mee vnto this deepe diftreffe,
but feeing women when one prayfeth them for their
deferts, deeme that he flattereth them to obteine his
defire, I am heere prefent to yeeld my felfe ˙to fuch
tryal, as your courtefie in this behalfe fhal require.
Yet will you commonly obieĉt this to fuch as ferue
you, and ftarue to winne your good wil, that hot loue
is foone colde : that the Bauin though it burne bright,
is but a blaze : that fcalding water if it ftand a while
tourneth almoft to Ice : that Pepper though it be hot
in the mouth, is colde in the Maw : that the faith of
men, though it fry in their words, it freefeth in their
workes : Which things (*Lucilla*) albeit they be fufficient
to reproue the lyghtneffe of fome one, yet can they
not conuince euery one of lewdnes : neither ought the
conftancie of all, to be brought in queftion through the
fubtiltie of a few. For although the worme entreth
almoft into euery wood, yet he eateth not the *Cedar*
tree. Though the ftone *Cylindrus* at euery thunder
clap, rowle from the hil, yet the pure fleeke ftone
mounteth at the noyfe : though the ruft fret the
hardeft fteele, yet doth it not eate into the Emeraulde :
though *Polypus* chaunge his hue, yet the *Salamander*
keepeth his coulour : though *Proteus* tranfforme himfelfe into euerie fhape : yet *Pigmalion* reteineth his
olde forme : though *Aeneas* were too fickle to *Dido*,
yet *Troylus* was too faithfull to *Crefsid :* though others
feeme counterfeit in their deedes, yet *Lucilla*, perfwade
your felfe, that *Euphues* will be alwayes currant in his

dealings. But as the true golde is tryed by the touch, [and] the pure flint by the ftroake of the yron, fo the loyall heart of the faithfull louer, is knowen by the tryall of his Ladie : of the which tryall (*Lucilla*) if you fhall accompt *Euphues* worthy, affure your felfe, he will be as readie to offer himfelfe a Sacrifice for your fweete fake, as your felfe fhall be willing to employe him in your feruice. Neither doth he defire to be trufted any way, vntil he fhal be tryed euery way : neither doth he craue credite at the firft, but a good coun-tenaunce, till time his defire fhall be made manifeft by his deferts. Thus not blinded by light affection, but dazeled with your rare perfection, and boldened by your exceeding courtefie : I haue vnfolded mine entire loue, defiring you hauing fo good leafure, to giue fo friendlye an aunfwere, as I may recciue com-forte, and you commendacion.

Lucilla, although fhe were contented to heare this defired difcourfe, yet did fhee feeme to bee fomewhat difpleafed. And truely I know not whether it be peculiar to that fexe to diffemble with thofe whom they moft defire, or whether by craft they haue learned outwardly to loath that, which inwardly they moft loue : yet wifely did fhe caft this in hir head, that if fhe fhould yeelde at the firft affault, he would thinke hir a light hufwife : if fhe fhould reiect him fcornfully a very haggard : minding therefore that he fhoulde neither take holde of hir promife, neither vnkinde-neffe of hir precifeneffe, fhe fed him indifferently, with hope and difpaire, reafon and affection, life and death. Yet in the ende arguing wittily vpon certeine queftions, they fel to fuch agreement, as poore *Philautus* would not haue agreed vnto if he had ben prefent, yet alwayes keeping the [her] body vndefiled. And thus fhe replyed :

Gentleman, as you may fufpect me of idleneffe in giuing eare to your talke, fo may you conuince me of lightneffe in aunfwering fuch

toyes: certes as you haue made mine earts glow at the
rehearfall of your loue, fo haue you galled my heart
with ye remembraunce of your folly. Though you
came to *Naples* as a ftraunger, yet were you welcome to my fathers houfe as a friend: And can you
then fo much tranfgreffe the bonds of honour (I
will not fay of honeftie,) as to folicite a fute more
fharpe to me then death? I haue hetherto God bee
thanked, lyued without fufpition of lewdeneffe, and
fhall I now incurre the daunger of fenfual libertie?
What hope can you haue to obteine my loue,
feeing yet I could neuer affoord you a good looke?
Do you therefore thinke me eafely entifed to the
bent of your bow, bicaufe I was eafely entreated to
liften to your late difcourfe? Or feeing mee (as finely
you glofe) to excell all other in beautie, did you deeme
that I would exceede all other in beaftlines? But yet
I am not angry *Euphues*, but in agonye: For who is
fhee that will frette or fume with one that loueth hir,
if this loue to delude me, be not diffembled. It is that
which caufeth me moft to feare, not that my beautie is
vnknown to my felf, but that commonly we poore
wenches are deluded through light beliefe, and ye men
are naturally enclined craftely to lead your lyfe.
When the Foxe preacheth, the Geefe perifh. The
Crocodile fhrowdeth greateft treafon vnder moft
pitiful teares: in a kiffing mouth there lyeth a galling
minde. You haue made fo large profer of your feruice,
and fo faire promifes of fidelytie, that were I not ouer
charie of mine honeftie, you woulde inueigle me to
fhake handes with chaftitie. But certes I will either
lead a virgins life in earth (though I lead Apes in hel)
or els follow thee rather then thy gifts: yet am I
neither fo precife to refufe thy profer, neither fo
peeuifh to difdain thy good wil: fo excellent alwayes
are the gifts which are made acceptable by the vertue
of ye giuer. I did at the firft entraunce difcerne thy
loue, but yet diffemble it. Thy wanton glaunces, thy
fcalding fighes, thy louing fignes caufed me to blufh

for fhame and to looke wanne for feare, leaft they
fhould be perceiued of any.   Thefe fubtill fhiftes, thefe
painted practifes (if I wer to be wonne) would foone
weane me from the teate of *Vefta* to the toyes of
Venus.   Befides this thy comly grace, thy rare qualyties, thy exquifite perfection, were able to moue a
minde halfe mortified to tranfgreffe the bonds of
maidenly modeftie.   But god fhield *Lucilla*, that thou
fhouldeft be fo careleffe of thine honour, as to commit
the ftate thereoff to a ftraunger.   Learne thou by me
Euphues to difpife things that be amiable, to forgoe
delightfull practifes, beleeue mee it is pietie to ab
fteine from pleafure.

Thou art not the firft that hath folicited this fute,
but the firft that goeth about to feduce me, neither
difcerneft thou more then other, but dareft more then
any, neither haft thou more art to difcouer thy me[a]ning,
but more heart to open thy minde.   But thou preferreft
me before thy lands, thy liuings, thy life : thou offereft
thy felfe a facrifice for my fecuritie, thou profereft me
the whole and only fouereignetie of thy feruice :
Truely I were very cruel and hard hearted, if I fhould
not loue thee : hard hearted albeit I am not, but
truly loue thee I cannot, whom I doubt to be my
louer.

Moreouer I haue not ben vfed to the court of
Cupide, wherin ther be more flights then ther be Hares
in *Athon*, then Bees in *Hybla*, then ftarres in heauen.
Befides this, the common people here in *Naples* are
not only both very fufpitious of other mens matters
and manners, but alfo very iealous ouer other mens
children and maidens, either therefore diffemble thy
fancie or defift from thy folly.

But why fhouldeft thou defift from the one, feeing
thou canft cunningly diffemble the other.   My father
is now gone to *Venice*, and as I am vncerteine of his returne, fo am I not priuy to the caufe of his trauayle : But
yet is he [he is] fo from hence, that he feeth me in his
abfence.   Knoweft thou not *Euphues*, that kinges haue

long armes, and rulers large reaches? neither let this
comfort thee, that at his departure he deputed thee in
Philautus place. Although my face caufe him to
miftruft my loyalty, yet my faith enforceth him to giue
me this liberty : though he be fufpitious of my faire
hiew, yet is he fecure of my firme honefty. But alas
Euphues, what truth can there be* found in a trauailer?
what ftay [truft] in a ft[r]aunger? whofe words and
bodyes both watch but for a winde, whofe feete are
euer fleeting, whofe faith plyghted on the fhoare, is
turned to periurye when they hoyfe [hoyft] fayle. Who
more traiterous to *Phillis* then *Demophoon.?* yet hee a
trauayler. Who more periured to *Dido* then *Aeneas?*
and he a ftraunger : both thefe Queenes, both they
Caytiffes. Who more falfe to *Ariadne* then *Thefeus?*
yet he a fayler. Who more fickle to *Medea* then
Iafon? yet he a ftarter : both thefe daughters to great
Princes, both they vnfaithfull of their promifes. Is it
then likely yat *Euphues* wil be faythfull to *Lucilla*, being
in *Naples* but a foiourner ? I haue not yet forgotten
the inuectiue (I can no [cannot] otherwyfe terme it)
which thou madeft againft beauty, fayinge, it was a
deceitful bayte with a deadly hooke, and a fweet
poyfon in a paynted pot. Canft thou then be fo
vnwife to fwallowe the bayte which will breede thy
bane ? To fwill the drinke that will expyre thy date ?
To defire the wight that will worke thy death ? But
it may be that with the Scorpion thou canft feede on
the earth, or with the Quayle and Roebucke, be fat
with poyfon : or with beautye liue in all brauerye. I
feare me thou haft the ftone *Contineus* about thee,
which is named of the contrarye, that though thou
pretende fayth in thy words, thou deuifeft fraude in
thy heart : yat though thou feeme to prefer loue, thou
art inflamed with luft. And what for that ? Though
thou haue eaten the feedes of Reckat [Rackat], which
breede incontinencie, yet haue I chewed the leafe
Creffe which mainteineth modeftie.

Though thou beare in thy bofom the hearb *Araxa,*

moſt noiſome to virginitie, yet haue I the ſtone that
groweth in the mount *Tmolus*, the vpholder of chaſtitie.
You may Gentleman accompt me for a colde Prophet,
thus haſtely to deuine of your diſpoſition : pardon me
Euphues, if in loue I caſt beyond the Moone, which
bringeth vs women to endles moane.   Although I my
ſelf were neuer burnt wherby I ſhould dread the fire,
yet the ſchorching of others in the flames of fancy,
warneth me to beware : Though I as yet neuer tryed
any faithles whereby I ſhould be feareful, yet haue I
read of many that haue ben periured, which cauſeth
me to be careful : though I am able to conuince none
by proofe, yet am I enforced to ſuſpect one vppon
probabylities.   Alas we ſilly foules which haue neither
wit to decypher the wiles of men, nor wiſdome to
diſſemble our affection, neither craft to traine in young
louers, neyther courage to withſtande their encounters,
neither diſcretion to diſcerne their dubling, neither
hard harts to reiect their complaints : we I ſay, are
ſoone enticed, beeing by nature ſimple, and eaſily
entangled, beeing apte to receiue the impreſſion of
loue.   But alas, it is both common and lamentable, to
behold ſimplicity intrapped by ſubtiltie, and thoſe that
haue moſt might, to be infected with moſt mallice.
The Spider weaueth a fine web to hang the Fly, the
Wolfe weareth a faire face to deuour the Lambe, the
Mirlin ſtriketh at the Partridge, the Eagle often
ſnappeth at the Fly, men are alwayes laying baites for
women, which are the weaker veſſels : but as yet I
could neuer heare man by ſuch ſnares to entrappe
man : For true it is that men themſelues haue by vſe
obſerued, yat it muſt be a harde Winter when one
Wolfe eateth another.   I haue read, that the Bull
being tyed to the Figge tree, looſeth his ſtrength, yat
the whole heard of Deare ſtand at the gaze, if they
ſmell a ſweete apple : that the *Dolphin* by the ſound
of Muſicke is brought to ye ſhoare.   And then no
meruaile it is yat if the fierce Bull be tamed with the
Fig tree, if that women being as weake as ſheepe, be

ouercome with a Figge : if the wilde Deare be caught with an apple, that the tame Damzell is wonne with a bloffome : if the fleete *Dolphin* be allured with harmony, that women bee entangled with the melody of mens fpeach, faire promifes and folemne proteflations. But folly it were for me to marke their mifchiefes, fith I am neither able, neither they willing to amende their manners : it becommeth me rather to fhew what our fexe fhould doe, then to open what yours doth.

And feeing I cannot by reafon reftraine your importunate fuite, I will by rygour done on my felfe, caufe you to refraine the meanes. I would to God *Ferardo* were in this point lyke to *Lyfander*, which woulde not fuffer his daughters to weare gorgeous apparell, faying, it would rather make them common then comely. I would it were in *Naples* a lawe, which was a cuftome in *Aegypt*, that women fhould alwayes goe bare foote to the intent they might keepe themfelues alwayes at home, that they fhold be euer like to the Snaile, which hath euer his houfe on his head. I meane fo to mortifie my felfe, that in fteede of filkes, I wil weare fackcloth : for Owches and Bracelletes, Leere and Caddys : for the Lute, vfe the Diftaffe : for the Penne, the Needle : for louers Sonettes, Dauids Pfalmes. But yet I am not fo fenceles altogether to reiect your feruice : which if I wer certeinly affured to proceede of a fimple mind, it fhold not receiue fo fimple a reward. And what greater tryall can I haue of thy fimplicitie and truth, then thine owne requeft which defireth a triall. I, but in the coldeft flint there is hot fire, the Bee that hath hunny in hir mouth, hath a fting in hir tayle : the tree that beareth the fweeteft fruite, hath a fower fap : yea, the wordes of men though they feeme fmooth as oyle : yet their heartes are as crooked as the ftalke of Iuie. I woulde not *Euphues* that thou fhouldeft condemne me of rigour, in that I feeke to affwage thy folly by reafon : but take this by the way, that although as yet I am difpofed to lyke of none ? yet whenfoeuer I fhall loue any, I wil not forget thee . in the meane feafon accompt me thy friend, for thy foe I will neuer be.

Euphues was brought into a great quandary, and as it were a colde fhiuering, to heare this newe kinde of kindneffe : fuch fweete meate, fuch fowre fauce : fuch fayre wordes, fuch fainte promifes : fuch hot loue, fuch colde defire : fuch certeine hope, fuch fodeine chaunge : and ftoode lyke one that had looked on *Medufaes* heade, and fo had beene tourned into a ftone.

Lucilla feeing him in this pitiful plight, and fearing he would take ftand if the lure were not caft out, toke him by the hand, and wringing him foftly, with a fmiling countenaunce began thus to comfort him.

Me thinks *Euphues* chaunging fo your colour, vpon the fodeine, you wil foone chaunge your coppie : is your minde on your meate ? a penny for your thought.

Miftres (quoth he) if you would by al my thoughts at that price ? I fhould neuer be wearye of thinking, but feeing it is too [fo] deere, reade it and take it for nothing.

It feemes to me (faid fhe) that you are in fome brown ftudy, what coulours you might beft weare for your Lady.

In deede *Lucilla* you leuel fhrewdly at my thought, by the ayme of your owne imagination, for you haue giuen vnto me a true loue[r]s knot wrought of chaunge-able Silke, and you deeme that I am deuifing how I might haue my coulours chaungeable alfo, that they might agree : But lette this with fuch toyes and deuifes paffe, if it pleafe you to commaunde me anye feruice I am heere ready to attend your [p]leafure. No feruice *Euphues*, but that you keepe filence, vntil I haue vttered my minde : and fecrecie when I haue vnfolded my meaning.

If I fhould offende in the one I were too bolde, if in the other too beaftly.

Well then *Euphues* (fayd fhee) fo it is, that for the hope that I conceiue of thy loyaltie, and the happie fucceffe that is like to enfue of this our loue, I am

content to yeelde thee the place in my heart which
thou defireſt and deferueſt aboue all other, which
confent in me if it may any wayes breede thy contentation, fure I am that it will euery way worke my
comfort. But as either thou tendereſt mine honour or
thine owne fafetie, vfe fuch fecrecie in this matter, that
my father haue no inckling heereoff, before I haue
framed his minde fit for our purpofe. And though
women haue fmall force to ouercome men by reafon,
yet haue they good fortune to vndermine them by
pollicie. The foft droppes of raine perce the hard
Marble, many ſtrokes ouerthrow the talleſt Oke, a filly
woman in time may make fuch a breach into a mans
heart, as hir teares may enter without refiſtaunce : then
doubt not, but I wil fo vndermine mine olde father, as
quickly I wil enioy my new friend. Tuſh *Philautus*
was liked for faſhion fake, but neuer loued for fancie
fake : and this I vowe by the faith of a Virgin, and by
the loue I beare thee, (for greater bands to confirme
my vow I haue not) that my father ſhall fooner martir
mee in the fire then marye mee to *Philautus*. No no,
Euphues, thou onely haſt wonne me by loue, and ſhalt
onely weare me by law : I force not *Philautus* his fury,
fo I may haue *Euphues* his friendſhip: neither wil I
prefer his poſſeſſions before thy perfon, neither eſteme
better of his lands, then of thy loue. *Ferardo* ſhal
fooner diſherite me of my patrimony, then diſhonour me
in breaking my promife ? It is not his great mannors,
but thy good manners, that ſhal make my mariage. In
token of which my fincere affeꞔtion, I giue thee my
hande in pawne, and my heart for euer to be thy
Lucilla. Vnto whom *Euphues* aunfwered in this
manner.
    If my tongue were able to vtter the ioyes that my
heart hath conceiued, I feare me though I be well
beloued, yet I ſhould hardly be beleeued. Ah my
Lucilla, how much am I bound to thee, which pre-
ferreſt mine vnworthineffe, before thy Fathers wrath :
my happineffe, before thine owne miffortune : my loue,

before thine owne life ? How might I excell thee in
curtefie, whom no mortall creature can exceed in conftancie ? I finde it now for a fetled truth, which earft
I accompted for a vaine talke, that the purple dye will
neuer ftaine, that the pure Cyuet will neuer loofe his
fauour, that the greene Laurell will neuer chaunge his
coulour, that beautie can neuer be blotted with difcourtefie. As touching fecrecie in this behalf, affure
thy felfe, that I will not fo much as tell it to my felfe.
Commaund *Euphues* to runne, to ride, to vndertake
any exployt be it neuer fo daungerous, to hazard himfelfe in any enterprife, be it neuer fo defperate. As
they wer thus pleafauntly conferring the one with the
other, *Liuia* (whom *Euphues* made his ftale) entered
into the Parlour, vnto whome *Lucilla* fpake in thefe
termes.

Doft thou not laugh *Liuia*, to fee my ghoftly father
keepe me heere fo long at fhrifte ? Truely (aunfwered
Liuia) me thinkes yat you fmile at fome pleafaunt
fhift, either he is flow in enquiring of your faults, or you
flacke in aunfwering of his queftions : and thus being
fupper time they all fate downe, *Lucilla* well pleafed,
no man better content then *Euphues*, who after his
repaft hauing no opportunitie to confer with his louer,
had fmall luft to continue with the gentlewomen any
longer, feeing therefore he could frame no meanes to
worke his delyght, he coyned an excufe to haften his
departure, promifing the next morning to trouble them
againe as a gueft more bold then welcome, although
in deede he thought himfelfe to be the better welcome,
in faying that he would come.

But as *Ferardo* went in poft, fo hee retourned in haft
hauing concluded with *Philautus*, that the mariage
fhould immediatly be confummated, which wrought
fuch a content in *Philautus*, that he was almoft in an
extafie through the extremitie of his paffions : fuch is
the fulneffe and force of pleafure, that ther is nothing
fo daungerous as the fruition, yet knowing that delayes
bring daungers, although hee nothing doubted of

Lucilia whome hee loued, yet feared he the fickleneffe of olde men, which is alwayes to be miftrufted.

Hee vrged therefore *Ferardo* to breake with his Daughter, who beeing willyng to haue the matche made, was content incontinentlye to procure the meanes : finding therefore his daughter at leafure, and hauing knowledge of hir former loue, fpake to hir as followeth.

Deere daughter as thou haft long time liued a maiden, fo now thou muft learne to be a Mother, and as I haue bene carefull to bring thee vp a Virgin, fo am I now defirous to make thee a Wife. Neither ought I in this matter to vfe any perfwafions, for that maidens commonly now a dayes are no fooner borne, but they beginne to bride it : neither to offer any great portions, for that thou knoweft thou fhalt enherite al my poffeffions. Mine onely care hath bene hetherto, to match thee with fuch an one, as fhoulde be of good wealth, able to mainteine thee : of great worfhip, able to compare with thee in birth : of honeft conditions, to deferue thy loue : and an *Italian* borne to enioy my landes. At the laft I haue found one aunfwerable to my defire, a Gentleman of great reuenewes, of a noble progenie, of honeft behauiour, of comly perfonage, borne and brought vp in *Naples*, *Philautus* (thy friend as I geffe) thy husband *Lucilla* if thou lyke it, neither canft thou diflike him, who wanteth nothing that fhould caufe thy liking, neither hath any thing that fhould breede thy loathing.

And furely I reioyce the more that thou fhalt bee linked to him in mariage, whom thou haft loued, as I heare beeing a maiden, neither can there any iarres kindle betweene them, wher the mindes be fo vnited, neither any iealoufie arife, where loue hath fo long bene fetled. Therefore *Lucilla*, to the ende the defire of either of you may now be accomplyfhed to the delyght of you both, I am heere come to finifhe the contract by giuing handes, which you haue already begunne betweene your felues by ioyning of hearts, that as

GOD doth witneſſe the one in your conſciences, ſo the world may teſtifie the other, by your conuerſations, and therefore *Lucilla*, make ſuch aunſwere to my requeſt, as may lyke me and ſatiſfie thy friende.

Lucilla abaſhed with this ſodaine ſpeach of hir father, yet boldened by the loue of hir friend, with a comly baſhfulneſſe, aunſwered him in this manner.

Reuerend ſir, the ſweeteneſſe that I haue found in the vndefyled eſtate of virginitie, cauſeth me to loath the ſower ſauce which is myxed with matrimony, and the quiet life which I haue tryed being a mayden, maketh me to ſhun the cares that are alwayes incident to a mother, neither am I ſo wedded to the world that I ſhould be moued with great poſſeſſions, neither ſo bewitched with wantonneſſe, that I ſhoulde be entyſed with any mans proportion, neither if I were ſo diſpoſed would I be ſo proude, to deſire one of noble progenie, or ſo preciſe to chooſe one onely in mine owne countrey, for that commonly theſe things happen alwayes to the contrary. Doe wee not ſee the noble to match with the baſe, the rich with the poore, the *Italian* oftentimes with the *Portingale*? As loue knoweth no lawes, ſo it regardeth no conditions : as the louer maketh no pawſe where he lyketh, ſo he maketh no conſcience of theſe idle ceremonies. In that *Philautus* is the man that threatneth ſuch kindeneſſe at my handes, and ſuche curteſie at yours, that he ſhoulde accompt me his wife before he wo[o]e mee, certeinly he is lyke for mee to make his rec[k]oning twice, bicauſe he reckoneth without his Hoſteſſe. And in this *Philautus* would either ſhew himſelfe of great wiſedome to perſwade, or me of great lyghtnes to be allured : although the Loadſtone draw yron, yet it cannot moue gold : though the Iette gather vp the lyght* ſtraw, yet can it not take vp the pureſteele. Although *Philautus* thinke himſelfe of vertue ſufficient to winne his louer, yet ſhall he not obteine *Lucilla*. I cannot but ſmyle to heare yat a maryage ſhould be ſolemnized, where neuer was any mention of aſſuring, and that the wooing ſhould be a daye after the wedding

Certes if when I looked merily on *Philautus* he deemed it in ye way of mariage, or if feeing me difpofed to ieft, he tooke me in good earneft: then fure hee might gather fome prefumption of my loue, but no promife. But me thinkes it is good reafon, that I fhoulde bee at mine owne brideall, and not giuen in the Church, before I knowe the Bridegroome. Therefore deere Father in mine opinion as ther can be no bargaine where both be not agreed, neither any Indentures fealed where the one will not confent : fo canne there be no contract where both be not content : no banes afked lawefully, where one of the parties forbiddeth them : no mariage made where no match was ment. But I wil hereafter frame my felf to be coy, feirg I am claimed for a wife bicaufe I haue bene curteous : and giue my felf to melancholy, feing I am ac-compted wonne in that I haue bene merry. And if euery Gentleman bee made of the mettall that *Philautus* is, then I feare I fhall be challenged of as many as I haue vfed to company with, and be a common wife to all thofe that haue commonly reforted hether.

My duetie therefore euer referued, I here on my knees forfweare *Philautus* for my husband, although I accept him for my friend, and feeing I fhal hardly be induced euer to match with any, I befech you if by your fatherly loue I fhall be compelled, that I may match with fuch a one as both I may loue and you may lyke.

Ferardo being a graue and wife Gentleman, although he were throughly angry, yet he diffembled his fury, to the ende he might by craft difcouer hir fancy, and whifpering *Philautus* in the eare (who ftoode as though he had a flea in his eare) defired him to kepe filence, vntil he had vndermined hir by fubtiltie, which *Philautus* hauing graunted, *Ferardo* began to fift his daughter with this deuice. *Lucilla*, thy coulour fheweth thee to bee in a great choler, and thy hotte wordes be-wray thy heauy wrath, but be patient, feing al my talke

was onely to trye thee: I am neither fo vnnaturall to
wreaſt thee againſt thine owne wil, neither fo malytious
to wedde thee to any againſt thine own lyking : for well
I know what iarres, what ieloufie, what ſtrife, what
ſtormes enfue, where the match is made rather by the
compulfion of the parents, then by the confent of the
parties : neither doe I like thee the leffe in that thou
lykeſt *Philautus* fo little, neither can *Philautus* loue thee
ye worfe in that thou loueſt thy felfe fo well, wiſhing
rather to ſtande to thy chaunce, then to the choyce of
any other. But this grieueth me moſt, that thou art
almoſt vowed to the vayne order of the veſtal virgins,
difpifing, or at the leaſt not defiring the facred bandes
of *Iunò*, hir bedde. If thy mother had bene of that
minde when ſhe was a mayden, thou haddeſt not nowe
bene borne, to be of this minde to be a virgin. Way
with thy felfe what ſlender profit they bring to the
common wealth, what ſlight pleafure to themfelues,
what great griefe to their parents, which ioy moſt in
their offpring, and defire moſt to enioy the noble and
bleffed name of a graundfather. Thou knoweſt that
the talleſt Aſh is cut down for fuell, bicaufe it beareth
no good fruite : that the Cow that giues no milke, is
brought to the ſlaughter : that the Drone that gathereth
no Honny is contemned : that the woman that maketh
hir felfe barren by not marrying, is accompted amonge
the Grecian Ladyes worfe then a carryon, as *Homer*
reporteth.

Therefore *Lucilla*, if thou haue any care to be a comfort to my hoary haires, or a commoditie to thy common
weale, frame thy felf to that honourable eſtate of Matrimony, which was fanctified in Paradife, allowed of [of]
the Patriarches, hallowed of the olde Prophets, and commended of al perfons. If thou lyke any, be not aſhamed
to tell it me, which onely am to exhort thee, yea and as
much as in me lyeth to commaunde thee, to loue one :
If he be bafe, thy bloud will make him noble : If
beggerly, thy goods ſhall make him wealthy : If a
ſtraunger thy freedome may [ſhall] enfraunchife him : If

he be young, he is the more fitter to be thy pheere: if he be olde, the lyker to thine aged father. For I had rather thou fhouldeft leade a lyfe to thine owne lyking in earthe, then to thy great torments, leade Apes in Hell. Be bolde therefore to make me partaker [partener] of thy defyre, which will be partaker of thy difeafe : yea, and a furtherer of thy delightes, as farre as either my friendes, or my landes, or my life will ftretch.

Lucilla perceiuing the drift of the olde Foxe hir father, waied with hir felf what was the beft to be done, at the laft not waying hir fathers ill will, but encouraged by loue, fhaped him an aunfwere which pleafed *Ferardo* but a lyttle, and pinched *Philautus* on the perfons fyde, on this manner.

Deere Father *Ferardo*, although I fee the bayte you laye to catch mee, yet I am content to fwallowe the hooke, neither are you more defirous to take mee napping, then I willing to confeffe my meaning. So it is that loue hath as well inuegled me as others, which make it as ftraunge as I. Neither doe I loue him fo meanely that I fhould be afhamed of his name, neither is his perfonage fo meane that I fhoulde loue him fhamefully : It is *Euphues* that lately a[r]riued here at *Naples*, that hath battered the bulwark of my breft, and fhal fhortly enter as conquerour into my bofome. What his wealth is, I neither knowe it nor way it: what his wyt is, all *Naples* doth know it and wonder at it : neyther haue I bene curious to enquire of his progenitors, for that I know fo noble a minde could take no original but from a noble man, for as no Bird can looke againft the Sunne but thofe that be bredde of the Eagle, neither any Hawke foare fo high as the broode of the Hobby, fo no wight can haue fuch excellent qualyties except he defcende of a noble race, neither be of fo high capacitie, vnleffe he iffue of a high progeny. And I hope *Philautus* will not be my foe, feeing I haue chofen his deere friend, neither you Father be difpleafed, in that *Fhilautus* is difplaced. You neede not mufe that I fhould fo fodeinely be entangled, loue giues no reafon of choyce, neither will it fuffer any repulfe. *Mirrha* was enamoured of hir naturall Father, *Biblis* of hir Brother, *Phædra* of hir fonne in lawe: If Nature canne no waye refift the furye of affeciton: how fhoulde it be ftayed by wife-dome?

Ferardo interrupting hir in the middle of hir difcourfe, although he were moued with inward grudge, yet he wifely repreffed his anger, knowing that fharp words would but fharpen hir froward will, and thus aunfwered hir briefely.

Lucilla, as I am not prefently to graunt my good wil, fo meane I not to reprehend thy choyce, yet wife-dome willeth me to pawfe, vntill I haue called what may happen to my remembraunce, and warneth thee to be circumfpect, leaft thy rafh conceipt bring a fharpe repentaunce. As for you *Philautus*, I would not haue you difpayre, feeing a woman doth oftentimes chaunge hir defyre. Vnto whome *Philautus* in few words made aunfwere.

Certeinely *Ferardo* I take the leffe griefe, in that I fee hir fo greedy after *Euphues*, and by fo much the more I am content to leaue my fute, by how much the more fhe feemeth to difdaine my feruice: but as for hope, bicaufe I would not by any meanes tafte one dramme thereoff, I wil abiure all places of hir abode, and loath hir company, whofe countenaunce I haue fo much loued: as for *Euphues*, and there ftaying his fpeach, he flang out of the dores and repairing to his lodging, vttered thefe words.

Ah moft diffembling wretch *Euphues*, O counterfayte companion, couldeft thou vnder the fhewe of a ftedfaft friende cloake the mallice of a mortall foe? vnder the coulour of fimplicitie, fhrowd the Image of deceipt? Is thy *Liuia*, tourned to my *Lucilla*? thy loue, to my louer: thy deuotion to my Saint? Is this the curtefie of *Athens*, the cauilling of fchollers, the crafte of *Grecians*? Couldeft thou not remember *Philautus*, that *Greece* is neuer withôut fome wily *Vliffes*, neuer

void of fome *Synon,* neuer to feeke of fome deceitful
fhifter? Is it not commonly faid of *Grecians,* that craft
commeth to them by kinde, that they learne to deceiue
in their cradell? Why then did his pretended curtefie
bewitch thee with fuch credulytie? fhall my good wil
be the caufe of his il wil? bicaufe I was content to be
his friend, thought he me meete to be made his foole?
I fee now that as the fifh *Scolopidus* in the floud *Araris,*
at the waxing of the Moone is as white as the driuen
fnow, and at the wayning as black as the burnt coale :
fo *Euphues,* which at the firft increafing of our familiaritie, was very zealous, is now at ye laft caft become moft
faithleffe. But why rather exclaime I not againft
Lucilla whofe wanton lookes caufed *Euphues* to violate
his plighted faith? Ah wretched wench, canft thou
be fo lyght of loue, as to chaunge with euery winde? fo
vnconftant as to prefer a new louer before thine [an] olde
friend? Ah well I wot that a new broome fweepeth
cleane, and a new garment maketh thee leaue off the
olde though it be fitter, and new Wine caufeth thee to
forfake the olde, though it be better : much lyke to the
men in the Iland *Scyrum,* which pull vp the olde tree
when they fee the young begin to fpring, and not vnlike
vnto the widow of *Lesbos,* which chaunged al hir old
golde for new Glaffe. Haue I ferued thee three yeares
faithfully, and am I ferued fo vnkindely? fhall the
fruite of my defire be tourned to difdaine? But vnleffe
Euphues had inueigled thee, thou hadft yet bene conftant : yea, but if *Euphues* had not feene thee willyng
to be wonne, he woulde neuer haue wo[o]ed thee : But
had not *Euphues* entifed thee with faire wordes, thou
wouldft neuer haue loued him : but hadft thou not
giuen him faire lookes, he would neuer haue liked thee :
I, but *Euphues* gaue the onfet : I, but *Lucilla* gaue the
occafion : I, but *Euphues* firft brake his minde : I, but
Lucilla firft bewrayed hir meaning. Tufh why goe I
about to excufe any of them, feeing I haue iuft caufe
to accufe them both. Neither ought I to difpute
which of them hath proferred me the greateft villany,

fith that either of them hath committed periury. Yet although they haue found me dull in perceiuing their falfehood, they fhall not finde me flacke in reuenging their folly. As for *Lucilla*, feing I meane altogether to forget hir, I meane alfo to forgiue hir, leaft in feeking meanes to be reuenged, mine olde defire be renewed.

Philautus hauing thus difcourfed with himfelfe, began to write to *Euphues* as followeth.

¶ Although hetherto *Euphues*, I haue fhrined thee in my heart for a truftie friende, I will fhunne thee heereafter as a trothleffe foe, and although I cannot fee in thee leffe wit then I was wont, yet doe I finde leffe honeftie. I perceiue at the laft (although being deceiued it be too late) that Mufke though it be fweet in ye fmel, is fowre in the fmacke : that the leafe of the *Cedar* tree, though it be faire to be feene, yet the firup depriueth fight, that friendfhip though it be plighted by fhaking the hand, yet it is fhaken off by fraud of the heart. But thou haft not much to boaft off, for as thou haft won a fickle Lady, fo haft thou loft a faithful friend. How canft thou be fecure of hir conftancie, when thou haft had fuch tryall of hir lyghtneffe ?

How canft thou affure thy felfe that fhe will bee faithfull to thee, which hath bene faithleffe to me ? Ah *Euphues*, let not my credulitie be an occafion heereafter for thee to practife the lyke crueltie. Remember this that yet there hath neuer bene any faythleffe to his friende, that hath not alfo bene fruiteleffe to his God. But I way the treacherie the leffe, in that it commeth from a *Grecian*, in whome is no trouth. Though I be to weake to wraftle for a reuenge, yet God who per-mitteth no guile to be guiltleffe, will fhortly requite this iniury : though *Philautus* haue no pollicie to vnder-mine thee, yet thine owne practifes will be fufficient to ouerthrow thee.

Couldeft thou *Euphues*, for the loue of a fruiteleffe plefure, violate the league of faithfull friendfhip ? Didft thou way more the entifing lookes of a lewde wench, then the entire loue of a loyall friend? If thou diddeſt determine with thy felfe at the firſt to be falfe, why diddeſt thou fweare to be true? If to be true, why art thou falfe? If thou waſt minded both falfely and forgedly to deceiue me, why didſt thou flatter and diffemble with me at the firſt? If to loue me, why doſt thou flinch at the laſt? If the facred bands of amitie did delight thee, why diddeſt thou breake them? If diflike thee, why diddeſt thou praife them? Doſt thou not know yat a perfect friend fhould be lyke the Glazeworme, which fhineth moſt bright in the darke? or lyke the pure Frankencenfe which fmelleth moſt fweet when it is in the fire? or at the leaſt not vnlike to the damafke Rofe, which is fweeter in the Still then on the ftalke? But thou *Euphues*, doſt rather refemble the Swallow which in the Summer creepeth vnder the eues of euery houfe, and in the Winter leaueth nothing but durt behinde hir: or the humble Bee, which hauing fucked hunny out of the fayre flower, doth leaue it and loath it: or the Spider which in the fineſt web doth hang the fayreſt Fly. Doſt thou thinke *Euphues* that thy craft in betraying mee, fhall any whit coole my courage in reuenging thy villany? or that a Gentleman of *Naples* will put vp fuch an iniury at the hands of a fcholler? And if I do, it is not for want of ſtrength to mainteine my iuſt quarrell, but of will which thinketh fcorne to gette fo vaine a conqueſt. I knowe that *Menelaus* for his ten yeares warre, endured ten yeares wo[e], that after al his ſtrife hee wan but a ſtrumpet, that for all his trauayle he reduced (I cannot fay reclaymed) but a ſtraggeler: which was as much in my iudgement, as to ſtriue for a broken glaffe, which is good for nothing. I wifh thee rather *Menelaus* care, then my felfe his conqueſt, that thou being deluded by *Lucilla*, maiſt rather know what it is to be deceiued, then I hauinge conquered thee, fhould proue what it were to bring backe a diffembler. Seeing therefore there can no greater reuenge lyght vppon thee, then that as thou

haſt reaped where an other hath ſowen, ſo an other
may threſh yat which thou haſt reaped.   I will pray
that thou maiſt be meſured vnto with the lyke meaſure
that thou haſt meaten vnto others : that [is,] as thou haſt
thought it no conſcience to betray mee, ſo others may
deeme it no diſhoneſtie to deceiue thee: that as
Lucilla made it a light matter to forſweare hir olde
friend *Philautus*, ſo ſhe may make it a mocke to forſake hir new pheere *Euphues*.   Which if it come to
paſſe, as it is lyke by my compaſſe, then ſhalt thou ſee
the troubles and feele the torments which thou haſt
already throwne into the heartes and eyes of others.

  Thus hoping ſhortly to ſee thee as hopeleſſe, as my
ſelfe is haples, I wiſh my wiſh, were as affeꝛtually ended,
as it is hartely looked for.   And ſo I leaue thee.

Thine once
Philautus.

  Philautus diſpatching a meſſenger with this letter
ſpeadely to *Euphues*, went into the fields to walk ther,
either to digeſt his choler, or chew vpon his melancholy.
But *Euphues* hauing reade the contents, was well content, ſetting his talke at naught, and anſwering his
taunts in theſe gibing termes.

I Remember *Philautus* how valyantly *Aiax* boaſted
in the feates of armes, yet *Vlyſſes* bare away the
armour : and it may be that though thou crake of
thine owne courage, thou maiſt eaſily loſe the conqueſt.
Doſt thou thinke *Euphues* ſuch a daſtarde, that hee is
not able to withſtande thy courage, or ſuch a dullarde
that he cannot diſcrye thy craft.   Alas good ſoule.   It
fayreth with thee as with the Hen, which when the
Puttocke hath caught hir Chekin beginneth to cackle,
and thou hauing loſt thy louer beginneſt to prattle.
Tuſh *Philautus*, I am in this point of *Euripides* his
minde, who thinkes it lawfull for the deſire of a kingdome to tranſgreſſe the bonds of honeſtie, and for the
loue of a Lady to violate and breake the bonds of

amitie. The friendfhip betweene man and man as it is
common fo is it of courfe: betweene man and woman, as
it is feldome fo is it fincere, the one proceedeth of the
fimilitude of manners, the other of ye fincerity of the
heart: if thou haddeft learned the firft point [part] of
hauking, thou wouldft haue learned to haue held faft,
or the firft noat of Defcant, thou wouldeft, haue kept
thy *Sol. Fa.* to thy felfe.

But thou canft blame me no more of folly in leauing
thee to loue *Lucilla*, then thou maift reproue him of
foolifhneffe that hauing a Sparrow in his hande letteth
hir goe to catch the Pheafant, or him of vnfkilfulneffe
that feing the Heron, leaueth to leuell his fhot at the
Stockdoue, or that woman of coyneffe, that hauing a
dead Rofe in hir bofome, throweth it away to gather
the frefh violet. Loue knoweth no lawes : Did not
Iupiter tranfforme himfelfe into the fhape of *Amphitrio*
to embrace *Alcmæna* ? Into the forme of a Swan to
enioy *Læda:* Into a Bull to beguile *Iò :* Into a fhowre
of golde to winne *Danae:* Did not *Neptune* chaunge
himfelfe into a Heyfer, a Ramme, a Floud, a *Dolphin*,
onely for the loue of thofe he lufted after? Did not
Apollo conuert himfelfe into a Shephearde, into a Birde,
into a Lyon : for the defire he had to heale his difeafe ?
If the Gods thought no fcorne to become beaftes, to
obteine their beft beloued, fhall *Euphues* be fo nice in
chaunging his coppie to gayne his Ladie ? No, no :
he that cannot diffemble in loue, is not worthy to liue.
I am of this minde, that both might and mallice,
deceyte and trecherye, all periurye, any impietie may
lawfully be committed in loue, which is lawleffe. In
that thou argueft *Lucilla* of lightneffe thy will hangs in
the light of thy witte : Doeft thou not know that the
weak ftomacke if it be cloyed with one dyet doth foone
furfet ? That the clownes Garlike cannot eafe the
courtiers difeafe fo wel as the pure Triacle ? that farre
fet and deere bought is good for Ladyes ? That
Euphues being a more dainty morfell then *Philautus*
ought better to be accepted ? Tufh *Philautus* fet thy

heart at reſt, for thy happe willeth thee to giue ouer all
hope both of my friendſhip, and hir loue : as for
reuenge thou art not ſo able to lende a blow as I to
ward it : neither more venterous to challenge the com-
batte, then I valiant to aunſwere the quarrell.   As
Lucilla was caught by fraude, ſo ſhal ſhe be kept by
force : and as thou waſt too ſimple to eſpie my crafte,
ſo I thinke thou wilt be too weake to withſtande my
courage : but* if thy reuenge ſtande onely vpon thy
wiſh, thou ſhalt neuer liue to ſee my woe, or to haue
thy wil, and ſo farewell.

                                    Euphues.

   This letter being diſpatched, *Euphues* ſent it, and
Philautus read it, who diſdayning thoſe proud termes,
diſdayned alſo to aunſwere them, being readie to ryde
with *Ferardo.*
   Euphues hauing for a ſpace abſented himſelfe from
the houſe of *Ferardo,* bicauſe he was at home, longed
ſore to ſee *Lucilla,* which nowe opportunitie offered
vnto him, *Ferardo* being gon again to *Venice* with
Philautus, but in this his abſence, one *Curio* a Gentleman of *Naples* of little wealth and leſſe wit, haunted
Lucilla hir company, and ſo enchaunted hir, that
Euphues was alſo caſt off with *Philautus,* which thing
being vnknown to *Euphues,* cauſed him the ſooner to
make his repayre to the preſence of his Lady, whome
he finding in hir muſes, began pleaſantly to ſalute in
this manner.
   Miſtreſſe *Lucilla,* although my long abſence might
breede your iuſt anger, (for that louers deſire nothing
ſo much as often meeting) yet I hope my preſence will
diſſolue your choler (for yat louers are ſoone pleaſed
when of their wiſhes they be fully poſſeſſed.)   My
abſence is the rather to be excuſed in yat your father
hath bene alwayes at home, whoſe frownes ſeemed to
threaten my ill fortune, and my preſence at this preſent
the better to be accepted, in that I haue made ſuch
ſpeedy repaire to your preſence.

Vnto whom *Lucilla* aunfwered with this glyeke.

Truely *Euphues* you haue mift the cufhion, for I was neither angry with your long abfence, neither am · I well pleafed at your prefence, the one gaue mee rather a good hope heereafter neuer to fee you, ye other giueth me a greater occafion to abhorre you.

Euphues being nipped on the head, with a pale countenaunce as though his foule had forfaken his body, replyèd as followeth.

If this fodaine chaunge *Lucilla*, proceed of any defert of mine, I am heere not only to aunfwere the fact, but alfo to make amends for my fault : if of any new motion or minde to forfake your new friend, I am rather to lament your inconftancie then reuenge it : but I hope that fuch hot loue cannot be fo foone colde, neither fuch fure faith be rewarded with fo fodeine forgetfulneffe.

Lucilla not afhamed to confeffe hir folly, aunfwered him with this frumpe.

Sir, whether your deferts or my defire haue wrought this chaunge, it will boote you lyttle to know, neither do I craue amends, neither feare reuenge : as for feruent loue, you know there is no fire fo hotte but it is quenched with water, neither affection fo ftrong but is weakened with reafon, let this fuffice thee, that thou knowe I care not for thee.

In deede (faid *Euphues*) to know the caufe of your alteracion would boote me lyttle, feing the effect taketh fuch force. I haue heard that women either loue entirely or hate deadly, and feeing you haue put me out of doubt of the one, I must needes perfwade my felfe of the other. This chaunge will caufe *Philautus* to laugh me to fcorne, and double thy lightneffe in tourning fo often. Such was the hope that I conceiued of thy conftancie, that I fpared not in all places to blaze thy loyaltie, but now my rafh conceipt wil proue me a lyer, and thee a lyght hufwife.

Nay (fayd *Lucilla*) now fhalt thou not laugh *Philautus*

to fcorne, feeing you haue both drunke of one cup . in mifery *Euphues* it is great comfort to haue a companion. I doubt not, but that you wil both confpire againft me to worke fome mifchiefe, although I nothing feare your malice, whofoeuer accompteth you a lyar for prayfing me, may alfo deeme you a lecher for beeing enamoured of mee : and whofoeuer iudgeth me lyght in forfaking of you, may thinke thee as lewd in louing of me : for thou that thoughteft it lawfull to deceiue thy friend, muft take no fcorne to be deceiued of thy foe.

Then I perceiue *Lucilla* (faid he) that I was made thy ftale, and *Philautus* thy laughing ftocke : whofe friendfhip (I muft confeffe in deede), I haue refufed to obteine thy fauour : and fithens an other hath won that we both haue loft, I am content for my parte, neither ought I to be grieued feeing thou art fickle.

Certes *Euphues* (faid *Lucilla*) you fpend your wind in waft, for your welcome is but fmall, and your cheere is like to be leffe, fancie giueth no refon of his [her] change neither will be controlled for any choice : this is therfore to warn you, that from henceforth you neither folicite this fute, neither offer any way your feruice : I haue chofen one (I muft needes confeffe) neither to be compared to *Philautus* in wealth nor to thee in wit, neither in birthe to the worft of you both, I thinke God gaue it me for a iuft plague for [in] renouncing *Philautus*, and choofing thee, and fithence I am an enfample to all women of lightneffe, I am like alfo to be a mirrour to them all of vnhappineffe, which il luck I muft take, by fo much the more patiently, by how much the more I acknowledge my felfe to haue deferued it worthely.

Well *Lucilla* (aunfwered *Euphues*) this cafe breedeth my forrow the more, in that it is fo fodeine, and by fo much the more I lament it, by how much ye leffe I looked for it. In that my welcome is fo colde, and my cheere fo fimple, it nothing toucheth me, feing your fury is fo hot and my miffortune fo great, that I am neither willing to receiue it, nor you to beftow it : if tract of time, or want of triall, had caufed this *Meta-morphofis*, my griefe had bene more tollerable, and your fleeting more excufable, but comming in a moment vndeferued, vnlooked for, vnthought off, it encreafeth my forrow and thy fhame.

Euphues (quoth fhee) you make a long Harueft for a lyttle corne, and angle for the fifh that is alreadie caught. *Curio*, yea, *Curio* is he that hath my loue at his pleafure, and fhall alfo haue my life at his com-maundement, and although you deme him vnworthy to enioye that, which earft you accompted no wight worthye to embrace, yet feeing I efteeme him more woorth then any, he is to be reputed as chiefe. The Wolfe chooĩeth him for hir make, that hath or doth endure moft trauayle for hir fake. *Venus* was content to take the blake Smith with his powlt foote. *Cor-nelia* heere in *Naples*, difdayned not to loue a rude Miller.

As for chaunging, did not *Helen* ye pearle of *Greece* thy countrywoman, firft take *Menelaus*, then *Thefeus* and laft of all *Paris*? If brute beafts giue vs enfam-ples that thofe are moft to be liked, of whome we are beft beloued, or if the Princeffe of beautie *Venus*, and hir heires *Helen* and *Cornelia*, fhewe that our affection ftandeth on our free will, then am I rather to be excufed then accufed. Therefore good *Euphues* be as merry as you may be, for time may fo turne that once again you may be.

Nay *Lucilla* (fayd he) my Harueft fhall ceafe, feeing others haue reaped my corne, for anglyng for the fifh that is already caught, that were but meere folly. But in my minde if you be a fifh you are either an Eele, which as foone as one hath hold on hir tayle, wil flip out of his hande, or els a Minnow which wil be nib-ling at euery baite, but neuer biting : But what fifh fo euer you be, you haue made both me and *Philautus* to fwallow a Gudgen.

If *.Curio* be the perfon, I would neither wifh thee a

G

greater plague, nor him a deadlyer poyſon.   I for my part thinke him worthy of thee, and thou vnworthie of him, for although he be in body deformed, in minde fooliſh, an innocent borne, a begger by miſſortune, yet doth he deſerue a better then thy ſelfe, whoſe corrupte manners haue ſtained thy heauenly hue, whoſe lyght behauior hath dimmed the lights of thy beautie, whoſe vnconſtant minde hath betrayed the innocencie of ſo many a Gentleman.

And in that you bring in the example of a Beaſt to confirme your follye, you ſhew therein your beaſtly diſpoſition, which is readye to follow ſuch beaſtlyneſſe. But *Venus* played falſe: and what for that? ſeeing hir lyghtneſſe ſerued for an example, I woulde wiſh thou mighteſt trye hir puniſhment for a reward, that beeing openly taken in an yron net, all the world might iudge whether thou be fiſh or fleſh? and certes in my minde no angle will hold thee, it muſt be a net. *Cornelia* loued a Miller and thou a miſer, can hir folly excuſe thy fault? *Helen* of *Greece* my country-woman borne, but thine by profeſſion, chaunged and rechaunged at hir pleaſure, I graunt.   Shall the lewdeneſſe of others animate thee in thy lyghtneſſe?   Why then doſt thou not haunt ye ſtewes, bicauſe *Lais* frequented them? why doſt thou not loue a bul, ſeing *Paſiphae* loued one? why art thou not enamoured of thy father, knowing that *Mirrha* was ſo incenſed?

Theſe are ſet down, that we viewing their incontinencie, ſhould flye the lyke impudencie, not follow the like exceſſe, neither can they excuſe thee of any inconſtancy.   Merry I will be as I may, but if I may hereafter as thou meaneſt, I will not, and therefore farewell *Lucilla*, the moſt inconſtant that euer was nurſed in *Naples*, farewel *Naples* the moſt curſed towne in all *Italy*, and women all farewell.

Euphues hauing thus giuen hir his laſt farewell, yet being ſolytary, began a freſh to recount his ſorrow on this manner.

Ah *Euphues* into what miſſortune art thou brought?

in what fodeine miferye art thou wrapped? it is lyke
to fare with thee as with the Eagle, which dyeth
neither for age, nor with fickeneffe, but with famine,
for although thy ftomake hunger, yet thy heart will
not fuffer thee to eate. And why fhouldeft thou
torment thy felfe for one in whome is neither fayth
nor feruencye? O the counterfayte loue of woemen.
Oh inconftaunt fexe. I haue loft *Philautus*, I haue
loft *Lucilla*: I haue loft that which I fhall hardlye
finde againe, a faithfull friende. A foolifh *Euphues*,
why diddeft thou leaue *Athens*, the nurfe of wifedome,
to inhabite *Naples* the nourifher of wantonneffe?
Had it not beene better for thee to haue eaten falt
with the Philofophers in *Greece*, then fugar with the
courtiers of *Italy*? But behold the courfe of youth,
which alwayes enclyneth to pleafure, I forfooke mine
olde companions to fearch for new friendes, I reiected
the graue and fatherly counfaile of *Eubulus*, to follow
the brainficke humor of mine owne will. I addicted
my felfe wholly to the feruice of woemen, to fpend my
life in the lappes of Ladyes, my lands in maintenance
of brauery, my wit in the vanities of idle Sonnettes.
I had thought that woemen had bene as we men, that
is true, faithfull, zealous, conftant, but I perceiue they
be rather woe vnto men, by their falfehoode, geloufie,
[and] inconftancye. I was halfe perfwaded that they
were made of the perfection of men, and would be
comforters, but nowe I fee they haue tafted of the
infection of the Serpent, and will bee corafiues: The
Phifition fayth, it is daungerous to minifter Phifick
vnto the pacient that hath a colde ftomacke and a
hotte lyuer, leaft in giuing warmth to the one, he
inflame the other: fo verely it is hard to deale with a
woman, whofe woordes feeme feruent, whofe heart is
congealed into hard yce, leaft trufting their outward
talke, he be betrayed with their inward trechery. I
will to *Athens*, there to toffe my bookes, no more in
Naples to liue with faire lookes. I will fo frame my
felf, as all youth heereafter fhal rather reioyce to fee

mine amendement, then be animated to follow my
former life.   Philofophy, Phifick, Diuinitie, fhal be
my ftudy.   O the hidden fecrets of Nature, ye
expreffe Image of morall vertues, the equall ballance
of Iuftice, the medicines to heale al difeafes, how
they begin to delight me.   The *Axiomaes* of *Ariftotle*,
the *Maxims* of *Iuftinian*, the *Aphorifmes* of *Galen*,
haue fodeinely made fuch a breach into my minde,
that I feeme onely to defire them, which did onely
earft deteft them.   If witte be employed in the
honeft ftudy of learning, what thing fo precious as
wit? if in the idle trade of loue, what thing more
peftilent then wit?

The proofe of late hath bene verified in me whome
nature hath endued with a lyttle witte, which I haue
abufed with an obftinate will : moft true it is that the
thing the better it is, the greater is the abufe, and that
there is nothing but through the malice of man may
be abufed.

Doth not the fire (an element fo neceffary that
without it man cannot liue) as well burne the houfe,
as burne in the houfe, if it be abufed?   Doth not
Tryacle as well poyfon as helpe, if it be taken out of
time?   Doth not wine, if it be immoderatly taken
kill the ftomack, enflame the Liuer, mifchiefe the
dronken?   Doth not Phificke deftroy if it be not well
tempered?   Doth not law accufe if it be not rightly
interpreted?   Doth not diuinitie condemne if it be
not faithfully conftrued?   Is not poyfon taken out of
the Hunnyfuckle by the Spider? venym out of the
Rofe by the Cancker? dunge out of the Maple tree
by the Scorpion?   Euen fo the greateft wickedneffe
is drawne out of the greateft wit, if it bee abufed
by wil, or entangled with the world, or inuegled with
women.

But feeing I fee mine owne impietie, I will endeauour my felfe to amende all that is paft, and to
bee a myrrour of Godlineffe hereafter.   The Rofe
though a lyttle it be eaten with the Canker yet beeing

diftilled yeeldeth fweet water : the yron though fretted with the ruft, yet being burnt in the fire fhineth brighter : and witte although it hath beene eaten with the canker of his owne conceite, and fretted with the ruft of vayne loue, yet beeing purified in the ftyll of wifdome, and tryed in the fire of zeale, will fhine bright and fmell fweete in the nofethrils of all young nouifes.

As therefore I gaue a farewell to *Lucilla*, a farewell to *Naples*, a farewell to women, fo nowe doe I giue a farewell to the worlde, meaning rather to macerate my felfe with melancholye, then pine in follye, rather choofing to dye in my ftudye amiddeft my bookes, then to court it in *Italy*, in ye company of ladyes.

Euphues hauing thus debated with himfelfe, went to his bed, ther either with fleepe to deceiue his fancye, or with mufing to renue his ill fortune, or recant his olde follyes.

But it happened immediatly *Ferardo* to returne home, who hearing this ftraunge euent, was not a lyttle amazed, and was nowe more readye to exhorte *Lucilla* from the loue of *Curio*, then before to the lyking of *Philautus*. Therefore in all hafte, with watrye eyes, and a woeful heart, began on this manner to reafon with his daughter.

Lucilla (daughter I am afhamed to call thee, feeing thou haft neither care of thy fathers tender affection, nor of thine owne credite) what fp[i]rite hath enchaunted thy fpirit, that euery minute thou altereft thy minde ? I had thought that my hoary haires fhould haue found comforte by thy golden lockes, and my rotten age great eafe by thy rype years. But alas I fee in thee neither wit to order thy doings, neither wii to frame thy felfe to difcretion, neither the nature of a childe, neither the nurture of a mayden, neither (I cannot without teares fpeake it) any regard of thine honour, neither any care of thine honeftie.

I am nowe enforced to remember thy mothers death, who I thinke was a Prophetefle in hir life, for oftentimes fhe woulde faye, that thou haddeft more beautie then was conuenient for one that fhoulde bee honeft, and more cockering then was meete for one that fhould be a Matrone.

Woulde I had neuer lyued to be fo olde, or thou to be fo obftinate, either woulde I hadde dyed in my youth in the court, or thou in thy cradle : I would to God that either I had neuer beene borne, or thou neuer bredde. Is this the comfort that the parent reapeth for all his care? Is obftinacye payed for obedyence, ftubberneneffe rendred for duetie, malycious defperatneffe, for filiall feare? I perceiue now that the wife painter faw more then the foolifh parent can, who paynted loue going downward, faying, it might well defcende, but afcende it coulde neuer. *Danaus* whome they report to be the father of fiftie children, had among them all, but one that difobeyed him in a thing moft difhoneft : but I that am father to one more then I would be, although one be al, haue that one moft difobedient to me in a requeft lawful and reafonable. If *Danaus* feeing but one of his daughters without awe, became himfelf without mercie, what fhal *Ferardo* do in this cafe, who hath one and all moft vnnaturall to him in a moft iuft caufe? Shall *Curio* enioy the fruite of my trauailes, poffeffe the benefite of my laboures, enherite the patrimony of mine aunceftors, who hath neither wife-dome to increafe them, nor witte to keepe them.

Wilt thou *Lucilla*, beftow thy felfe on fuch an one, as hath neither comelyneffe in his bodye, nor knowledge in his minde, nor credite in his countrey. Oh I would thou hadft either bene euer faithfull to *Philautus*, or neuer faithleffe to *Euphues*, or would thou wouldeft be moft fickle to *Curio*. As thy beautie hath made thee the blaze of *Italy*, fo wil thy lightneffe make thee the bye word of the worlde. O *Lucilla*, *Lucilla*, would thou wert leffe faire or more fortunate, either of leffe honour, or greater honeftie : either better minded, or foone buryed.

Shall thine olde father lyue to fee thee match with a young foole ? fhall my kinde heart be rewarded with fuch vnkinde hate ? Ah *Lucilla*, thou knoweft not the care of a father, nor the duetie of a childe, and as farre art thou from pietie as I from crueltie.

Nature will not permit me to difherit my daughter, and yet it will fuffer thee to difhonour thy father. Affe&tion caufeth me to wifh thy lyfe, and fhall it entice thee to procure my death ? It is mine onely comfort to fee thee flourifh in thy youth, and is it thine to fee me fade in mine age ? to conclude I defire to liue to fee thee profper, and thou to fee me perifh. But why caft I the effe&te of this vnnatural-neffe in thy teeth, feeing I my felfe was the caufe ? I made thee a wanton, and thou haft made me a foole : I brought thee vp like a cockney, and thou haft handled me like a cockefcombe. (I fpeake it to mine owne fhame,) I made more of thee then became a Father, and thou leffe of me then befeemed a childe. And fhall my louing care be caufe of thy wicked crueltie ? Yea, yea, I am not the firft that hath bene too carefull, nor the laft that fhall bee handeled fo vnkindely : It is common to fee fathers too fonde, and children too frowarde. Well *Lucilla*, the teares which thou feeft trickle downe my cheekes, and my droppes of bloude (which thou canft not fee) that fal from my heart, enforce mee to make an ende of my talke, and if thou haue any duetie of a childe, or care of a friende, or courtefie of a ftraunger, or feelyng of a Chriftian, or humanitie of a reafonable creature, then releafe thy father of griefe, and acquite thy felfe of vngratefulneffe : Otherwife thou fhalt but haften my death, and encreafe thine owne defame : Which if thou doe, the gaine is mine, and the loffe thine, and both infinite.

Lucilla either fo bewitched that fhe could not relent, or fo wicked that fhe would not yeelde to hir Fathers requeft, aunfwered him on this manner.

Deere Father, as you would haue me to fhewe the

duetie of a childe, fo ought you to fhewe the care of a
Parent, for as the one ftandeth in obedience fo the
other is grounded vpon refon.   You would haue me
as I owe duetie to you to leaue *Curio,* and I defire
you as you owe mee any loue that you fuffer me to
enioy him.   If you accufe me of vnnaturalnes in that
I yeeld not to your requeft, I am alfo to condempne
you of vnkindneffe, in that you graunt not my
peticion.

You obieᵭt I know not what to *Curio,* but it is the
eye of the mafter that fatteth the horfe, and the loue
of the woeman, that maketh the man.   To giue
reafon for fancie were to weigh the fire, and meafure
the winde.   If therefore my delight be the caufe of
your death, I thinke my forrow woulde be an occafion
of your folace.   And if you be angry bicaufe I am
pleafed, certes I deeme you would be content if I
were deceafed : which if it be fo that my pleafure
breed your paine, and mine annoy your ioye, I may
well fay that you are an vnkinde father, and I an
vnfortunate childe.   But good father either content
your felfe with my choice, or lette mee ftande to the
maine chaunce, otherwife the griefe will be mine
and the fault yours, and both vntollerable [intollerable].

Ferardo feeing his daughter, to haue neither regarde
of hir owne* honour nor his requeft, conceyued fuch
an inward griefe that in fhort fpace he dyed, leauing
Lucilla the onely heire of his lands, and *Curio* to
poffeffe them, but what ende came of hir, feing it is
nothing incident to the hiftory of *Euphues,* it were
fuperfluous to infert it, and fo incredible that all
women would rather wonder at it then beleeue it,
which euent beeing fo ftraunge, I had rather leaue
them in a mufe what it fhould be, then in a maze in
telling what it was.

Philautus hauing intellygence of *Euphues* his fucceffe, and the falfehoode of *Lucilia,* although he began
to reioyce at the miferie of his fellow, yet feeing hir
fickleneffe, coulde not but lament hir folly, and pitie

his friends misfortune. Thinking that the lyghtneffe of *Lucilla* enticed *Euphues* to fo great lyking.

Euphues and *Philautus* hauing conference between themfelues, cafting difcourtefie in thee teeth each of the other, but chiefely noting difloyaltie in the demeanor of *Lucilla*, after much talke renewed their old friendfhip both abandoning *Lucilla*, as moft abhominable. *Philautus* was earneft to haue *Euphues* tarye in *Naples*, and *Euphues* defirous to haue *Philautus* to *Athens*, but the one was fo addicted to the court, the other fo wedded to the vniuerfitie, that each refufed the offer of the other, yet this they agreed betweene themfelues, that though their bodies were by diftance of place feuered, yet the coniunction of their mindes fhould neither be feperated by ye length of time nor alienated by change of foyle, I for my part faid *Euphues*, to confirme this league, giue thee my hande and my heart, and fo likewife did *Philautus*, and fo fhaking handes, they bidde each other farewell.

Euphues, to the intent he might bridle the ouer-lafhing affections of *Philautus*, conuayed into his ftudie a certeine pamphlet which he termed a cooling carde for *Philautus*, yet generally to be applyed to all louers, which I haue inferted as followeth.

¶ *A cooling Carde for Philautus
and all fond louers.*

Vfing with my felfe beeing idle, howe I
might be wel employed (friende *Philautus*)
I coulde finde nothing either more fit to
continue our friendfhippe, or of greater
force to diffolue our folly, then to write a
remedy for that, which many iudge paft cure, for loue
(*Philautus*) with the which I haue bene fo tormented,
that I haue loft my time, thou fo troubled that thou
haft forgot reafon, both fo mangled with repulfe,
inueigled by deceit, and almoft murthered by difdaine,
that I can neither remember our miferies without
griefe, nor redreffe our mifhaps without grones.   How
wantonly, yea, and how willingly haue we abufed our
golden time, and mifpent our gotten treafure?   How
curious were we to pleafe our Lady, how careleffe to
difpleafe our Lorde?   Howe deuout in feruing our
Goddeffe, how defperate in forgetting our God?   Ah
my *Philautus*, if the wafting of our money might not
dehort vs, yet the wounding of our mindes fhould
deterre vs, if reafon might nothing perfwade vs to
wifdome, yet fhame fhould prouoke vs to wit.   If
Lucilla reade this trifle, fhee will ftraight proclaime
Euphues for a traytour, and feing me turne my tippet,
will either fhut me out for a Wrangler, or caft mee off
for a Wiredrawer : either conuince me of malyce in
bewraying their fleightes, or condemne me of mifchiefe in arming young men againft fleeting minions.
And what then?   Though *Curio* bee as hot as a
toaft, yet *Euphues* is as colde as [a] clocke, though hee
bee a cocke of the game, yet *Euphues* is content to
bee crauen and crye creake, though *Curio* be olde
huddle and twang, *ipfe*, he, yet *Euphues* had rather
fhrinke in the wetting then waft in the wearing.   I
know *Curio* to be fteele to the backe, ftanderd bearer
to *Venus* camp, fworne to the crew, true to ye crowne,
knight marfhall to *Cupid*, and heyre apparaunt to his

kingdome. But by that time that he hath eaten but one bufhell of falt with *Lucilla*, he fhall taft tenne quarters of forrow in his loue, then fhall he finde for euery pynte of Hunny a gallon of Gall : for euerye dramme of pleafure, an ounce of payne : for euery inch of myrth, an ell of moane. And yet *Philautus*, if there be any man in difpaire to obteyne his purpofe, or fo obftinate in his opinion, that hauing loft his freedome by folly would alfo lofe his life for loue, let him repaire hether, and hee fhall reape fuch profite, as will either quench his flames, or affwage his fury, either caufe him to renounce his Ladye as moft pernitious, or redeeme his libertie as moft precious. Come therefore to me al ye louers that haue bene deceiued by fancy, the glaffe of peftilence, or deluded by woemen, the gate to perdition, be as earneft to feeke a medicine, as you were eager to runne into a mifchiefe, the earth bringeth forth as well Endiue to delight the people, as Hemlocke to endaunger the patient, as wel the Rofe to diftil, as the Nettle to fting, as wel the Bee to giue Hunny, as the Spyder to yeeld poyfon.

If my lewde lyfe Gentlemen haue giuen you offence, let my good counfaile make amends, if by my folly any be allured to luft, let them by my repentance be drawne to continency. *Achilles* fpeare could as wel heale as hurt, the fcorpion though he fting, yet he ftints the paine, though the hearb *Nerius* poyfon the Sheepe, yet is a remedy to man againft poyfon, though I haue infeéted fome by example, yet I hope I fhall comfort many by repentaunce. Whatfoeuer I fpeake to men, the fame alfo I fpeake to women, I meane not to run with the Hare and holde with the Hounde, to carye fire in the one hand and water in the other, neither to flatter men as altogether faultleffe, neither to fall out with woemen as altogether guiltie, for as I am not minded to picke a thanke with the one, fo am I not determined to picke a quarrell with the other, if women be not peruerfe they fhall reape profite, by

remedye of pleafure. If *Phillis* were nowe to take
counfayle fhee would not be fo foolyfh to hang hir
felfe, neither *Dido* fo fonde to dye for *Aeneas*, neither
Pafiphae fo monftrous to loue a Bull, nor *Phædra* fo
vnnaturall to bee enamoured of hir fonne.

This is therefore to admonifh all young Imps and
nouifes in loue, not to blow the coales of fancy with
defire, but to quench them with difdayne. When
loue tickleth thee, decline it, leaft it ftifle thee : rather
faft then furfette, rather ftarue then ftriue to exceede.
Though the beginning of loue bring delight, the ende
bringeth deftruction. For as the firft draught of wine
doth comfort the ftomacke, the feconde enflame the
lyuer, the thirde fume into the heade, fo the firft fippe
of loue is pleafant, the feconde perilous, the thirde
peftilent. If thou perceiue thy felfe to be entifed with
their wanton glaunces, or allured with their wicked
guiles, either enchaunted with their beautie, or
enamoured with their brauery, enter with thy felfe
into this meditation.

What fhall I gaine if I obteine my purpofe ? nay
rather what fhal I loofe in winning my pleafure ? If
my Lady yeeld to be my louer, is it not likely fhe will
be an others lemman ? and if fhe be a modeft matrone,
my labour is loft. This therefore remaineth, that either
I muft pine in cares or perifh with curfes.

If fhe be chaft then is fhe coye ? if lyght, then is fhe
impudent, if a graue matrone, who can woe hir ? if a
lewde minion, who woulde wedde hir ? if one of the
Veftall Virgins, they haue vowed virginitie, if one of
Venus court, they haue vowed difhoneftye. If I loue
one that is faire, it will kindle geloufie, if one that is
foule, it wil conuert me into phrenfie. If fertile to
beare children my care is increafed, if barren my curfe
is augmented. If honeft I fhall feare hir death, if
immodeft I fhall be weary of hir life.

To what ende then fhall I liue in loue, feeing
alwayes it is a life more to be feared then death? for
all my time wafted in fighes and worne in fobbes,

for all my treafure fpente on Iewells, and fpylte in
iolytye, what recompence fhall I reape befides repentaunce? What other reward fhall I haue then
reproch? What other folace then endles fhame?
But happely thou wylt fay, if I refufe their curtefie, I
fhall be accompted a Mecocke, a Milkfop, taunted
and retaunted with check and checkmate, flowted and
reflowted with intollerable glee.

Alas fond foole, art thou fo pinned to their fleeues
yat thou regardeft more their babble then thine own
bliffe, more their frumpes then thine owne welfare?
Wilt thou refemble the kinde Spaniel, which the more
he is beaten the fonder he is, or the foolifh Gieffe,
which wil neuer away? Doft thou not know that
woemen deeme none valyaunt vnleffe he be too
venterous? That they accompt one a daftard if he
be not defperate, a pynch penny if he be not
prodygall, if filent a fotte, if full of wordes a foole?
Peruerfly doe they alwayes thinke of their louers and
talke of them fcornefully, iudging all to be clownes
which be no courtiers, and al to be pinglers that be
not courfers.

Seeing therfore the very bloffome of loue is fower,
the budde cannot be fweete: In time preuent
daunger, leaft vntimely thou runne into a thoufande
perills.

Search the wound while it is greene, too late
commeth the falue when the fore feftereth, and the
medicine bringeth double care, when the maladye is
paft cure.

Beware of delayes. What leffe then the grayne of
Muftardfeed, in time almoft what thing is greater
then the ftalke thereoff. The flender twigge groweth
to a ftately tree, and that which with the hande might
eafely haue bene pulled vp, wil hardly with the axe be
hewen downe. The leaft fparke if it be not quenched
will burft into a flame, the leaft Moath in time eateth
the thickeft cloath, and I haue read that in a fhorte
fpace, there was a Towne in *Spayne* vndermined with

Connyes, in *Theſſalia* with Mowles, with Frogges in *Fraunce*, in *Africa* with Flyes. If theſe ſilly Wormes in tracte of time ouerthrowe ſo ſtatelye Townes, how much more will Loue, which creepeth ſecretly into the minde, (as the ruſt doth into the yron and is not perceiued) confume the body, yea, and confound the ſoule. Defer not from houre to day, from day to month, from month to yeare, and alwayes remaine in miſery.

He that to day is not willyng, will to morrow bee more wilful. But alas it is no leſſe common then lamentable to behold the tottering eſtate of louers, who thinke by delayes to preuent daungers, with Oyle to quench fire, with ſmoake to clear the eye ſight. They flatter themſelues with a fainting farewell, deferring euer vntil to morrow, when as their morrow doth alwayes increaſe their ſorrow. Let neither their amiable countenaunces, neither their painted proteſtacions, neither their deceitfull promiſes allure thee to delayes.

Thinke this with thy ſelfe, that the ſweete ſongs of *Calipſo*, were ſubtill ſnares to entice *Vliſſes*, yat the Crab then catcheth the Oyſter, when the Sun ſhineth, that *Hiena* when ſhe ſpeaketh lyke a man, deuiſeth moſt miſchiefe, that women when they be moſt pleaſaunt, pretend moſt trecherie [miſchiefe].

Follow *Alexander* which hearing the commendation and ſingular comelineſſe of the wife of *Darius*, ſo couragiouſly withſtood the aſſaults of fancie, that hee would not ſo much as take a view of hir beautie. Imitate *Cyrus*, a king endued with ſuch continencie, that hee loathed to looke on the heauenly hue of *Panthea*, and when *Araſpus* tolde him that ſhe excelled al mortall wights in amiable ſhewe, by ſo much the more (ſayd *Cyrus*) I ought to abſtaine [refraine] from hir ſight, for if I followe thy counſaile in going to hir, it may be, I ſhall deſire to continue with hir, and by my lyght affection, neglect my ſerious affaires. Learne of *Romulus* to refraine [abſtaine] from wine, be it neuer ſo delycate: of *Ageſilaus* to diſpiſe coſtly apparell, be

it neuer fo curious : of *Diogenes* to deteft women be
they neuer fo comely.   Hee that toucheth Pitch
fhall bee defiled, the fore eye infecteth the founde,
the focietie with women breedeth fecuritie in the
foule, and maketh all the fences fenceleffe.   Moreouer
take this counfaile as an Article of thy Creede, which
I meane to follow as the chiefe argument of my faith,
that Idleneffe is the onely nourfe and nourifher of fenfual appetite, ye fole maintenaunce of youthful [youthly]
affeCtion, the firft fhaft that *Cupid* fhooteth in the hot
liuer of a heedeleffe louer.   I would to god I were
not able to finde this for a truth by mine owne tryal,
and I would the example of others idleneffe had caufed
me rather to auoyde that fault, then experience of mine
owne folly.

How diffolute haue I bene in ftriuing againft good
counfaile ? how refolute in ftanding in mine own conceipt ? how forward to wickedneffe, how frowarde to
wifdome ? how wantonne with too much cockering ?
how wayward in hearing correCtion.   Neither was I
much vnlyke thefe Abbaie lubbers in my lyfe (though
farre vnlyke them in beliefe) which laboured till they
were colde, eat till they fweat, and lay in bed til their
boanes aked.   Heeroff commeth it Gentlemen that
loue creepeth into the minde by priuie craft, and
keepeth his holde by maine courage.

The man beeing idle, the minde is apte to all
vncleaneneffe, the minde being voyde of exercife, the
man is voyde of honeftie.   Doth not the ruft fret the
hardeft yron, if it be not vfed ?   Doth not the Moathe
eate the fineft garment, if it be not worne ?   Doth
not Moffe grow on the fmootheft ftone if it be not
ftirred ?   Doth not impietie infeCt the wifeft wit, if it
be giuen to idleneffe ?   Is not the ftanding water
fooner frofen then the running ftreame ?   Is not he
yat fitteth more fubieCt to fleepe then he that
walketh ?   Doth not common experience make th:s
common vnto vs that the fatteft ground bringeth
foorth nothing but weedes if it be not well tilled ?

That the fharpeft wit enclyneth onely to wickedneffe, if it be not exerciled? Is it not true which
Seneca reporteth, that as too much bending breaketh
the bowe, fo too much remiffion fpoyleth the minde.
Befides this immoderate fleepe, immodeft play,
vnfatiable fwilling of wine, doth fo weaken the fences,
and bewitch the foule, that before we feele the motion
of loue, we are refolued into luft. Efchew Idleneffe
my *Philautus*, fo fhalt thou eafely vnbende the bow
and quench the brandes of *Cupide.* Loue giues place
to labour, labour and thou fhalt neuer loue. *Cupide*
is a craftie childe, following thofe at an ynch that
ftudie pleafure, and flying thofe fwiftly that take
paines.

Bende thy minde to the Lawe whereby thou mayeft
haue vnderftanding of olde and auntient cuftomes,
defend thy Clyents, enrich thy cofers, and cary credite
in thy Countrey.

If Law feeme loathfome vnto thee, fearche the
fecrets of Phyficke, whereby thou mayft know the
hidden natures of hearbes, whereby thou mayft gather
profite to thy purfe, and pleafure to thy minde.

What can be more exquifite in humaine affaires,
then for euery feuer be it neuer fo hot, for euery palfie
be it neuer fo cold, for euery infection, be it neuer fo
ftraunge, to giue a remedy? The old verfe ftandeth
as yet in his old vertue. That *Galen* giueth goods,
Iuftinian honors.

If thou be fo nice, that thou canft no way brooke
the practife of Phificke, or fo vnwife, that thou wilt
not beat thy braines about the inftitutes of the Law,
conferre all thy ftudie, all thy time, all thy treafure to
the atteining of ye facred and fincere knowledge of
diuinitie. By this maift thou bridle thine incontinencie, raine thy affections, reftraine thy luft.
Heere fhalt thou behold as it were in a glaffe, that al
the glory of man is as the graffe, that all things vnder
Heauen, are but vaine, that our lyfe is but a fhadow,
a warfare, a pilgrimage, a vapor, a bubble, a blaft: of

fuch fhortneffe, that *Dauid* faith, it is but a fpan long: of fuch fharpnes, that *Iob* noteth it replenifhed with al miferies, of fuch vncerteinetie, that we are no fooner borne but we are fubieĉt to death, the one foote no fooner on the ground, but the other ready to flip into the graue. Heere fhalt thou finde eafe for thy burden of finne, comfort for thy confcience pined with vanitie, mercie for thine offences by the Martirdome of thy fweete Sauiour.

By this thou fhalt be able to inftruĉt thofe that be weake, to confute thofe that be obftinate, to confound thofe that bee erronious, to confirme the faithfull, to comfort the defperate, to cut off the prefumpt[u]ous, to faue thine owne foule by thy fure faith, and edifie the hearts of many by thy found doĉtrine. If this feeme to ftraight a diet for thy ftraying [straunge] difeafe, or too holy a profeffion, for fo hollow a perfon, then employe thy felfe to marcial feates, to iuftes, to turneyes, yea, to al tormentes rather then to loyter in loue, and fpende thy lyfe in the lappes of Ladyes : what more monftrous can there be, then to fee a younge man abufe thofe giftes to his owne fhame, which God hath giuen him for his owne preferment? What greater infamy, then to conferre the fharpe witte to the mak-ing of lewde Sonettes, to the idolatrous worfhypping of their Ladyes, to the vaine delyghtes of fancye, to all kinde of vice as it were againft kinde and courfe of Nature? Is it not folly to fhewe witte to woemen which are neither able nor willing to receiue fruite thereoff? Doeft thou not knowe that the tree *Siluacenda* beareth no fruite in *Pharo*? That the *Perfian* trees in *Rhodes* doe onely waxe greene, but neuer bring foorth apple. That *Amonius* and *Nardus* will onely growe in *India*. *Balfamum* onely in *Syria*, that in *Rhodes* no Eagle will build hir neaft, no Owle lyue in *Creete*, no wit fpring in the will of women? Mortifie therefore thy affeĉtions, and force not Nature againft Nature to ftriue in vaine. Goe into the Contrey, looke to thy groundes, yoke thine Oxen, follow the

H

Plough, graft thy trees, beholde thy cattell, and deuife
with thy felfe, howe the encreafe of them may encreafe
thy profite. In *Autumne* pull thine apples, in Summer
ply thy harueft, in the Springe trimme thy Gardens, in
the Winter thy woodes, and thus beginninge to delyght
to be a good hufband, thou fhalt begin to deteft to be
in loue with an idle hufwife, when profite fhall beginne
to fill thy purfe with golde, then pleafure fhall haue
no force to defile thy minde with loue. For honeft
recreation after thy toyle, vfe hunting or haukeing,
either rowfe the Deere, or vnpearch the Phefant, fo
fhalt thou roote out the remembraunce of thy former
loue, and repent thee of thy foolifhe luft. And
although thy fweete hearte binde thee by othe alwaye
to holde a candle at hir fhrine, and to offer thy deuotion to thine owne deftruction, yet goe, runne, flye
into the Country, neither water thou thy plants, in that
thou departeft from thy Pygges nye, neither ftande
in a mammering whether it be beft to depart or not,
but by howe much the more thou art vnwilling to goe,
by fo much the more haften thy fteppes, neither faine
for thy felfe any fleeueleffe excufe, whereby thou
maift tarrye. Neither lette rayne nor thunder, neither
lightening nor tempeft ftay thy iourney, and recken
not with thy felfe how many myles thou haft gone,
that fheweth wearines, but how many thou haft to go,
that proueth manlyneffe. But foolifh and franticke
louers, will deeme my precepts hard, and efteeme my
perfwafions haggarde : I muft of force confeffe, that
it is a corafiue to the ftomake of a louer, but a comfort to a godly lyuer, to runne through a thoufande
pikes to efcape ten thoufand perills. Sowre potions
bring founde health, fharp purgations make fhort
difeafes, and the medicine the more bitter it is, the
more better it is in working. To heale the body we
trye Phificke, fearch cunninge, proue forcery, venture
through fire and water, leauing nothing vnfought that
may be gotten for money, be it neuer fo much, or
procured by any meanes be they neuer fo vnlawfull.

How much more ought we to hazard all things for
the fauegard of minde, and quiet of confcience? And
certes eafier will the remedy be, when the reafon is
efpyed : doe you not knowe the nature of women
which is grounded onely vpon extremities? Doe
they thinke any man to delyght in them, vnleffe
he doate on them? Any to be zealous except they
bee iealous? Any to be feruent in cafe he be not
furious? If he be cleanelye, then terme they him
proude, if meane in apparell a flouen, if talle a lungis,
if fhort, a dwarfe, if bolde, blunt : if fhamefaft, a
cowarde : Infomuch as they haue neither meane in
their frumps, nor meafure in their folly. But at the
firft the Oxe weyldeth not the yoke, nor the Colt the
fnaffle, nor the louer good counfel, yet time caufeth
the one to bend his neck, the other to open his
mouth, and fhoulde enforce the thirde to yeelde his
right to reafon. Laye before thine eyes the flightes
and deceits of thy Lady, hir fnathching in ieft and
keeping in earneft, hir periury, hir impietie, the
countenance fhe fheweth to thee of courfe, the loue
fhe beareth to others of zeale, hir open malice, hir
diffembled mifchiefe.

O I woulde in repeating their vices thou couldeft
be as eloquent as in remembring them thou oughteft
to bee penitent : be fhe neuer fo comely call hir
counterfaite, bee fhe neuer fo ftraight thinke hir
cro[o]ked. And wreft all partes of hir body to the
worft, be fhe neuer fo worthy. If fhee be well fette,
then call hir a Boffe, if flender, a Hafill twygge, if
Nutbrowne, as blacke as a coale, if well couloured, a
paynted wall, if fhee bee pleafaunt, then is fhee a
wanton, if fullenne, a clowne, if honeft, then is fhee
coye, if impudent a harlot.

Search euery vaine and finewe of their difpofition,
if fhe haue no fight in defcante, defire hir to chaunt
it, if no cunning to daunce requeft hir to trippe it, if
no fkill in muficke, profer hir the Lute, if an ill gate,
then walke with hir, if rude in fpeach, talke with hir,

if fhee be gagge toothed, tell hir fome merry ieft, to make hir laughe, if pinke eyed, fome dolefull Hiftorye to caufe hir weepe, in the one hir grinning will fhew hir deformed, in the other hir whyning like a Pigge halfe rofted.

It is a world to fee howe commonly we are blinded with the collufions of women, and more entifed by their ornaments beeing artificiall, then their proportion beeinge naturall. I loath almoft to thincke on their oyntments and appoticary drugges, the fleeking of their faces, and all their flibber fawces, whiche bring quefineffe to the ftomacke, and difquiet to the minde.

Take from them their perywigges, their paintings, their Iewells, their rowles, their boulftrings, and thou fhalt foone perceiue that a woman is the leaft parte of hir felfe. When they be once robbed of their robes, then wil they appeare fo odious, fo vgly, fo monftrous, that thou wilt rather think them ferpents then faints, and fo like Hags, that thou wilt feare rather to be enchaunted then enamoured. Looke in their clofettes, and there fhalt thou finde an Appoticaryes fhop of fweete confeótions, a furgions boxe of fundry falues, a Pedlers packe of newe fangles. Befides all this their fhadowes, their fpots, their lawnes, their leefekyes, their ruffes, their rings : Shew them rather Cardinalls curtifans, then modeft Matrons, and more carnally affeóted, then moued in confcience. If euery one of thefe things feuerally be not of force to moue thee, yet all of them ioyntly fhould mortifie thee.

Moreouer to make thee the more ftronger to ftriue againft thefe *Syrenes*, and more fubtil to deceiue thefe tame Serpents, my counfaile is that thou haue more ftrings to thy bow then one, it is fafe riding at two ankers, a fire deuided in twayne burneth flower, a fountaine running into many ryuers is of leffe force, the minde enamoured on two women is leffe affeóted with defire, and leffe infeóted with difpaire, one loue expelleth an other, and the remembraunce of the latter quencheth the concupifcence of the firſt.

Yet if thou be fo weake being bewitched with their wiles that thou haſt neither will to efchue, nor wit to auoyd their company, if thou be either fo wicked that thou wilt not, or fo wedded that thou canſt not abſtein from their glaunces, yet at the leaſte diſſemble thy griefe. If thou be as hot as ye mount *Aetna*, faine thy felfe as colde as the hil *Caucafus*, cary two faces in one hood, couer thy flaming fancie with fained aſhes, ſhew thy felfe found when thou art rotten, let thy hewe be merry, when thy heart is melancholy, beare a pleafaunt countenaunce with a pined confcience, a painted ſheath with a leaden dagger : Thus diſſembling thy griefe, thou maiſt recure thy difeafe. Loue creepeth in by ſtealth, and by ſtealth ſlideth away.

If ſhe breake promife with thee in the night, or abſent hir felfe in the day, feeme thou careleſſe, and then will ſhe be carefull, if thou languiſh, then wil ſhe be lauiſh of hir honour, yea and of the other ſtrange beaſt hir honeſtie. Stande thou on thy pantuffles, and ſhee will vayle bonnet ? lye thou aloofe and ſhe wil ceaze on the lure, if thou paſſe by hir dore, and bee called backe, either feeme deafe and not to heare, or defperate and not to care. Fly the places, the parlours, the portals, wherein thou haſt bene conuerfant with thy lady, yea *Philautus* ſhunne the ſtreet where *Lucilla* doth dwell ? leaſt the fight of hir window renue the fumme of thy forrow.

Yet although I would haue thee precife, in keeping thefe precepts, yet would I haue thee to auoyd folly-tarineſſe, that breedes melancholy ; melancholy, mad-neſſe ; madneſſe, mifchiefe and vtter defolation : haue euer fome faithful pheere, with whom thou maiſt communicate thy councells, fome *Pilades* to encourage *Oreſtes*, fome *Damon* to releafe *Pithias*, fome *Scipio* to recure *Lælius*. *Phillis* in wandring the woodes, hanged hir felfe. *Aſiarchus* forfaking companye, ſpoyled himfelfe with his owne bodkin. *Biarus* a

Romaine more wife then fortunate, being alone
deftroyed himfelfe with a potfherd. Beware [of] folitarineffe. But although I would haue thee vfe company
for thy recreation, yet woulde I haue thee alwayes to
leaue the companye of thofe that accompany thy
Lady, yea, if fhe haue any iewell of thine in hir
cuftodie, rather loofe it then goe for it, leaft in feeking
to recouer a trifle, thou renewe thine olde trouble.
Be not curious to curle thy haire, nor carefull to be
neat in thine apparel, be not prodigal of thy golde,
nor precife in thy going, be not lyke the Englifhman,
which preferreth euery ftraunge fafhion before the
vfe of his countrey, be thou diffolute, leaft thy Lady
thinke thee foolifh in framing thy felfe to euerye
fafhion for hir fake. Beleeue not their othes and
folempne proteftations, their exorcifmes and coniurations, their teares which they haue at commaundement,
their alluring lookes, their treading on the toe, their
vnfauery toyes.

Let euery one loath his Ladye, and bee afhamed to
be hir feruaunt. It is riches and eafe that nourifheth
affection, it is play, wine and wantonneffe, that feedeth
a louer as fat as a foole, refraine from all fuch meates,
as fhall prouoke thine appetite to luft, and all fuch
meanes as may allure thy minde to folly. Take
cleere water for ftrong wine, browne breade for
fine manchet, beefe and brewys, for Quailes and
Partridge : for eafe labour, for pleafure paine : for
furfetting, hunger : for fleepe watching : for the fellowfhip of Ladies, the company of Philofophers. If thou
faye to mee, Phifition heale thy felfe. I aunfwere,
that I am meetly well purged of that difeafe, and yet
was I neuer more willyng to cure my felfe then to
comfort my friend. And feeing the caufe that made
in mee fo colde a deuotion, fhould make in thee alfo
as frofen a defire, I hope thou wilt be as ready to
prouide a falue as thou waft haftie in feeking a fore.
And yet *Philautus*, I would not that al women fhould
take pepper in the nofe, in that I haue difclofed the

ıegerdemaines of a fewe, for well I know none will winch except fhe bee gawlded, neither any be offended vnleffe fhe be guiltie. Therfore I earneftly defire thee, that thou fhew this coolyng carde to none, except thou fhew alfo this my defence to them all. For although I way nothing the ill will of light hufwiues, yet would I be loath to lofe the good wil of honeft matrones. Thus being ready to goe to *Athens*, and ready there to entertein thee whenfoeuer thou fhalt repaire thether. I bidde thee farewell, and fly women.

Thine euer,
Euphues.

To the graue Matrones,
and honeft Maidens
of Italy.

Entlewomen, bicaufe I wold neither be miftaken of purpofe, neither mifconftrued of malice, leaft either the fimple fhould fufpeĉt me of folly, or the fubtile condemne me of blafphemy againft the noble fexe of women, I thought good that this my faith fhould be fet downe to finde fauour with the one, and confute the cauills of the other Beleeue me Gentlewomen, although I haue bene bold to inuay againft many, yet am I not fo brutifh to enuie them all, though I feeme not fo game-fome as *Ariftippus* to play with *Lais*, yet am I not fo dogged as *Diogenes* to abhorre all Ladies, neither would I, you fhould thinke me fo foolifh (although of late I haue ben very fantafticall) that for the lyght behauiour of a few I fhould cal in queftion the demeanour of all. I know that as there hath bene an vnchaft *Helen* in *Greece*, fo ther hath ben alfo a chaft *Penelope*, as ther hath ben a prodigious *Pafiphae*, fo

there hath bene a godly *Theocrita*, though many haue
defired to be beloued, as *Iupiter* loued *Alcmæna*, yet
fome haue wifhed to be embraced, as *Phrigius* embraced *Pieria*, as ther hath reigned a wicked *Iezabel*,
fo hath ther ruled a deuout *Debora*, though many
haue bene as fickle as *Lucilla*, yet hath there many
bene as faithful as *Lucretia*. Whatfoeuer therfore I
haue fpoken of the fplene againft the flights and
fubtilties of women, I hope ther is none wil miflike it,
if fhe be honeft, neither care I if any doe, if fhe be an
harlot. The fower Crabbe hath the fhew of an Apple
as well as the fweet Pippin, the blacke Rauen the
fhape of a bird, as wel as the white Swan, ye lewd
wight, the name of a woman as wel as the honeft
Matrone. There is great difference between the
ftanding puddle and the running ftreame, yet both
water : great oddes between the Adamant and the
Pommice, yet both ftones, a great diftinction to be
put betweene *Vitrum* and the Chriftall, yet both
glaffe : great contrarietie betweene *Lais* and *Lucretia*,
yet both women. Seeing therefore one may loue the
cleere Conduit water, though he loath the muddie
ditch, and weare the precious Diamonde, though he
difpife the ragged bricke, I thinke one may alfo with
fafe confcience reuerence the modeft fex of honeft
maidens, though he forfweare the lewd fort of
vnchaft minions. *Vlyffes* though he detefted *Calipfo*
with hir fugred voice, yet he embraced *Penelope* with
hir rude diftaffe. Though *Euphues* abhorre ye beautie
of *Lucilla*, yet wil he not abfteine from the company
of a graue mayden. Though the teares of the Hart
be falt, yet the teares of the Bore be fweete : though
the teares of fome women be counterfayte to deceiue,
yet the teares of many be currant to trye their loue.

I for my part will honour thofe alwayes that bee
honeft, and worfhip them in my life whom I fhall
know to be worthy in their liuinge : neither can I
promife fuch precifeneffe that I fhall neuer be caught
againe with the bayte of beautye, for although the

falfehood of *Lucilla* haue caufed me to forfake my
wonted dotage, yet ye faith of fome Lady may caufe
me once againe to fall into mine olde difeafe. For as
ye fire ftone in *Liguria* though it be quenched with
milke, yet again it* is kindled with water, or as the
rootes of *Auchufa* [*Anchufa*], though it be hardned with
water, yet it* is* againe [it is] made foft with Oyle, fo the
heart of *Euphues* enflamed earft with loue, although it
bee cooled with the deceites of *Lucilla*, yet will it againe
flame with the loyaltie of fome honeft Ladye, and
though it be hardned with the water of wilyneffe, yet
will it be molyfied with the Oyle of wifedome. I
prefume therefore fo much vppon the difcreation of
you Gentlewoemen that you will not thinke the worfe
of mee in that I haue thought fo ill of fome women, or
loue me the worfe in yat I loath fome fo much. For
this is my faith, that fome one Rofe will be blafted in
the bud, fome other neuer fall from the ftalke : that
the Oke will foone be eaten with the worme, the
Walnut tree neuer: that fome women wil eafily be
entifed to folly, fome other neuer allured to vanitie :
You ought therefore no more to bee agrieued with
that whiche I haue faide, then the Mint Maifter to fee
the coyner hanged, or the true fubiect the falfe traytour
araigned, or the honeft man the theefe condemned.

And fo farewell.

You haue heard (Gentlemen) how foone the hotte
defire of *Euphues* was turned into a cold deuotion, not
that fancy caufed him to chaunge, but that the fickleneffe of *Lucilla* enforced him to alter his minde.
Hauing therefore determined with himfelfe neuer againe
to be entangled with fuch fonde delyghts, according to
the appointment made with *Philautus*, he immedyatly
repayred to *Athens*, there to followe his owne priuate
ftudy : And calling to minde his former loofeneffe, and
how in his youth he had mifpent his time, he thought to

giue a Caueat to al parents, how they might bring [up] their
children vp*in vertue, and a commaundement to al youth,
how they fhould frame themfelues to their fathers in-
ftructions : in which is plainly to be feene, what wit can
and will doe, if it bee well imployed, which difcourfe
followinge, although it bring leffe pleafure to
your youthfull mindes then his firft [dif]courfe,
yet will it bring more profite : in the
one being conteyned the race
of a louer, in the other
the reafons of a
Philofo-
pher.

Euphues and his Ephœbus.

T is commonly faid, yet doe I thinke it a common lye, that experience is the miftreffe of fooles, for in my opinion they be moft fooles that want it. Neyther am I one of the leaft that haue tried this true, neither he onely that heretofore thought it to be falfe. I haue ben heere a ftudent of great welth, of fome wit, of no fmall acquaintance, yet haue I learned that by Experience, that I fhould hardly haue feene by learning. I haue thorowly fifted the difpofition of youth, wherein I haue founde more branne then meale, more dowe then leauen, more rage then reafon. Hee that hath beene burned knoweth the force of the fire, he that hath beene ftong, remembreth the fmart of the Scorpion, hee that hath endured the brunts of fancy, knoweth beft how to efchew the broiles of affe6tion. Let therefore my counfayle be of fuch authority as it may commaund you to be fober, your conuerfation of fuch integritie, as it may encourage mee to go forward in that which I haue taken in hand: the whole effe6t fhall be to fet downe a young man fo abfolute, as yat nothing may be added to his further perfe6tion. And although *Plato* hath bene fo curious in his common weale, *Ariftotle* fo precife in his happye man, *Tullie* fo pure in his Orator, that we may wel wifh to fee them, but neuer haue any hope to enioy them, yet fhal my young Impe be fuch an one as fhal be perfe6t euery way and yet common, if diligence and induftry be imployed to the atteining of fuch perfe6tion. But I would not haue young men flow to follow my precepts, or idle to deferre ye time lyke faint *George*, who is euer on horfebacke, yet neuer rydeth.

If my counfell fhal feeme rigorous to fathers to inftru6t their children, or heauy for youth to follow

their parents wil : let them both remember that the
Eſtrich diſgeſteth harde yron to preſerue his health,
that the ſouldier lyeth in his harneſſe to atchi[e]ue conqueſt, that the ſicke pacient ſwalloweth bitter pilles to
be eaſed of his griefe, that youth ſhoulde endure ſharpe
ſtormes to finde reliefe.

I my ſelfe had bene happie if I had bene vnfortunate, wealthy if left meanely, better learned if I had
bene better liued : we haue an olde (prouerbe) youth
wil haue his courſe. Ah Gentlemen, it is a courſe
which we ought to make a courſe accompt off, re-
plenyſhed with more miſeries then old age, with
more ſinnes then common cutthroats, with more
calamityes then the date of *Priamus* : we are no
ſooner out of the ſhell but wee reſemble the *Cocyx*
which deſtroyeth it ſelfe thorowe ſelfe will, or the
Pellican which perceth a wounde in hir owne breaſt :
we are either leade with a vaine glorye of our proper
perſonage, or with ſelfe loue of our ſharpe capacitie,
either entangled with beautie, or ſeduced by idle
paſtimes, either witcht with vycious company of others,
or inuegled with our owne conceits : of all theſe things
I may the bolder ſpeake, hauing tryed it true to mine
owne trouble.

To the intent therefore that all young Gentlemen
might ſhunne my former looſeneſſe, I haue ſet it down,
and that all might followe my future life, I meane
heere to ſhewe what fathers ſhoulde doe, what children
ſhoulde followe, deſiring them both not reieᶜt it bycauſe
it proceedeth from one which hath beene lewde, no
more then if they would negleᶜt the golde bicauſe it
lyeth in the durtye earth, or the pure wine for that
it commeth out of a [the] homelye preſſe, or the precious
ſtone *Aetites* whiche is founde in the filthy neaſtes of
the Eagle, or the precious gemme *Dacromtes* [*Draconites*] that is euer taken out of the heade of the poyſoned
Dragon, but to my [our] purpoſe.

¶ *That the childe ſhouldbe [be] true borne,*
no baſtarde.

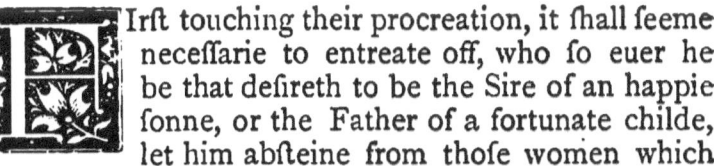Irſt touching their procreation, it ſhall ſeeme
neceſſarie to entreate off, who ſo euer he
be that deſireth to be the Sire of an happie ·
ſonne, or the Father of a fortunate childe,
let him abſteine from thoſe women which
be either baſe of birth, or bare of honeſtie : for if the
mother be noted of incontinencie, or the father of
vice, the childe wil either during life, be infeċted with
the like crime, or the trecheries of his parents, as
ignomy to him wil be caſt in his teeth : For we commonly cal thoſe vnhappie children which haue ſprong
from vnhoneſt parents. It is therfore a great treaſure
to the faiher, and tranquilitie to the minde of the
childe, to haue yat libertie, which both nature, law,
and reaſon hath ſet down. The guiltie conſcience of
a father that hath troden awry, cauſeth him to thinke
and ſuſpeċt yat his father alſo went not right, wherby
his owne behauiour is as it were a witneſſe, of his
owne baſeneſſe : euen as thoſe yat come of a noble
progenie boaſt of their gentrie. Heerevppon it came
that *Diophantus*, *Themiſtocles* his ſonne, would often
and that openly ſay in a great multitude, that whatſoeuer he ſhould ſeeme to requeſt of the *Athenians*, he
ſhould be ſure alſo to obteine, for ſaith he, whatſoeuer
I wil, that wil my mother, and what my mother ſaith
my father ſootheth, and what my father deſireth, that
the *Athenians* will graunt moſt willingly. The bolde
courage of the *Lacedemonians* is to be praiſed, which
ſet a fine on the heade of *Archidamus* their king, for
that he had maried a woman of a ſmal perſonage,
ſaying he minded to begette Queenes, not Kings to
ſucceede him. Lette vs not omitte that which our
aunceſtours were wont preciſely to keepe, that men
ſhould either be ſober, or drinke lyttle wine, that
would haue ſober and diſcreet children, for that the

fact of the father woulde be figured in the Infant. *Diogenes* therefore feeing a young man either ouercome with drincke or bereaued of his wittes, cryed with a loude voice, Youth, youth, thou hadſt a dronken Father. And thus much for procreation, now how the lyfe ſhould be ledde I will ſhewe briefely.

¶ *How the lyfe of a young man,* *ſhould be ledde.*

HERE are three things which cauſe per-fection in man, Nature, Reaſon, Vſe. Reaſon I call diſcipline, Vſe, Exerciſe, if anye one of theſe braunches want, cer-teinely the Tree of Vertue muſt needes wither. For Nature without Diſcipline is of ſmall force, and Diſcipline without Nature more feeble : if exerciſe or ſtudie be voyd of any of theſe it auayleth nothing. For as in tilling of the ground and huſ-bandry, there is firſt choſen a fertill ſoyle, then a cunning ſower, then good ſeede, euen ſo muſt we compare Nature to the fatte earth, the expert huſ-bandman to the Schoolemaſter, the faculties and ſciences to the pure ſeedes. If this order had not bene in our predeceſſors, *Pithagoras, Socrates, Plato,* and who ſo euer was renowmed in *Greece,* for the glorie of wifedome, they had neuer bene eterniſhed for wiſe men, neither canoniſed as it were for Sainēts, among thoſe that ſtudie Sciences. It is therefore a moſt euident ſigne of Gods ſingular fauour towards him that is endued with al theſe qualities without the* leaſt* of* the which, man is moſt miſerable. But if ther be any one that thinketh wit not neceſſary to the obteining of wifedome, after he hath gotten the waye to vertue by Induſtrie and Exerciſe, he is an Hereticke in my opinion, touching the true fayth of learning, for if Nature play not hir part in vaine is labour, and as I ſaid before, if ſtudie be not imployed, in vain is Nature. Sloth tourneth the edge of wit, Studie fharpeneth the minde, a thing be it neuer fo eafie is harde to the (idle) a thing be it neuer fo hard, is eafie to the wit well employed. And moft playnly we may fee in many things the efficacie of induftrie and labour.

The lyttle droppes of rayne pearceth hard Marble, yron with often handling is worne to nothing. Befides this, Induftrie fheweth hir felfe in other things, the fertill foyle if it be neuer tilled, doth waxe barren, and that which is moft noble by nature, is made moft vyle by neglygence. What tree if it be not topped beareth any fruite? What Vine if it be not proyned, bringeth foorth Grapes? Is not the ftrength of the bodye tourned too weakeneffe with too much delycacie, were not *Milo* his armes brawnefallen for want of wraft-lyng? Moreouer by labour the fierce Vnicorne is tamed, the wildeft Fawlchon is reclaimed, the greateft bulwarke is facked. It was well aunfwered of that man of *Theffalie,* who beeing demaunded, who among the *Theffalians* were reputed moft vile, thofe fayde hee that lyue at quyet and eafe, neuer giuing themfelues to martiall affaires : but what fhoulde one vfe many words in a thing already proued. It is Cuftome, Vfe, and Exercife, that bring a young man to Vertue, and Vertue to his perfeĉtion. *Lycurgus* the lawgiuer of the *Spartans* did nourifh two Whelpes both of one fire and one damme : But after a fundry manner, for the one he framed to hunt, and the other to lye alwayes in the chimneyes ende at the porredge pot, afterward calling the *Lacedemonians* into one affembly he faide : To the atteining of vertue ye *Lacedemonians,* Educa-tion, Induftrie, and Exercife, is the moft nobleft meanes, the truth of which I will make manifeft vnto you by tryal, then bringing forth the whelpes, and fetting downe there a pot and a Hare, the one ran at the Hare, the other to the porredge pot, the *Lacede-monians* fcarce vnderftanding this miftery, he faid : both of thefe be of one fire and one damme, but you fee how Education altereth Nature.

¶ Of the education
of youth.

T is moſt neceſſary and moſt naturall in mine opinion, that the mother of the childe be alſo the nurſe, both for the entire loue ſhe beareth to the babe, and the great deſire ſhe hath to haue it well nouriſhed : for is there any one more meete to bring vp the infant then ſhe that bore it ? or will any be ſo carefull for it, as ſhe that bredde it ? For as the throbs and throwes in child birth wrought hir paine, ſo the ſmiling countenaunce of the Infant increaſeth hir pleaſure, the hired nurſe is not vnlike to the hired ſeruaunt which not for good wil but gaine not for loue of the man but the deſire of the mony, accomplyſheth his dayes worke. Moreouer Nature in this poynt enforceth the Mother to nourſe hir owne childe, which hath giuen vnto euerye Beaſt milke to ſuccour hir owne, and me thinketh Nature to be a moſt prouident foreſeer and prouider for the ſame, which hath giuen vnto a woman two pappes, that if ſhee coulde conceiue two, ſhe might haue wherewith alſo to nouriſh twaine, and that by ſucking of the mothers breaſts there might be a greater loue both ot the mother towardes the childe, and the childe towards . the mother, which is very lykely to come to paſſe, for we ſee commonly thoſe that eate and drinke and liue together, to be more zealous one to the other, then thoſe that meete ſeldome, is not the name of a mother moſt ſweete ? If it be, why is halfe that title beſtowed on a woeman which neuer felt the paines in conceiuing, neither can conceyue the like pleaſure in nourſing [nouriſhing] as the mother doth? Is the earth called the mo·ther of all things onely bicauſe it bringeth forth ? No, but bicauſe it nouriſheth thoſe things that ſpringe out of it, whatſoeuer is bred in ye ſea, is fed in the ſea, no plant, no tree, no hearbe commeth out of the ground that is

not moyſtened and as it were nourſed of the moyſture and mylke of the earth : the Lyoneſſe nurſeth
hir whelps, the Rauen cheriſheth hir byrdes, the
Viper hir broode, and ſhal a woman caſt away hir
babe ?

I accompt it caſt away which in the ſwath clouts
is caſt aſide, and lyttle care can the Mother haue,
which can ſuffer ſuch crueltie : and can it be tearmed
with any other title then cruelty, the infant yet looking
redde of the mother, the mother yet breathing through
the torments of hir trauaile, the child crying for helpe
which is ſaid to moue wilde beaſtes, euen in the ſelfe
ſaid moment it is borne, or the nexte minute, to deliuer
to a ſtraunge nurſe, which perhappes is neither wholе-
ſome in body, neither honeſt in manners, whiche
eſteemeth more thy argent although a trifle, then thy
tender infant thy greateſt treaſure ? Is it not neceſſarye
and requiſite that the babe be nurſſed with that true
accuſtomed iuyce, and cheriſhed with his wonted
heate, and not fedde with counterfaite dyet ? Wheate
throwne into a ſtrange grounde tourneth to a contrary
graine, the vine tranſlated into an other ſoyle changeth
his kinde. A ſlyp pulled fro the ſtalke wythereth, the
young childe as it were ſlypped from the paps of his
Motner, either chaungeth his nature or altereth his
diſpoſition. It is pretely ſayd of *Horace*, a newe veſſel
will long time ſauour of that liquor that is firſt powred
into it, and the infant will euer ſmel of the nurſes
manners hauing taſted of hir milke. Therefore let the
Mother as often as ſhe ſhall beholde thoſe two foun-
taynes of milke, as it were of their owne accorde
flowing and ſwelling with liquor, remember that ſhe
is admoniſhed of nature, yea, commaunded of duetie,
to cheriſh hir owne childe, with hir owne teates,
otherwiſe when the babe ſhall now begin to tattle
and call hir Mamma, with what face can ſhe heare it
of his mouth, vnto whom ſhe hath denyed Mamma ?
It is not milke onely yat encreaſeth the ſtrength or
augmenteth the body, but the naturall heate and

I

agreement of the mothers body with the childes, it craueth the fame accuftomed moyfture that before it receiued in ye bowels by the which the tender partes were bound and knit together by the which it encreafed and was fuccoured in the body.

Certes I am of that minde, that the witte and dif-pofition is altered and chaunged by the mylke, as the moyfture and fap of the earth, doth chaunge the nature of that tree or plant that it nourifheth. Wherefore the common bye word of the common people feemeth to be grounded vpon good experience, which is: This fellow hath fucked mifchiefe euen from the teate of his nurfe. The *Grecians* when they faw any one fluttifhly fedde, they would fay euen as nurffes: whereby they noted the great · diflyking they had of their fulfome feedinge: the *Etimologie* of mother among ye *Grecians* may aptly be applyed to thofe mothers which vnna-turally deal with their children, they call it *Meter a meterine*, that is mother of not making much off, or of not nourifhing, heereoff it commeth that the fonne doth not with deepe defire loue his mother, neither with duetie obeye hir, his naturall affeChion being as it were deuided and dyftraught into twaine, a mother and a nurfe: heereoff it proceedeth that the Mother beareth but a colde kindneffe towards hir childe, when fhe fhall fee the nature of hir [the] nurfe in the nurture of hir [the] childe. The cheefeft way to learning is, if there be a mutual loue and feruent defire betweene the teacher and him that is taught, then verely the greateft furtheraunce to education is, if the Mother nouryfh the childe, and the childe fucke the Mother, that there bee as it were a relation and reciprocall order of affeChion. Yet if the Mother either for the euill habit of hir body or the weakeneffe of hir pappes, cannot though fhe would nurfe hir infant, then let hir prouide fuch a one as fhall be of a good compleChion, of honeft condition, carefull to tender the childe, louing, to fee well to it, willing to take paines, dilligent in tending and prouiding all things neceffary, and as lyke both in the liniaments of the body and dyfpofition of the minde to the mother as may bee. Let hyr forflow no occafion that may bring the childe to quyetneffe and cleanlyneffe, for as the parts of a childe as foone as it is borne, are framed and fafhioned of the midwife, that in all points it may be ftreight and comely, fo the manners of the childe at the firft are to be looked vnto that nothing difcommend the minde, that no crooked behauiour, or vndecent demeanour be found in the man.

Young and tender age is eafely framed to manners, and hardly are thofe things mollyfied which are hard. For as the fteele is imprinted in the foft waxe, fo learning is engrauen in ye minde of an young Impe. *Plato* that diuine Philofopher admonifhed all nurffes and weaners of youth, that they fhould not be too bufie to tell them fonde fables or filthy tales, leaft at theyr entraunce into the worlde they fhoulde bee contaminated with vnfeemely behauiour, vnto the which *Phocilides* the Poet doth pithely allude, faying: Whileft that the childe is young, let him be inftruéted in vertue and lytterature.

Moreouer they are to be trayned vp in the language of their country, to pronounce aptly and diftinétly without ftammering euery word and fillable of their natiue fpeach, and to be kept from barbarous talke, as the fhip from rockes: leaft being affeéted with their barbarifme, they be infeéted alfo with their vncleane conuerfation.

It is an olde Prouerbe that if one dwell the next doore to a cre[e]ple he will learne to hault, if one bee conuerfant with an hipocrit, he wil foone endeuour to diffemble. When this young infant fhall grow in yeares and be of that ripeneffe that he can conceiue learning, infomuch that he is to be committed to the tuityon of fome tutour, all dillygence is to be had to fearch fuch a one as fhall neither be vnlearned, neither ill lyued, neither a lyght perfon.

A gentleman that hath honeft and difcreet feruants

dyfpofeth them to the encrcafe of his Segnioryes, one
he appointeth ftewarde of his courtes, an other
ouerfeer of his landes, one his factor in far countries
for his merchaundize, an other puruayour for his cates
at home.   But if among all his feruaunts he fhal efpy
one, either filthy in his talke or foolifh in his behauior,
either without wit or voyde of honeftye, either an
vnthrift or a wittall, him he fets not as a furuayour
and ouerfeer of his manors, but a fuperuifour of hys
childrens conditions and manners, to him he committeth ye guiding and tuition of his fons, which is by
his proper nature a flaue a knaue by condition, a
beaft in behauior.   And fooner will they beftow an
hundreth crownes to haue a horfe well broken, then a
childe well taught, wherein I cannot but maruell to
fee them fo carefull to encreafe their poffeffions, when
they be fo careleffe to haue them wife that fhould
inherite them.

A good and difcreete fchoolemafter fhould be fuch
an one as *Phœnix* was the inftructor of *Achilles*, whom
Pelleus (as *Homer* reporteth) appoynted to that ende
that he fhould be vnto *Achilles* not onely a teacher of
learning, but an enfample of good lyuing.   But that is
moft principally to be looked for, and moft diligently
to be forefeene, that fuch tutors be fought out for the
education of a young childe, whofe lyfe hath neuer bene
ftayned with difhoneftie, whofe good name hath neuer
bene called vnto queftion, whofe manners hath ben
irreprehenfible before the world.   As hufbandmen
hedge in their trees, fo fhould good fchoolemafters with
good manners hedge in the wit and difpofition of the
fcholler, whereby the bloffomes of learning may the
fooner encreafe to a budde.

Many parents are in this to be miflyked, which
hauing neither tryal of his honeftie, nor experience of
his learning to whome they commit the childe to be
taught, without any deepe or due confideration put
them to one either ignoraunt or obftinate, the which if
they themfelues fhall doe of ignoraunce the folly can-

not be excufed, if of obftinacie, their lewdneffe is to bee abhorred.

Some fathers are ouercome with ye flatterie of thofe fooles which profeffe outwardly great knowledge, and fhew a certeine kinde of diffembling finceritie in their lyfe, others at the entreating of their familiar friends are content to commit their fonnes to one, without either fubftaunce of honeftie or fhadow of learning. By which their vndifcreet dealing, they are like thofe ficke men which reiect the expert and cunning Phifition, and at the requeft of their friendes admitte the heedeleffe practifer, which daungereth the patient, and bringeth the bodye to his bane: Or not vnlyke vnto thofe, which at the inftaunt and importunate fute of their acquaintaunce refufe a cunning Pilot, and chufe an vnfkilfull Marriner, which hazardeth the fhip and themfelues in the calmeft Sea.

Good God can there be any that hath the name of a father which will efteeme more the fancie of his friend then the nurture of his fonne? It was not in vayne that *Crates* would often fay, that if it were lawfull euen in the market place he would cry out: Whether runne you fathers, which haue all your carke and care to multiplye your wealth, nothing regarding your children vnto whom you muft leaue all. In this they refemble him which is very curious about the fhoe, and hath no care of the foote. Befides this there be [are] many Fathers fo inflamed with the loue of wealth, that they be as it were incenfed with hate agaynft their children : which *Ariftippus* feeing in an olde mifer did partlye note it, this olde mifer afking of *Ariftippus* what he woulde take to teache and bring vp his fonne, he aunfwered a thoufand groates : a thoufand groats, God fhield aunfwered this olde huddle, I can haue two feruaunts of yat price. Vnto whom he made aunfwere, thou fhalt haue two feruaunts and one fon, and whether wilt thou fell? Is it not abfurd to haue fo great a care of the right hande of

the childe to cut his meat, that if he handle his knife
in the left hand we rebuke him feuerely, and to be
fecure of his nourture in difcipline and learning? But
what doe happen vnto thofe parents, that bring vp
their children like wantons.

When their fonnes fhal grow to mans eftate, difdayning now to be corrected, ftubborne to obey,
giuing themfelues to vayne pleafures, and vnfeemelye
paftimes, then with the foolifh trowants they begin to
waxe wife and to repent them of their former follye,
when their fonnes fhall infinuate themfelues in the
company of flatterers, (a kinde of men more perillous
to youth then any kinde of beaftes.) When they fhall
haunt harlottes, frequent tauerns, be curious in their
attyre, coftlye in their dyet, careleffe in their behauiour,
when they fhall either bee common Dicers with
Gamefters, either wanton dalliers with Ladies, either
fpend al their thrift on wine, or al their wealth on
women : then the Father curffeth his owne fecuritie,
and lamenteth too late his childes miffortune, then
the one accufeth his Sire as it were of malyce, that
hee woulde not bring him vp in learning, and himfelfe of mifchiefe, that he gaue not his minde to good
letters. If thefe youthes had bene trayned vppe in
the company of any Philofopher, they would neuer
haue ben fo diffolute in their life, or fo refolute in their
own conceipts.

It is good nurture that leadeth to vertue, and difcreete demeanour that playneth the path to felicitie.
If one haue either the giftes of Fortune, as greate
riches, or of Nature, as feemely perfonage, he is to be
difpifed in refpect of learning.   To be a noble man it
is moft excellent, but that is our aunceftours, as *Vliffes*
fayde to *Aiax*, as for our nobilytie, our ftocke, our
kindred, and whatfoeuer we our felues haue not done,
I fcarcely accompt ours.   Riches are precious, but
Fortune ruleth the roft, which oftentimes taketh away
all from them that haue much, and giueth them more
that had nothing, glory is a thing worthy to be followed,

but as it is gotten with great trauaile, fo is it loft in a fmall time.

Beautie is fuch a thing as we commonly preferre before all things, yet it fadeth before we perceiue it to flourifh : health is that which all men defire, yet euer fubieft to any difeafe : ftrength is to be wifhed for, yet is it either abated with an ague, or taken away with age : whofoeuer therefore boafteth of force, is too beaftly, feeing hee is in that qualytie not to be compared with beaftes, as the Lyon, the Bull, the Elephant.

It is vertue, yea vertue Gentlemen, that maketh gentlemen : that maketh the poore rich, the bafe borne noble, the fubieft a fouereigne, the deformed beautiful, the ficke whole, the weake ftrong, the moft miferable moft happy. There are two principall and peculiar gifts in the nature of man, Knowledge and Reafon : the one commaundeth, the other obeyeth : thefe things neither the whirling wheele of Fortune can chaunge, neither the deceitful cauilling of world-lings feperate, neither fickeneffe abate, neither age abolifh.

It is onely Knowledge, which worne with yeares waxeth young, and when all things are cut away with the Cicle [fickle] of Time, Knowledge flourifheth fo high that Time cannot reach it. Warre taketh all things with it euen as the whirlepoole, yet muft it leaue learning behinde it, wherefore it was wifely aunfwered in my opinion of *Stilpo* the Philofopher, for when *Demetrius* wonne the Citie, and made it euen to the ground leauing nothing ftanding, he demaunded of *Stilpo* whether he had loft any thing of his in this great fpoyle : vnto whom he aunfwered, no verely, for warre getteth no fpoyle of vertue.

Vnto the lyke fence may the aunfwere of *Socrates* be applyed, when *Gorgias* afked him whether hee thought the *Perfian* king happy or not : I knowe not faide hee, howe much vertue or difcipline he hath, for happineffe doth not confift in the gifts of fortune, but

in ye grace of vertue.   But as there is nothing more
conuenient then inftruction for youth, fo would I haue
them nurtured in fuch a place as is renowmed for learning, voyde of incorrupt [corrupt] manners, vndefiled
with vice, that feeing no vaine delyghtes, they may the
more eafily abftein from lycencious defires, they that
ftudy to pleafe the multytude are fure to difpleafe the
wife, they that feeme to flatter rude people with their
rude pretences, leuell at great honour hauing no ayme
at honefty.   When I was heere a ftudent in *Athens*, it
was thought a great commendation for a young fcholler
to make an Oration extempore, but certeinely in my
iudgement it is vtterly to be condemned, for whatfoeuer is done rafhly is done alfo rawly, he that taketh
vpon him to fpeake without premeditation, knoweth
neither howe to beginne, nor where to ende, but
falling into a vayne of babling, vttereth thefe thinges
which with modeftye he fhould haue concealed, and
forgetteth thofe things that before he had conceiued.
An Oration either penned, either premeditated, keepeth
it felfe within the bonds [bands] of *Decorum*, I haue read
that *Pericles* being at fundrye times called of the
people to pleade, would alwayes anfwere that he was
not ready : euen after the fame manner *Demofthenes*
being fent for to declaime amiddeft the multitude,
ftayd and faide, I am not yet prouided.

And in his inuectiue againft *Mydas*, he feemeth to
praife the profitableneffe of premeditation, I confeffe
faith he, yee *Athenians*, that I haue ftudied and confidered deepely with my felfe what to fpeake, for I
were a fotte if without due confideration had of thofe
things that are to be fpoken, I fhould haue talked
vnaduifedly.   But I fpeake this not to this ende to
condemne the exercife of the wit, but that I woulde
not haue any young fcholler openly to exercife it, but
when he fhall grow both in age and eloquence, infomuch as hee fhall through great vfe and good memory
bee able aptly to conceiue and readily to vtter any
thing then this faying, extempore bringeth an admira-

tion and delight to the auditory, and finguler praife and commendation to the Orator. For as he yat hath long time ben fettered with chaynes, being releafed, halteth through the force of his former yrons, fo he that hath bene vfed to a ftrickt kinde of pleading, when he fhal talke extempore wil fauour of his former penning. But if any fhal vfe it as it were a precept for youth to tatle extempore, he wil in time bring them to an immoderate kinde of humilytie. A certeine Painter brought *Appelles* the counterfaite of a · face in a table, faying : loe *Appelles*, I drew this euen now, whervnto he replyed. If thou hadſt ben filent, I would haue iudged this picture to haue been framed of the fodein. I meruaile yat in this time thou couldeſt not paint many more of thefe. But return we again, as I would haue tragical and ſtately ſtile fhunned, fo would I haue yat abiect and bafe phrafe efcheued, for this fwelling kind of talk hath little modeſty, the other nothing moueth.

Befides this, to haue the Oration al one in euery part, neither adorned with fine figures, neither fprinkled with choyfe phrafes, bringeth tedioufneffe to the hearers, and argueth the fpeaker of little learning and leffe eloquence. He fhoulde moreouer talke of many matters, not alwayes harp vpon one ſtring, he that alwayes fingeth one note without defkant [Defcant] breedeth no delight, he yat alwayes playeth one part, bringeth lothfomeneffe to the eare. It is varietie that mooueth the minde of al men, and one thing faid twice (as we fay commonly) deferueth a trudge. *Homer* woulde faye, that it loathed him to repeat any thing again though it were neuer fo pleafaunt or profitable. Though the Rofe be fweet yet being tyed with the Vyolet the fmel is more fragraunte, though meat nouriſh, yet hauing good fauor it prouoketh the* appetite. The fayreſt nofegay is made of many flowers, the fineſt picture of fundry colours, ye wholfomeſt medicine of diuers hearbs : wherefore it behoueth youth with all induſtry to fearch

not onely the hard queftions of the Philofophers, but
alfo the fine cafes of ye Lawyers, not only the quirks
and quiddities of the *Logicians*, but alfo to haue a fight
in the numbers of the *Arithmetricians*, the Tryangles
and Circles of the *Geometricians*, the Spheere and
Globe of the *Aftrologians*, the notes and crochets of
the Mufitions, the odd conceits of the Poets, the
fimples of the Phifitions, and in all things, to the ende
that when they fhal be willed to talke of any of them,
they may be ignorant in nothing.   He that hath a
garden plot doth afwel fow the Pothearb as the Mar-
gerom, as wel the Leeke as the Lylly, as wel ye
wholfome Ifoppe, as the faire Carnation, the which he
doth to the intent he may haue wholefome hearbs as
wel to nourifh his inward parts as fweet flowers to
plefe his outward defire, as wel fruitfull plants to
refrefh his fences, as faire fhewes to pleafe his fight.
Euen fo whofoeuer that hath a fharpe and capable
witte, let him as well giue his mind to facred knowledge of diuinitie, as to the profound ftudy of
Philofophye, that by his wit he may not onely reape
pleafure but profit, not only contentation in minde
but quietneffe in confcience.   I will proceede in the
education.

   I would haue them firft of all to followe Philofophy,
as moft auncient, yea, moft excellent, for as it is
pleafaunt to paffe thorowe many faire cities, but moft
pleafant to dwell in the fayreft : euen fo to read many
Hiftories and artés it is pleafaunt, but as it were to
lodge with Philofophy moft profitable.

   It was pretely faide of *Byon* the Philofopher : Euen
as when the wo[o]ers could not haue the companye of
Penelope, they ranne to hir handmaydens : fo they that
cannot atteine to the knowledge of Philofophy, apply
their mindes to things moft vile and contemptible.
Wherefore we muft preferre Philofophy, as the onely
Princeffe of al Sciences, and other arts as wayting
Maydes.   For the curing and keeping in temper of
the body, man by his induftry hath found two things,

Phificke and Exercife, the one cureth ficknefſe, the other preferueth the body in temper: but ther is nothing that may heale difeafes or cure the woundes of the minde but onely Phylofophy. By this fhal we learne what is honeft, what difhoneft, what is right, what is wrong, and yat I may in one word fay what may be faid, what is to be knowen, what is to be a voyded: how we ought to obey our parents, reuerence our elders, entertein ftrangers, honour Magiftrates, loue our friends, liue with our wiues, vfe our feruauntes. Howe wee fhoulde worfhippe God, bee duetifull to our Fathers, ftand in awe of our fuperiours, obey lawes, giue place to Officers, how we may choofe friends, nurture our children and that which is moft noble, how we fhould neither be too proude in profperitie, neither penfiue in aduerfitie, neither like beaftes ouer-come with anger. And heere I cannot but lament *Athens*, which hauing ben alwayes ye nurfe of Philofo-phers, doth now nourifh only the name of Philofophy. For to fpeak plainly of the diforder of *Athens*, who doth not fe it, and forrow at it? fuch playing at dice, fuch quaffing of drink, fuch daliaunce with women, fuch dauncing, that in my opinion there is no quaffer in *Flaunders* fo giuen to tipplyng, no Courtier in *Italy* fo giuen to ryot, no creature in the world fo mifled, as a ftudent in *Athens*. Such a confufion of degrees, that the fcholler knoweth not his dutie to the Bachelor, nor the Bachelor to the Mafter, nor the Mafter to the Doctor. Such corruption of manners, contempt of Magiftrates, fuch open finnes, fuch priuie villanye, fuch quarrelling in the ftreets, fuch fubtile practifes in chambers, as maketh my heart to melt with forrow to thinke of it, and fhould caufe your mindes Gentlemen to bee penitent to remember it.

Moreouer, who doth know a fcholler by his habite? Is there any hat of fo vnfeemely a fafhion, anye dublet of fo long a wafte, any hofe fo fhort, any attyre, either fo coftly or fo courtly, either fo ftraunge in making, or fo monftrous in wearing, that is not worn of a

Scholler? haue they not now in fteede of black cloth
blacke veluet, in fteede of courfe fackecloth, fine
filke? Be they not more lyke courtiers then fchollers,
more like ftage-players then ftudents, more like
ruffians of *Naples* then difputers in *Athens*? I would
to god they did not imitate al other nations in ye vice
of the minde, as they doe in the attire of their body,
for certeinely as there is no nation whofe fafhion in
apparel they do not vfe, fo there is no wickedneffe
publyfhed in anye place, that they do not practife. I
thinke that in *Sodom* and *Gomora*, there was neuer
more filthineffe, neuer more pride in *Rome*, more
poyfoning in *Italy*, more lying in *Crete*, more priuie
fpoylyng in *Spayne*, more Idolatry in *Aegypt*, then is
at this day in *Athens*, neuer fuch fects among the
Heathens, fuch fchifmes amongft the *Turkes*, fuch
mifbeliefe among ye Infidels, as is now among
Schollers.

But there not many in *Athens* which thinke there is
no God, no redemption, no refurrection?

What fhame is this gentlemen, that a place fo
renowmed for good learning fhould be fo fhamed for
ill lyuing? that where grace doth abounde, finne
fhoulde fo fuperabound? that where the greateft profeffion of knowledge is, ther fhould alfo be the leaft
practifing of honeftie. I haue read of many Vniuerfities, as of *Padua* in *Italy*, *Paris* in *Fraunce*, *Witten-
berge* in *Germany*, in *England* of *Oxford* and *Cambridge*,
which if they were halfe fo ill as *Athens* they were too
too bad, and as I haue heard, as they be, they be
ftarke naught.

But I can fpeake the leffe againft them for that I
was neuer in them, yet can I not choofe but be
agrieued, that by report, I am enforced rather to
accufe them of vanitie, then excufe them any way.
Ah Gentlemen what is to be looked for, nay, what is
not to be feared, when the temple of *Vefta* wher
virgins fhould liue is like the ftewes fraught with
ftrumpets, when the alter, wher nothing but fanctitie

and holyneffe fhould be vfed, is polluted with vncleanneffe, when the vniuerfities of Chriftendome which
fhould be the eyes, the lights, the leauen, the falt, the
feafoning of the world, are dimmed with blinde concupifcence, put out with pride, and haue loft their
fauour with impietie.

Is it not become a bye word amongft the common-
people, that they had rather fende their children to the
carte, then to the Vniuerfitie, being induced fo to fay,
for the abufe that reigneth in the Vniuerfities, who
fending their fonnes to atteine knowledge, find them
little better learned, but a great deale worfe liued,
then when they went, and not onely vnthrifts of their
money, but alfo banckerouts of good manners : was
not this the caufe that caufed a fimple woman in
Greece, to exclayme againft *Athens*, faying : The
Mafter and the Scholler, the Tutor and the Pupil
be both agreed, for the one careth not how lyttle
payne he taketh for his mony, the other how lyttle
learning.

I perceiue that in *Athens* ther be no chaunglyngs :
when of olde it was fayd to a *Lacedemonian*, that all
the *Grecians* knewe honeftie, but not one practifed it.
When *Panthœnea* wer celebrated at *Athens*, an olde
man gooing to take a place was mockingly reiected,
at the laft comming among the *Lacedemonians*, all the
youth gaue him place, which the *Athenians* liked
wel off, then one of the *Spartans* cryed out : Verily
the *Athenians* know what fhould be done, but they
neuer doe it.

When one of the *Lacedemonians* had ben for a
certeine time in *Athens*, feeing nothing but dauncing,
dicing, banquetting, furfetting, and licentious behauiour, retourning home, he was afked how all
things ftoode in *Athens*, to whom hee aunfwered all
things are honeft ther, meaning that the *Athenians*
accompted all things good, and nothing badde.

How fuch abufes fhould or might be redreffed in al
Vniuerfities efpecially in *Athens*, if I were of authoritie

to commaund, it fhould foone be feene, or of credite
to perfwade thofe yat haue the dealings with them, it
fhould foone be fhowne.

And vntill I fee better reformation in *Athens*, my
young *Ephœbus* fhall not be nourtured in *Athens*, I
haue fpoken all this, that you Gentlemen might fee
how the Philofophers in *Athens* practife nothing leffe
then Philofophie, what fcholler is he that is fo zealous
at his booke as *Chrifippus*, who had not his mayd
Meliffa thruft meate in his mouth, had perifhed with
famine, beeing alwaye ftudying? Who fo watchfull as
Ariftotle, who going to bed would haue a bal of braffe
in his hande, that if hee fhoulde bee taken in a flumber
it might fall and awake him? No, no, the times are
chaunged as *Ouid* faith, and we are chaunged in the
times, let vs endeauour euery one to amend one, and
we fhall all foone be amended, let vs giue no occafion
of reproch and we fhall more eafely beare the burden
of falfe reportes, and as wee fee by learning what we
fhould doe, fo lette vs doe as we learne, then fhall
Athens flourifh, then fhal the ftudents be had in great
reputacion, then fhall learning haue his hire, and euery
good fcholler his hope. But returne we once againe
to *Philo*.

Ther is amongft men a trifold kinde of life, Actiue,
which is about ciuill function and adminiftration of
the common weale. Speculatiue, which is continuall
meditation and ftudie. The thirde a lyfe ledde, moft
commonly a lewde lyfe, an idle and vaine life, the life
that the *Epicures* accompt their whole felicitie, a
voluptuous lyfe replenifhed with all kinde of vanitie,
if this actiue life be without philofophie, it is an idle
life, or at the leaft a life euill imployed which is worfe:
if the contemplatiue lyfe be feperated from the Actiue,
it is moft vnprofitable.

I woulde therefore haue my youth, fo to beftowe his
ftudie, as he may be both exercifed in the common
weale to common profite, and well employed priuately
for his owne perfection, fo as by his ftudie the rule

he fhal beare may be directed, and by his gouernment his ftudie may be increafed : in this manner did *Pericles* deale in ciuill affaires, after this fort did *Architas* [the] *Tarentine, Dion* the *Syracufian,* the *Theban Epaminondas* gouerne their cities.

For the exercife of the body it is neceffary alfo fomwhat be added, that is, that the child fhould be at fuch times, permitted to recreate himfelfe, when his minde is ouercome with ftudye, leaft dullyng himfelfe with ouermuch Induftrie he become vnfitte afterwarde to conceiue redily, befides this, it wil caufe an apt compofition and yat natural ftrength yat it before reteined. A good compofition of the body, layeth a good foundation of olde age, for as in the fayre Summer wee prepare [repaire] all thinges neceffarye for the colde winter, fo good manners in youth and lawful exercifes be as it were victualls and nourifhments for age, yet are their labours and paftimes fo to be tempered, that they weaken not their bodyes more by play, then otherwife they fhould haue done by ftudie, and fo to be vfed that they addict not themfelues more to ye exercife of the limmes then the following of learninge : the greateft enimyes to difcipline, as *Plato* recompteth, are labours and fleepe. It is alfo requifite that he be expert in marcyall affayres, in fhootinge, in dartinge, that hee hauke and hunte for his honeft paftime and recreation, and if after thefe paftimes hee fhall feeme fecure, nothing regardinge his bookes, I woulde not haue him fcourged with ftripes, but threatened with wordes, not dulled with blowes, lyke feruaunts, the which the more they are beaten the better they beare it, and the leffe they care for it, for children of good difpofition are either incited by praife to goe forward, or fhamed by difprayfe to commit the like offence : thofe of obftinate and blockifh behauiour, are neither with wordes to be perfwaded, neither with ftripes to bee corrected. They muft nowe be taunted with fharpe rebukes, ftraight wayes admonifhed with fayre wordes, now threatned a payment, by and by promifed a reward, and dealt withal as nurffes do with the [their] babes, whom after they haue made to cry they profer the teate, but diligent heede muft be taken that he be not praifed aboue meafure, leaft ftanding too much in his own conceit, he become alfo obftinate in his owne opinions. I haue knowne many fathers whofe great loue towards their fonnes hath ben the caufe in time that they loued them not, for when they fee a fharpe wit in their fonne to conceiue, for the defire they haue that he fhould outrunne his fellowes, they loaden him with continual exercife, which is the onely caufe that he finketh vnder his burden, and giueth ouer in the plaine fielde. Plantes are nourifhed with little raine, yet drowned with much : euen fo the mind with indifferent labour waxeth more perfeĉt, with [ouer-] much ftudye it is made fruitleffe. We muft confider that all our life is deuided into remiffion and ftudy.

As there is watchinge, fo is there fleepe : as there is warre, fo is there peace : as there is winter, fo is there Summer : as there be many working dayes, fo is there alfo many holy-dayes : and if I may fpeak al in one worde, eafe is the fauce of labour, which is plainly to be feene, not onely in lyuing thinges, but alfo in thinges without life. Wee vnbend the bowe that wee maye the better bend him, we vnloofe the Harpe, that we may the fooner tune him, the body is kept in health as well with fafting as eating, the minde healed with eafe, as wel as with labour : thofe parents are in mind to be mifliked which commit the whole care of theyr childe to the cuftodye of a hyrelinge, neither afkinge neither knowing howe their children profite in learning. For if the father were defirous to examine his fonne in that which he hath learned, the mafter would be more carefull what he did teach. But feeing the father careleffe what they learne, he is alfo fecure what he teacheth : that notable faying of the horfe-keeper may [be] here bee* applyed, which faid, nothing did fo fatte the horfe as the eye of the king. More-ouer I would haue the memorye of children continually

to be exercyfed, which is the greateft furtheraunce to
learninge that can be.

For this caufe they fayned in their olde fables,
memory to be the mother of perfection. Children
are to be chaftifed if they fhal vfe any filthy or
vnfeemely talk, for as *Democrates* faith, the worde is
the fhadowe of the worke : they muft be curteous in
their behauiour, lowly in their fpeach, not difdayning
their cockmates or refraining their company : they
muft not liue wantonly, neither fpeake impudently,
neither angry without caufe, neither quarellous without
colour. A young man beeing peruerfe in nature and
proud in words and manners, gaue *Socrates* a fpurne,
who being moued by his fellowes to giue him an other,
if fayde *Socrates* an Affe had kycked mee, would you alfo
haue mee to kick him againe, the greateft wifedome in
Socrates in compreffing [fuppreffing] his anger is worthy
great commendation. *Architas* [ye] *Tarentine*, returning
from war and finding his ground ouergrowen with
weeds, and turned vp with Mowles, fent for his
farmour, vnto whome hee fayde, if I were not angry
I would make thee repent thy ill hufbandry. *Plato*
hauing a feruaunt whofe bliffe was in filling of his
belly, feeing him on a .time idle and vnhoneft in behauiour, faid, out of my fight, for I am incenfed with
anger.

Although thefe enfamples [examples] be hard to imitate, yet fhould euery man do his endeuour to repreffe
that hot and heady humor which he is by nature fubiect
vnto. To be filent and difcreete in companye, though
many thinke it a thing of no great wayght or import-
aunce, yet is it moft requifite for a young man and moft
neceffary for my *Ephœbus.* It neuer hath bene hurtfull to any to holde his peace, to fpeake, damage to
many : what fo is kept in filence is hufht, but whatfoeuer is babled out, cannot againe be recalled. We
may fee the cunning and curious work of Nature,
which hath barred and hedged nothing in fo ftrongly
as the tongue, with two rowes of teeth, and therewith

K

two lips, befid[e]s fhe hath placed it farre from the heart, that it fhoulde not vtter that which the heart had conceiued, this alfo fhoulde caufe· vs to be filent, feeinge thofe that vfe much talke, though they fpeake truely are neuer beleeued. Wyne therefore is to be refrained, which is termed to be the glaffe of the minde, and it is an old Prouerbe, Whatfoeuer is in the heart of the fober man, is in the mouth of the drunckarde. *Bias* holdinge his tongue at a feaft, was tearmed there of a tatler to be a foole, who faid, is there any wife man that can hold his tongue amidft the wine? vnto whom *Bias* anfwered, there is no foole that can.

A certeine Gentleman heere in *Athens*, inuited the Kings Legats to a coftly and fumptuous feaft, wher alfo he affembled many Philofophers, and talking of diuers matters, both of the common weale and learning, onely *Zeno* faid nothing. Then the ambaffadors faid, what fhall we fhewe of thee O *Zeno* to the king. Nothing aunfwered he, but that there is an olde man in *Athens* that amiddeft the pottes could hold his peace. *Anacharfis* fupping with *Solon*, was founde a fleepe, hauing his right hande before his mouth, his left vpon his priuities, wherby was noted that ye tongue fhould bee rayned with the ftrongeft brydie. *Zeno* bicaufe hee woulde not be enforced to reueale any thing againft his will by torments, bit of his tongue and fpit it in the face of the tyrant.

Nowe when children fhall by wifdome and vfe refrayne from ouer-much tatling, let them alfo be admonifhed that when they fhall fpeake, they fpeake nothing but truth : to lye is a vice moft deteftable, not to be fuffered in a flaue, much leffe in a fonne. But the greateft thing is yet behinde, whether that thofe are to bee admytted as cockemates with children whiche loue them entirely, or whether they be to be banifhed from them.

When as I fee many fathers more cruell to their children then carefull of them, which thinke it not

neceffarye to haue thofe about them, that moft tender
them, then I am halfe as it were in a doubte to giue
counfayle. But when I call to my remembraunce,
Socrates, *Plato*, *Xenophon*, *Efchines*, *Sœbetes*, and all
thofe that fo much commend the loue of men, which
haue alfo brought vp many to great rule, reafon, and
pietie, then I am encouraged to imitate thofe whofe ·
excellencie doth warrant my precepts to be perfeft [true].
If any fhall loue the childe for his comely countenaunce, him would I haue to be banifhed as a moft
daungerous and infeftious beaft, if he fhall loue him
for his fathers fake or for his own good qualities, him
would I haue to be with him alwayes, as fuperuifour of
his manners : fuch hath it bene in times paft, the loue
of one *Athenian* to the other, and of one *Lacedemonian*
to the other.

But hauing faide almoft fufficient for the education
of a childe, I wil fpeake two words, how he fhould be
trayned when he groweth in yeares. I cannot but
miflyke the Nature of diuers Parents which appoynt
ouerfeers and tutors for their children in their tender
age, and fuffer them when they come to be young
men, to haue the bridle in their owne hande, knowing
not that age requireth rather a harde fnaffle, then a
pleafaunt bit, and is fooner allured to wickednes then
childehoode.

Who knoweth not the efcapes of children, as they
are fmal fo they are foone amended? either with
threats they are to be remedied, or with faire promifes
to bee rewarded. But the finnes and faults of young
men are almoft or altogether intollerable, which giue
themfelues to be delicate in their dyet, prodigall in
their expence, vfing dicing, dauncing, dronkennes,
deflowring of virgins, abufing wiues, committing adul-
teries, and accounting al things honeft, that are moft
deteftable. Heere therefore muft be vfed a due
regarde that their luft may be repreffed, their ryot
abated, their courage cooled : for harde it is to fee a
young man to be Mafter of himfelfe, which yeeldeth

himfelfe as it were a bond flaue to fonde and ouerlafhing
affections.   Wife Parents ought to take good heede,
efpecially at this time, yat they frame their fonnes to
modeftie, either by threats or by rewards, either by faire
promifes or feuere practifes, either fhewing the miferies
of thofe that haue ben ouercome with wildneffe, or ye
happineffe of them that haue conteined [contented]
themfelues, within the bandes of reafon : thefe two are
as it wer the enfignes of vertue, the hope of honour,
the feare of punifhment.   But chiefly parents muft
caufe their youths to abandon the focietie of thofe
which are noted of euill liuing and lewde behauiour,
which *Pithagoras* feemed fomwhat obfcurely to note in
thefe his fayings.

Firft, that one fhould abftein from the taft of thofe
things that haue blacke tayles : That is we muft not
vfe the company of thofe whofe corrupt manners doe
as it were make their lyfe blacke.   Not to goe aboue
the ballaunce, that is to reuerence Iuftice, neither for
feare or. flatterie to leane vnto any one partially.
Not to lye in idleneffe, that is, that floth fhoulde be
abhorred.   That we fhould not fhake euery man by
ye hand : That is, we fhould not contract friendfhippe
with all.   Not to weare a ftraight ring : that is, that
we fhoulde leade our lyfe, fo as wee neede not to fetter
it with chaynes.   Not to bring fire to a flaughter :
that is, we muft not prouoke any that is furious with
words.   Not to eate our heartes : that is, that wee
fhoulde not vexe our felues with thoughts, confume our
bodies with fighes, with fobs, or with care to pine our
carcaffes.   To abfteine from beanes, that is, not to
meddle in ciuile affaires or bufineffe of the common
weale, for in the old times the election of Magiftrates
was made by the pullyng of beanes.   Not to put
our meat in *Scapio :* that is, we fhould not fpeake of
manners or vertue, to thofe whofe mindes are [be] infected with vice.

Not to retire when we are come to the ende of
our race : that is, when we are at the poynt of death

we fhould not be oppreffed with griefe, but willingly
yeeld to Nature. But I will retourne to my former
precepts : that is, that young men fhoulde be kept
from the company of thofe that are wicked, efpecially
from the fight of ye flatterer. For I fay now as I
haue often times before fayde, that there is no kinde
of beaft fo noyfome as the flatterer, nothing that will
fooner confume both the fonne and the father and all
honeft friendes.

When the Father exhorteth the fonne to fobrietie, the
flatterer prouoketh him to Wine : when the Father
weaneth [warneth] them to continencie, the flatterer allureth them [him] to luft : when the Father admonifheth
them to thrifte, the flatterer haleth them to prodigaly-
tie, when the Father incourageth them to labour, the
flatterer layeth a cufhion vnder his elbowe, to fleepe,
bidding them [him] to eate, drinke, and to be merry, for
that the lyfe of man is foone gone, and but as a fhort
fhaddowe, and feeing that we haue but a while to lyue,
who woulde lyue [doe] lyke a feruant ? They faye that
now their fathers be olde, and doate through age like
Saturnus.

Heeroff it commeth that young men giuing not
only attentiue eare but ready coyne to flatterers, fall
into fuch miffortune : heereoff it proceedeth that they
haunt the ftewes, mary before they be wife, and dye
before they thriue. Thefe be the beaftes which liue
by the trenchers of young Gentlemen, and confume
the treafures of their reuenewes, thefe be they that
footh young youths in al their fayings, that vphold
them in al their doings, with a yea, or a nay, thefe be
they that are at euery becke, at euery nod, freemen by
fortune, flaues by free will.

Wherfore if ther be any Fathers that would haue
his children nurtured and brought vp in honeftie, let
him expell thefe Panthers which haue a fweete fmel,
but a deuouring minde : yet would I not haue parents
altogether precife, or too feuere in correction, but lette
them with mildeneffe forgiue light offences, and

remember that they themfelues haue ben young : as ye Phifition by minglyng bitter poyfons with fweete lyquor, bringeth health to the body, fo the father with fharpe rebukes, fefoned with louing lookes caufeth a redreffe and amendement in his childe. But if the Father bee throughly angry vppon good occafion, let him not continue his rage, for I had rather he fhould be foone angry then hard to be pleafed, for when the fonne fhall perceiue that the Father hath conceiued rather a hate then a heat agaynſt him, hee becommeth defperate, neither regarding his fathers ire, neither his owne duetie.

Some lyght faults lette them diffemble as though they knew them not, and feeing them let them not feeme to fee them, and hearing them, lette them not feeme to heare. We can eafely forget ye offences ot our friendes be they neuer fo great, and fhall wee not forgiue the efcapes of our children be they neuer fo fmall? Wee beare oftentimes with our feruaunts, and fhal we not fometimes with our fonnes : the faireſt Iennet is ruled as well with the wande as with the fpurre, the wildeſt child is as foone corrected with a word as with a weapon. If thy fonne be fo ſtubburne obſtinately to rebel againſt thee, or fo wilful to per-feuer in his wickedneffe, yat neither for feare of punifhment, neither for hope ot reward, he is any way to be reclaymed, then feeke out fome mariage fit for his degree, which is the fureſt bond of youth, and the ſtrongeſt chayne to fetter affections yat can be found. Yet let his wife be fuch a one as is neither much more noble in birth or far more richer in goods, but according to the wife faying : choofe one euery way, as neere as may be equal in both : for they that do defire great dowryes do rather mary themfelues to the wealth then to their wife. But to returne to the matter, it is moſt requifite that fathers both by their difcreete counfayle, and alfo their honeſt conuerfation, be an example of imitation to their children, yat they feing in their parents, as it were in a glaffe, the per-

fection of manners, they may be encouraged by their
vpright liuing to practife the like pietie. For if a
father rebuke his child of fwearing, and he himfelfe a
blafphemor, doth he not fee that in detecting his fons
vice, hee alfo noteth his owne ? If the father counfaile
the fonne to refrayne wine as moft vnwholfome, and
drinke himfelfe immoderately, doth hee not as well
reproue his owne folly, as rebuke his fonnes? Age
alway ought to be a myrrour for youth, for where olde
age is impudent, there certeinly youth muft needes be
fhamelefe, where the aged haue no refpect of their
honorable and gray haires, there the young gallants
haue little regard of their honeft behauiour : and in
one worde to conclude al, wher age is paft grauity
ther youth is paft grace. The fum of al wherwith I
would haue my *Ephœbus* endued, and how I would
haue him inftructed, fhal briefly appeare in this
following. Firft, that he be of honeft parents, nurfed
of his mother, brought vp in fuch a place as is
incorrupt, both for the ayre and manners with fuch a
perfon as is vndefiled, of great zeale, of profound
knowledge, of abfolute perfection, yat be inftructed in
Philofophy, whereby he may atteine learning, and
haue in al fciences a fmacke, whereby he may readily
difpute of any thing. That his body be kept in his
pure ftrength by honeft exercife, his wit and memory
by diligent ftudy.

That he abandon al allurements of vice, and continually encline to vertue, which if it fhall as it may
come to paffe, then do I hope that if euer *Platoes*
common weale fhal flourifh, that my *Ephœbus* fhall
bee a citizen, yat if *Ariftotle* fined any happy man it
wil be my childe, if *Tully* confeffe any to be an
abfolute Orator, it will be my young youth. I am
heere therefore gentlemen to exhort you, that with all
induftry you apply your minds to the ftudy of
Philofophy, that as you profeffe your felues ftudents,
fo you may be ftudents, that as you difdaine not the
name of a fcholler, fo you wil not be found voyd of

the duety of fchollers, let not your mindes be caryed
away with vaine delights, as with trauailing into farre
and ftraunge countries wher you fhal fee more wickedneffe then learn vertue and wit. Neither with coftly
attyre of the newe cut, the *Dutch* hat, the *French* hofe,
the *Spanifh* rapier, ye *Italian* hilt, and I know not
what?

Caft not your eyes on the beauty of women, leaft
ye caft away your hearts with folly, let not that fond
loue, wherewith youth fatteth himfelfe as fatte as a
foole infect you, for as a finewe being cut though it be
healed, there wil alwayes remaine a fcarre, or as fine
lynnen ftayned with blacke ynke, though it bee
wafhed neuer fo often, will haue an yron Mowle: fo
the minde once mangled or maymed with loue,
though it be neuer fo well cured with reafon, or
cooled by wifedome, yet there wil appeare a fcarre,
by the which one may geffe the minde hath ben
perced, and a blemmifh whereby one may iudge the
heart hath ben ftayned.

Refraine from dicing, which was the onely caufe
that *Pyreus* was ftriken to the heart, and from dauncing which was the meanes that loft *Iohn Baptifts*
heade: I am not he that will difallowe honeft recreation, although I deteft the abufes, I fpeake boldely
vnto you bicaufe I my felfe know you: what *Athens*
hath ben, what *Athens* is, what *Athens* fhal be, I can
geffe. Let not euery Inne and Alehoufe in *Athens* be
as it were your chamber, frequent not thofe ordinary
tables wher either for the defire of delicate cates, or
the meetinge of youthfull companions, yee both fpend
your money vainely and your time idly, imitate him
in life whom ye [you feeme to] honour for his learning.
Ariftotle who was neuer feene in the company of thofe
that idly beftowed their time.

There is nothing more fwifter then time, nothing
more fweeter: wee haue not as *Seneca* faith little time
to liue, but we leefe muche, neither haue we a fhort
life by Nature, but we make it fhorter by naughtyneffe,

our life is long if we know how to vfe it. Follow *Appelles* that cunning and wife Painter, which would lette no day paffe ouer his head, without a lyne, without fome labour. It was pretely fayde of *Hefiodas*, lette vs endeauour by reafon to excell beaftes, feeinge beafts by nature excell men, although ftrick[t]ely taken it be not fo, for that man is endewed with a foule, yet · taken touching their perfeótion of fences in their kind it is moft certeine. Doth not the Lyon for ftrength, the Turtle for loue, the Ante for labour excell man? Doth not the Eagle fee cleerer, the Vulter fmel better, the Mowle heare lyghtlyer? Let vs therefore endeauour to excell in vertue, feeing in qualyties of ye body we are inferiour to beaftes. And heere I am moft earneftly to exhort yoú to modefty in your behauiour, to duetye to your elders, to dylligence in your ftudyes. I was of late in *Italy*, where mine eares gloed, and my heart was galled to heare the abufes that reygne in *Athens*: I cannot tell whether thofe things fprang by the lewde and lying lippes of the ignoraunt, which are alwayes enimyes to learning, or by the reports of fuch as faw them and forrowed at them. It was openly reported of an olde man in *Naples* that there was more lightneffe in *Athens* then in all *Italy*, more wanton youths of fchollers, then in all *Europe* befids, more Papifts, more *Atheifts*, more feóts, more fchi[f]mes, then in all the Monarch[i]es in the world, which thinges although I thincke they be not true, yet can I not but lament that they fhoulde be deemed to be true, and I feare me they be not altogether falfe, ther can no great fmoke arife, but there muft be fome fire, no great reporte without great fufpition. Frame therefore your lyues to fuch integritie, your ftudyes to atteininge of fuch perfeótion, that neither the might of the ftronge, neyther the mallyce of the weake, neither the fwifte reportes of the ignoraunt be able to fpotte you wyth difhoneftie, or note you of vngodlyneffe. The greateft harme that you can doe vnto the enuious, is to doo well, the

greateſt coraſiue that you can giue vnto the ignoraunte,
is to proſper in knowledge, the greateſt comforte that
you can beſtowe on your parents is to lyue well and
learne well, the greateſt commoditie that you can
yeelde vnto your Countrey, is with wiſedome to beſtowe
that talent, that by grace was giuen you.

And here I cannot chooſe but giue you that counſel
that an olde man in *Naples* gaue mee moſt wiſely,
although I had then neither grace to followe it, neyther
will to giue eare to it, deſiring you not to reieᶜt it
bicauſe I did once diſpiſe it. It was this [thus] as
I can remember word for word.†

Deſcende into your owne conſciences, conſider with
your ſelues the great difference between ſtaring and
ſtarke blynde, witte and wiſedome, loue and luſt :
Be merry but with modeſtie, be ſober but not too*
ſullen : be valiaunt, but not too venterous : let your
attire be comely, but not too coſtly : your dyet wholeſome, but not exceſſiue : vſe paſtime as the word
importeth, to paſſe ye time in honeſt recreation :
miſtruſt no man without cauſe, neither be ye credulous
without proofe : be not lyght to follow euery mans
opinion, neither obſtinate to ſtand in your owne conceipts : ſerue God, feare God, loue God, and God will
bleſſe you, as either your hearts can wiſh, or your
friends deſire.

This was his graue and godly aduiſe, whoſe counſel
I would haue you all to follow, frequent leᶜtures, vſe
diſputacions openly, negleᶜt not your priuate ſtudies,
let not degrees be giuen for loue but for learning, not
for mony, but for knowledge, and bicauſe you ſhall
bee the better incouraged to follow my counſell, I wil
be as it were an example my ſelfe, deſiring you al to
imitate me.

Euphues hauing ended his diſcourſe, and finiſhed
thoſe precepts which he thought neceſſary for the
inſtruᶜtion of youth, gaue his minde to the continual
ſtudie of Philoſophie, inſomuch as he became publique

Reader in the Vniuerfitie, with fuch commendation as neuer any before him, in the which he continued for the fpace of tenne yeares, only fearching out the fecrets of Nature and the hidden mifteries of philofophy, and hauing collećted into three volumes his lećtures, thought for the profite of young fchollers to fette them foorth in print, which if he had done, I would alfo in this his *Anatomie* haue inferted, but he altering his determination, fell into this difcourfe with himfelfe.

Why *Euphues*, art thou fo addićted to the ftudie of the Heathen that thou haft forgotten thy God in heauen? fhal thy wit be rather employed to the atteining of humaine wifedome then diuine knowledge? Is *Ariftotle* more deare to thee with his bookes, then Chrift with his bloud? What comfort canft thou finde in Philofophy for thy guiltie confcience? What hope of the refurrećtion? What glad tidings of the Gofpell?

Confider with thy felfe that thou art a gentleman, yea, and a Gentile, and if thou neglećt thy calling thou art worfe then a *Iewe*. Moft miferable is the eftate of thofe Gentlemen, which thinke it a blemmifh to their aunceftours, and a blot to their owne gentrie, to read or praćtize Diuinitie. They thinke it now fufficient for their felicitie to ryde well vppon a great horfe, to hawke, to hunt, to haue a fmacke in Philofophie, neither thinking of the beginning of wifedome, neither the ende, which is Chrift: onely they accompt diuinitie moft contemptible, which is and ought to be moft notable. Without this there is no Lawyer be he neuer fo eloquent, no Phifition be he neuer fo excelent, no Philofopher bee hee neuer fo learned, no King, no Keyfar, be he neuer fo royall in birth, fo polytique in peace, fo expert in warre, fo valyaunt in proweffe, but he is to be detefted and abhorred. Farewell therefore the fine and filed phrafes of *Cicero*, the pleafaunt *Eligues* of *Ouid*, the depth and profound knowledge of *Ariftotle.* Farewell Rhethoricke, fare-

well Philofophie, farewel all learning which is not
fprong from the bowells of the holy Bible.

In this learning fhal we finde milke for the weake
and marrow for the ftrong, in this fhall we fee how
the ignoraunt may be inftruĉted, the obftinate confuted, the penitent comforted, the wicked punifhed,
the godly preferued. Oh I would Gentlemen would
fome times fequefter themfelues from their owne
delights, and employ their wits in fearching thefe
heauenly and diuine mifteries. It is common yea
and lamentable to fee that if a young youth, haue the
giftes of Nature, as a fharpe wit, or of Fortune, as
fufficient wealth to mainteine them, he employeth the
one, in the vayne inuentions of loue, the other in the
vile brauerie of pride : the one in the paffions of his
minde and prayfes of his Lady, the other in furnifhing of
his body and furthering of his luft. Heeroff it commeth
that fuch vaine ditties, fuch idle fonnets, fuch enticing
fongs, are fet foorth to the gaze of the world and griefe
of the godly. I my felfe know none fo ill as my felfe,
who in times paft haue bene fo fuperfticioufly addiĉted,
yat I thought no Heauen to ye Paradife of loue, no
Angel to be compared to my Lady, but as repentaunce
hath caufed me to leaue and loath fuch vaine delights,
fo wifdome hath opened vnto me, the perfeĉt gate to
eternall lyfe.

Befides this I my felfe haue thought that in Diuinitie
there could be no eloquence, which I might imitate,
no pleafaunt inuention which I might follow, no
delycate phrafe that might delight me, but now I fee
that in the facred knowledge of Gods will, the onely
eloquence, the true and perfeĉt phrafe, the teftimonie
of faluation doth abide, and feeing without this all
learning is ignoraunce, al wifdome more folly, all witte
plaine bluntnes, al Iuftice iniquitie, al eloquence barbarifme, al beautie deformitie. I will fpend all the
remainder of my life in ftudying the olde Teftament,
wherin is prefigured the comming of my Sauiour, and
the new teftament, wherin my Chrift doth fuffer for

my finnes, and is crucified for my redemption, whofe
bitter agonyes fhould caft euery good chriftian into a
eeuering ague to remember his anguifh, whofe
fweating of water and bloud fhould caufe euery deuout
and zealous Catholique to fhedde teares of repentaunce,
in remembraunce of his torments.

Euphues hauing difcourfed this with himfelfe, did
immediately abandon all lyght company, all the difputations in fchooles, all Philofophie [Schooles of Phi-
lofophie], and gaue himfelfe to the touchftone of holineffe in diuinitie, accompting all other things as moft
vyle and contemptible.

¶ *Euphues to the Gentlemen fchollers
in Athens.*

He Merchant that trauaileth for gain, the
hufbandman that toyleth for increafe, ye
lawier that pleadeth for gold, the crafts
man that feeketh to lyue by his labour, al
thefe after they haue fatted themfelues with
fufficient, either take their eafe, or leffe payne then
they were accuftomed. *Hippomanes* ceafed to runne
when he had gotten the goale. *Hercules* to labour,
when he had obteined the victorie. *Mercurie* to pipe
when he had caft *Argus* in a flumber. Euery action
hath his ende, and then we leaue to fweat when we
haue founde the fweete. The Ant though fhe toyle
in Summer, yet in Winter fhee leueth to trauaile.
The Bee though fhe delight to fuck the faire flower,
yet is fhe at laft cloyed with Honny. The Spider that
weaueth the fineft threede ceafeth at the laft when fhe
hath finifhed hir webbe. But in the action and ftudy
of the mind (Gentlemen) it is farre otherwife, for hee
that tafteth the fweet of learning endureth all the
fower of labour. He that feeketh the depth of knowledge : is as it were in a *Laborinth*, in the which ye
farther he goeth, the farther he is from the end : or
like ye bird in the limebufh, which the more fhe

ſtriueth to get out, ye faſter ſhe ſticketh in. And
certeinly it may be ſaid of learning, as it was fained of
Nectar the drinke of the Gods, the which the more it
was dronk, the more it would ouerflow the brim of the
cup, neither is it farre vnlike the ſtone that groweth in
the riuer of *Caria,* the which the more it is cut the
more it encreaſeth. And it fareth with him that
followeth it as with him that hath the dropſie, who the
more he drinketh the more he thirſteth. Therefore in
my minde the ſtudent is at leſſe eaſe then the Oxe yat
draweth, or the Aſſe that caryeth his burthen, who
neither at the boord when others eate is voyd of
labour, neither in his bed when others ſleepe is without
[voyd of] meditation. But as in manuary craftes though
they be all good, yet that is accompted moſt noble
that is moſt neceſſary, ſo in the actions and ſtudyes of
the minde, although they be all worthy, yet that
deſerueth greateſt praiſe which bringeth greateſt profit.
And ſo we commonly do make beſt accompt of that
which doth vs moſt good. Wee eſteeme better of the
Phiſition that miniſtreth the potion, then of the
Apothecary yat ſelleth the drugs. How much more
ought we with al diligence, ſtudy, and induſtry, [to]
ſpend our ſhort pilgrimage in the ſeeking out of our
ſaluation. Vaine is Philoſophy, vaine in Phiſick, vaine
is Law, vaine is al learning without yat taſt of diuine
knowledge. I was determined to write notes of
philoſophy, which had ben to feede you fat with folly,
yet yat I might ſeeme neither idle, neither you euil
imployed, I haue heere ſet downe a briefe diſcourſe
which of late I haue had with an hereticke which kept
me from idlenes, and may if you read it deterre you
from hereſie. It was with an *Atheyſt,* a man in my
opinion monſtrous, yet tractable to be perſwaded. By
this ſhal you ſee ye abſurde dotage of him that
thinketh ther is no god, or an vnſufficient god, yet
heere ſhall you finde the ſumme of faith which
iuſtifieth onely in Chriſt, the weakneſſe of the lawe the
ſtrength of the goſpel, and the knowledge of gods

wil. Heere fhall ye finde hope if you be in difpaire, comfort if ye be diftreffed, if ye thirft drinke, meate if ye hungur, if ye feare *Mofes* who faith without you fulfil the lawe you fhall perifh. Beholde Chrift, which faith, I haue ouercommen the lawe. And yat in thefe defperate dayes wherein fo many fectes are fowen, and in the wayning of the world, wherein fo many falfe Chrifts are come, you might haue a certeintie of your faluati-on, I meane to fet downe the touchftone whervnto e-uery one ought to truft, and by the which euery one fhoulde trie himfelfe, which if you fol-low, I doubt not but that as you haue proued learned Philofophers, you will alfo proceede excellent diuines, which God graunt.

¶ EVPHVES AND ATHEOS.

**THEOS.** I am gladde *Euphues* that I haue founde thee at leafure, partly yat we might be merry, and partly that I might bee perfwaded in a thing that much troubled my confcience. It is concerning God. There be many that are of this minde, that there is a God whom they tearme the creator of all thinges, a God whom they cal the fonne, the redeemer of the world, a God whom they name the holye Ghoft the worker of all things, the comforter, the fpirite, and yet are they of this opinion alfo, that they be but one God, coequal in power, coeternall, incomprehenfible, and yet a Trinity in perfon. I for my part although I am not fo credulous to beleeue their curious opinions, yet am I defirous to heare the reafons yat fhoulde driue them into fuch fond and franticke imaginations. For as I knowe nothing to be fo abfurde which fome of the Philofophers haue not defended, fo thinke I nothing fo erronious which fome of our Catholikes haue not mainteined. If there were as diuers dreame, a God that woulde reuenge the oppreffion of the widdowes and fatherleffe, that would rewarde the zeale of the mercifull, pitie the poore, and pardon the penitent, then woulde the people either ftand in greater awe, or owe more loue towards their God. I remember *Tully* difputing of the nature of Gods, bringeth *Dionifius* as a fcoffer of fuch vaine and deuifed Deities, who feeing *Aefculapius* with a long bearde of golde, and *Appollo* his father beardleffe, played the Barber and fhaued it from him, faying, it was not decent that the fonne fhould haue a beard and the father none. Seeing alfo *Iupiter* with an ornament of golde, tooke it from him iefting thus, in Summer this aray is too heauy, in

Winter too colde, heere I leaue one of wollen both warmer for the cold and lyghter for the heate. He comming alfo into the Temple wher certeine of the gods with golden gifts ſtretched out their hands, tooke them al away, ſaying: Who will be ſo mad as to refuſe thinges ſo gently offered: Doſt thou not ſee *Euphues* what ſmall accompt he made of their gods, for at the laſt ſailing into his countrey with a proſperous winde, hee laughing ſayd, loe ſee you not my Maſters, howe well the Gods reward our Sacriledge. I coulde rehearſe infinite opinions of excellent men who in this pointe holde on my ſide, but eſpecially *Protagoras* [*Pithagoras*]. And in my iudgement, if there bee any God, it is the worlde wherein we liue, that is the onely God, what can we beholde more noble then the world, more faire, more beautifull, more glorious? what more maieſticall to the ſight, or more conſtant in ſubſtance? But this by the way *Euphues*, I haue greater and more forcible arguments to confirme my opinion, and to confute the errors of thoſe that imagine that there is a God. But firſt I woulde gladlye heare thee ſhape an aunſwere to that which I haue ſaid, for wel I know yat thou art not onely one of thoſe which beleeue that there is a God, but of them alſo which are ſo preciſe in honouring him, that they bee ſcarce wiſe in helping themſelues.

Euphues. If my hope *(Atheos)* were not better to conuert thee, then my happe was heere to conferre with thee, my heart would breake for griefe, whiche beginneth freſhly to bleede for ſorrow, thou haſt ſtroken me into ſuch a ſheuering and cold terror at the rehearſinge of this thy monſtrous opinion, that I looke euery minute when the grounde ſhould open to ſwallow thee vp, and that GOD which thou knoweſt not, ſhoulde with thunder from heauen, ſtrike thee to hell. Was there euer *Barbarian* ſo ſenceleſſe, euer miſcreaunt ſo barbarous, that did not acknowledge a liuing and euerlaſting *Iehouah*? I cannot but tremble at the remembraunce of his Maieſtie, and doſt thou

L

make it a mockerie ?   O iniquitie of times, O corruption of manners, O blafphemie againſt the heauens.
The Heathen man faith, yea that *Tully* whom thou
thy felfe alleadgeſt, that there is no nation fo barbarous,
no kinde of people fo fauage, in whom reſteth not this
perfwaſion that there is a God, and euen they that in
other parts of their lyfe feeme very lyttle to differ from
brute beaſts, doe continally keepe a certeine feede of
Religion, fo throughly hath this common principle
poſſeſſed al mens mindes, and fo faſt it ſticketh in all
mens bowells.   Yea, Idolatrie it felfe is fufficient
proofe of this perfwaſion, for we fee how willingly man
abaſeth himfelf to honour other creatures, to doe
homage to ſtockes, to goe on pilgrimage to Images, if
therefore man rather then he would [wil] haue no God,
doe worſhip a ſtone : how much more art thou duller
then a ſtone, which goeſt againſt the opinion of all
men.

Plato a Philofopher would often fay, there is one
whom we may cal God omnipotent, glorious, immortall, vnto whofe fimilitude we that creepe heere on the
earth haue our foules framed, what can be faid more
of a Heathen, yea, what more of a Chriſtian ?

Ariſtotle when hee could not finde out by the
fecrecie of Nature, the caufe of the ebbing and
flowing of the Sea, cryed out with a lowd voyce. O
thing of things haue mercy vppon me.

Cleanthes alleadged foure caufes, which might induce man to acknowledge a God, the firſt by the forefeeing of things to come, the fecond by the infinite
commodities which we daily reape, as by the temperature of the ayre, the fatneſſe of the earth, the
fruitefulneſſe of trees, plants, and hearbes, the aboundaunce of all things that maye either ferue for the
neceſſitie of many, or the fuperfluitie of a few, the
thirde by the terror that the minde of man is ſtroken
into, by lyghtenings, thunderings, tempeſts, hayles,
fnowe, earthquakes, peſtilence, by the ſtraunge and
terrible fights which caufe vs to tremble, as the

rayning of bloud, the fi[e]rie impreffions in the Element, the ouerflowing of floudes in the earth, the prodigious fhapes and vnnaturall formes of men, of beaftes, of birdes, of fifhes, of all creatures, the appearing of blafing Comettes, which euer prognofticate fome ftraunge mutation, the fight of two Sunnes which happened in the Confulfhippe of *Tuditanus* and *Aquilius*, with thefe things mortall men being afrighted, are inforced to acknowledge an immortal and omnipotent god. The fourth by the equalytie in mouing in the heuen, the courfe of the Sunne, the order of the ftars, the beautifulneffe of the Element, ye fight wheroff might fufficiently induce vs to beleeue they proceede not by chaunce, by nature, or deftenie, but by the eternal and diuine purpofe of fome omnipotent Deitie. Heereoff it came that when the Philofophers could giue no reafon by Nature, they would fay there is one aboue Nature, an other would call him the firft mouer, an other the ayder of Nature, and fo foorth.

But why goe I about in a thing fo manifeft to vfe proofes fo manifolde. If thou deny the truth, who can proue it, if thou deny that blacke is blacke, who can by reafon reproue thee, when thou oppofeft thy felf againft reafon, thou knoweft that manifeft truthes are not to be proued but beleeued, and that he that denyeth the principles of any Arte, is not to be con- futed by arguments, but to be left to his owne folly. But I haue a better opinion of thee, and therefore I meane not to trifle with Philofophy, but to trye this by the touchftone of the Scriptures. Wee reade in the fecond of *Exodus*, that when *Mofes* defired of God to knowe what he fhoulde name him to the children of *Ifrael* : hee aunfwered thou fhalt faye, I am that I am. Againe, he that is hath fent me vnto you. The Lord euen your God, he is God in the heauen aboue, and in the Earth beneath. I am the firft, and the laft I am. I am the Lord, and there is none other befides me. Againe, I am the Lorde, and there is none other. I haue created the lyght and made darkeneſſe, making peace and framing euill. If thou deſire to vnderſtand what God is, thou ſhalt heare, he is euen a conſuming fire, the Lord [God] of reuenge, the God of iudgement, the lyuing God, the ſearcher of the reynes, he that made all things of nothing, *Alpha* and *Omega*, the beginning, and yet without beginning : the ende, and yet euerlaſting. One at whoſe breath the mountaines ſhall ſhake, whoſe feat is the loftie *Cherubins*, whoſe foote-ſtoole is the earth. Inuiſible, yet ſeeing all things, a iealous God, a louing God, miraculous in all points, in no part monſtrous. Beſides this, thou ſhalt well vnder-ſtande that hee is ſuch a God as will puniſh him who-ſoeuer hee bee that blaſphemeth his name, for holy is the Lord. It is written, bring out the blaſphemer without the tents, and let al thoſe that heard him, lay their hands vpon his head, and let all the people ſtone him. He that blaſphemeth the name of the Lorde, ſhall dye the death. Such a iealous God, that whoſoeuer committeth Idolatrye with ſtraunge GODS, hee will ſtrike with terrible plagues. Tourne not to Idolls, neither make Gods with handes, I am the Lord your God. Thou ſhalt make no Image which the Lorde thy God abhorreth. Thou ſhalt haue no new God, neither worſhippe any ſtraunge Idoll. For all the Gods of the Gentiles are diuells.

My ſons keepe your ſelues from Images, the wor-ſhipping of Idolls is the cauſe of all euill, the begin-ning and the ende. Curſed be that man that en-graueth any Images, it is an abhomination before the Lorde. They ſhall be confounded that worſhip grauen Images, or glorie in Idolls. I will not giue my glory to an other nor my praiſes to grauen Images.

If all theſe teſtimonies of the Scriptures can not make thee to acknowledge a lyuing GOD, harken what they ſay of ſuch as be altogether incredulous. Euery vnbeleeuer ſhall dye in his incredulite. Wo be to thoſe that be looſe in heart, they beleeue there is no God, and therefore they fhall not be protected of him. The wrath of the Lorde fhall kindle againſt an vnbeleeuing Nation. If ye beleeue not, you fhal not endure. He that beleeueth, fhall not be dampned. He that beleeueth not, is iudged already. The portion of the vnbeleeuers fhall be in the lake that burneth with fire and brimſtone, which is the fecond death.

If thou feele in thy felfe *Atheos,* any ſpark of grace, pray vnto the Lord and he will caufe it to flame, if thou haue no feeling of faith, yet pray, and the Lord wil giue aboundaunce, for as he is a terrible God, whofe voyce is lyke the ruſhing of many waters, fo is hee a mercifull God, whofe wordes are as foft as Oyle. Though he breath fire out of his noſtrels againſt finners, yet is he milde to thofe that afke forgiueneffe. But if thou be obſtinate, that feing thou wilt not fee, and knowing thou wilt not acknowledge, then fhal thy heart be hardened with *Pharao,* and grace fhal be taken away from thee with *Saul.*

Thus faith the Lorde, who fo beleeueth not fhall periſh, heauen and earth fhall paffe, but the worde of the Lord fhall endure for euer.

Submit thy felfe before the throne of his Maieſty, and his mercy fhall faue thee. Honour the Lorde and it fhall be well with thee. Befid[e]s him feare no ſtrange God. Honour the Lord with al thy foule. Offer vnto God the facrifice of praife. Be not like the Hipocrit[e]s which honour God with their lyppes, but be farre from him with their hearts, neither like the foole which faith in his heart, there is no God.

But if thou wilt ſtill perfeuer in thine obſtinacie, thine end fhalbe worfe then thy beginning, the Lord, yea thy Sauiour, fhall come to be thy Iudge, when thou fhalt behold him come in glory, with Millions of Angels and Archangels, when thou fhalt fee him ap-peare in thundringes and lyghtninges and flaſhinges of Fyre, when the mountaines fhall melt, and the heauens be wrapped vp lyke a fcrowle, when al the

earth fhall tremble, with what face wilt thou beholde
his glory, that denyeft his Godhead? Howe canft
thou abide his prefence that beleeueft not his effence?
What hope canft thou haue to be faued which diddeft
neuer acknowledge any to be thy Sauiour? Then
fhall it bee faide [vn]to thee and to all thofe of thy fect,
(vnleffe ye repent) Depart all ye workers of iniquitie,
there fhalbe weeping and gnafhing of teeth When you
fhall fee *Abraham, Ifaac* and *Iacob,* and all the
Prophets in the kingdome of God, and ye to be
thruft out : You fhall conceiue heate and bring foorth
wood, your owne confciences fhall confume you like
fire. Heere doeft thou fee *Atheos* the threatnings
againft vnbeleeuers, and the punifhment prepared for
mifcreants. What better or founder proofe canft
thou haue that there is a God, then thine owne confcience, which is vnto thee a thoufand witneffes?
Confider with thy felfe that thy foule is immortall,
made to the Image of the Almightye God : be not
curious to enquire of God, but carefull to beleeue,
neither bee thou defperate if thou fee thy finnes
abounde, but faithfull to obteine mercye, for the
Lorde will faue thee bicaufe it is his pleafure. Search
therefore the Scriptures, for they teftifie of him.

Atheos. Truely *Euphues* you haue faide fomewhat,
but you goe about contrarye to the cuftomes of
fchooles, which mee thinckes you fhould dilygently
obferue, being a profeffed Philofopher : for when I
demaunde by what reafon men are induced to acknowledge a God, you confirme it by courfe of Scripture,
as who fhould fay there were not a relation betwene
GOD and the Scripture, bicaufe as the olde fathers
define, without Scripture there were no GOD, no
Scripture without a GOD. Whofoeuer therefore denyeth a Godhead, denieth alfo the Scriptures which
teftifie of him. This is in my opinion *abfurdum per
abfurdius,* to proue one abfurditie by an other.

If thou canft as fubftantially by reafon proue thy
authoritie of Scriptures to be true, as thou haft proued

by Scriptures there is a God, then will I wyllyngly
with thee both beleue the Scriptures, and worfhippe
thy GOD. I haue heard that *Antiochus* commaunded
all the copyes of the Teftament to be burnt, from
whence therefore haue we thefe newe bookes, I thinke
thou wilt not fay by reuelation, therefore goe forward.

𝕰𝔲𝔭𝔥𝔲𝔢𝔰. I haue read of the milke of a Tygreffe, that
the more falt there is throwne into it, the frefher it is,
and it may be that [either] thou haft either* eaten of that
milke, or that thou art the whelpe of that monfter, for
the more reafons that are beate[n] into thy head, the
more vnreafonable thou feemeft to bee, the greater my
authorities are, the leffer is thy beleefe. As touching
the authoritie of Scriptures although there be many
arguments which do proue yea and enforce the
wicked to confeffe that the Scriptures came from God,
yet by none other meane then by the fecreat teftimony
of the holy Ghoft our heartes are truely perfwaded that
it is God which fpeaketh in the lawe, in the Prophetes,
in the Gofpell, the orderly difpofition of the wifedome
of God, the doctrine fauoring nothing of earthlyneffe,
the godly agreement of all partes among themfelues,
and efpecially the bafeneffe of contemptible words
vttering the high mifteries of the hauenly kingedome,
are fecond helpes to eftablifh the Scriptures.

Moreouer the antiquitie of the Scripture, wher as
the bookes of other Religions are later then the books
of *Mofes*, which yet doth not himfelfe inuent a newe
God, but fetteth foorth to the *Ifraelites* the God of
their fathers. Whereas *Mofes* doth not hide the
fhame of *Leuy* his father, nor the mourning of *Aaron*
his brother, and of *Marie* his fifter, nor doth aduaunce
his owne children : The fame are arguments that in
his booke is nothing fayned by man. Alfo the
myracles yat happened as well at the publyfhing of
the lawe as in all the reft of time are infallible proofes
that the fcriptures proceeded from the mouth of God.
Alfo where as *Mofes* fpeaking in the perfon of *Iacob*,
affigneth gouernement to the Tribe of *Iuda*, and

where he telleth before of the calling of the Gentiles,
whereof the one came to paffe foure hundreth yeares
after, the other almoſt two thouſande yeares, thefe are
arguments that it is GOD himfelfe that fpeaketh in
the bookes of *Moſes*.

Whereas *Eſay* telleth before of the captiuitie of the
Iewes and their reſtoringe by *Cyrus* (whiche was
borne an hundreth yeares after the death of *Eſay*)
and whereas *Ieremy* before the people were led awaye,
apointeth their exile to continue three ſcore and ten
years.   Whereas *Ieremy* and *Ezechiel* being farre
diſtant in places the one from the other, do agree in
all their ſayinges.   Where *Daniel* telleth of thinges to
come fixe hundreth yeares after.   Thefe are moſt
certeine prooues to eſtabliſh the auĉthoritie of the
books of the Prophets, the ſimplicitie of the ſpeach of
the firſt three Euaungeliſts, conteining heauenly miſte-
ries, the praife of *Iohn*, thundering from an [on] high
with weightie ſentences, the heauenly maieſtie ſhining
in the wrytings of *Peter* and *Paul*, the ſodayn calling
of *Mathew* from the receipt of cuſtome, the calling of
Peter and *Iohn* from their fiſher boates to the preach-
ing of the Goſpell, the conuerſion and calling of *Paul*
being an enimy to the Apoſtleſhip, are ſignes of the
holy Ghoſt ſpeaking in them.   The conſent of ſo
many ages, of ſo ſundry nations, and of ſo dyuers
mindes, in embracing the Scriptures, and the rare
godlyneffe of ſome, ought to eſtabliſh the authoritie
theroff amongſt vs.   Alfo the bloud of ſo many
Martyrs which for ye confeffion theroff haue ſuffered
death, with a conſtant and ſober zeale, are vndoubted
teſtimonies of the trueth and authoritie of the
Scriptures.

The myracles that *Moſes* recounteth are ſufficient
to perſwade vs that God, yea, the God of hoaſtes, ſet
downe the Scriptures.   For this that he was caryed in
a cloude vpp into the mountaine : that there euen
vntill the fortith day he continued without the company of men.   That in the very publiſhing of the law

his face dyd fhyne as it were befette with Sunne
beames, that lyghteninges flafhed round about, that
Thunder and noyfes were each where hearde in the
ayre, that a Trompette fownded being not fownded
with any mouth of man. That the entry of the
Tabernacle by a clowd fet betweene was kept from
the fight of the people, that his authoritie was fo
miraculoufly reuenged with the horrible deftruction of
Chorah, Dathan, and *Abiron,* and all that wicked
faction, that the Rocke ftroken with a rod, did by
and by poure forth a riuer, that at his prayer it rained
Manna from heauen. Did not God heerein commend
him from heauen as an vndoubted Prophet? Now as
touching the tyranny of *Antiochus,* which commaunded
all the bookes to be burned, herein Gods finguler
prouidence is feene, which hath alwaies kept his woord
both from ye mightie that they could neuer extinguifh
the fame, and from the malitious that they could neuer
diminifh it. Ther were diuers copyes which God of
his great goodneffe kept from the bloudy proclamation
of *Antiochus,* and by and by followed the tranflating
of them into *Greek,* that they might be publifhed vnto
the whole worlde. The *Hebrew* tongue lay not onely
vnefteemed but almoft vnknowne, and furely had it
not bene gods wil to haue his religion prouided for, it
had altogether perifhed.

Thou feeft *Atheos* how the Scriptures come from
the mouth of God, and are written by the finger of
the Holy Ghoft, in the confciences of all the faithful.
But if thou be fo curious to afke other queftions, or fo
quarrellous to ftriue againft the truth, I muft aunfwere
thee as an an olde father aunfwered a young foole,
which needes woulde know what God did before hee
made Heauen, to whome he faide, hell, for fuch curious
inquifitors of gods fecrets, whofe wifedome is not to be
comprehended, for who is he yat can meafure the
winde, or way the fire, or attain vnto the vnfearchable
iudgements of the Lorde.

Befides this where the holy Ghoft hath ceafed to

fette downe, there ought we to ceafe to enquire, feeing we haue the fufficiencie of our faluation conteined in holy Scripture. It were an abfurditie in fchooles, if one being vrged with a place in *Ariſtotle* could finde none other ſhift to auoyde a blancke, then in doubting whether *Ariſtotle* fpake ſuch words or no. Shal it then be tollerable to deny the Scriptures hauing no other colour to auoyd an inconuenience, but by doubting whether they proceede from the holy Ghoſt? But that ſuch doubts ariſe among many in our age, the reafon is their little faith, not the infufficient proofe of the caufe.

Thou maiſt as well demaund how I proue white to be white, or blacke b[l]acke, and why it ſhould be called white rather then greene. Such groſſe queſtions are to be aunſwered with ſlender reaſons, and ſuch idle heads ſhould be ſcoffed with adle aunſweres. He that hath no motion of god in his minde, no feeling of the ſpirite, no taſte of heauenly things, no remorce in conſcience, no ſparke of zeale, is rather to be confounded by torments, then reaſons, for it is an euident and infallible ſigne that the holy ghoſt hath not ſealed his conſcience, whereby hee might crye, *Abba Father*, I could alledge Scripture to proue that the godly ſhould refrayne from the company of the wicked, which although thou wilt not beleeue, yet will it condempne thee. Sainct *Paul* ſaith, I deſire you bretheren that you abſteine from the company of thoſe that walke inordinately. Againe, my ſonne, if ſinners ſhall flatter thee giue no eare vnto them, flye from the euill, and euills ſhall flye from thee.

And ſurely wer it not to confute thy deteſtable hereſie, and bring thee if it might be to ſome taſt of the holy Ghoſt, I would abandon all place of thy abode, for I thinke the grounde accurſed whereon thou ſtandeſt: Thy opinions are ſo monſtrous that I cannot tel whether thou wilt caſt a doubt alſo whether thou haue a foule or no, which if thou doe, I meane not to waſt winde in prouing that, which thine infi-

delytie will not permit thee to beleeue, for if thou haſt
as yet felt no taſt of the ſpirit working in thee, then
ſure I am that to proue the immortalytie of the ſoule
were booteleſſe, if thou haue a ſecret feelyng, then it
were needeleſſe. And God graunt thee that glowing
and ſting in conſcience, that thy ſoule may witneſſe to
thy ſelfe that ther is a liuing god, and thy heart ſhed
drops of blood as a token of repentaunce, in that thou
haſt denied that God, and ſo I commit thee to God,
and that which I cannot doe with any perſwaſion I
will not leaue to attempt with my prayer.

𝔄𝔱𝔥𝔢𝔬𝔰. Nay ſtaye a while good *Euphues,* and leaue
not him perplexed with feare, whome thou maiſt make
perfect by fayth : for nowe I am brought into ſuch a
double and doubtfull diſtreſſe that I know not how to
tourne me,* if I beleeue not the ſcriptures, then ſhall I
be damned for vnbeliefe, if I beleeue them, then [ſhall] I
ſhal* be confounded for my wicked life. I know the
whole courſe of ye Bible, which if I ſhould beleue,
then muſt I alſo beleue that I am an abiect. For
thus ſaith *Heli* to his ſonnes. If man ſinne againſt
man, God can forgiue it, if againſt God, who ſhall
intreate for him ? He that ſinneth is of the diuell, the
rewarde of ſin is death, thou ſhalt not ſuffer the wicked
to liue : take all the Princes of the people and hang
them vp againſt the Sunne on Iybbets, that my anger
may bee tourned from *Iſrael,* theſe ſayings of holy
Scripture, cauſe me to tremble and ſhake in euery
ſinew. Againe this ſaith the holy Bible, now ſhall the
ſcourge fal vpon thee for thou haſt ſinned, behold I
ſet a curſe before you to day, if you ſhall not harken
to the commaundements of the Lord, al they that
haue forſaken ye Lord ſhall be confounded. Further-
more, where threats are poured out againſt ſinners my
heart bleedeth in my belly to remember them.

I will come vnto you in iudgement ſaith the Lord,
and I wil be a ſwift and a ſeuere witneſſe, offenders,
adulterers, and thoſe that haue committed periury,
and reteined the duetie[s] of hirelyngs, oppreſſed the

Widdowes, mifufed the ftraunger, and thofe that haue
not feared me the Lorde of hoafts.   Out of his mouth
fhal come a two edged fword.   Behold I come quickly,
and bring my reward with me, which is to yeeld euery
one according to his deferts.

Great is the day of the Lord and terrible, and who
is he that may abide him?   What fhal I then do
when the Lord fhall arife to iudge, and when he fhall
demaund what fhal I aunfwere?   Befides this, the
names that in holy fcripture are attributed to God,
bring a terror to my guiltie confcience.   He is faid to
be a terrible God, a God of reuenge, whofe voyce is
lyke the thunder, whofe breath maketh all the corners
of the Earth to fhake and tremble.   Thefe things
Euphues, teftifie vnto my confcience, that if ther be a
god, he is the god of the righteous, and one that will
confound the wicked.   Whether therefore fhal I go,
or who may auoyd the day of vengeaunce to come?
If I go to heauen, that is his feat : if into the earth,
that is his foot-ftoole : if into the depth, ther he is
alfo?   Who can fhrowd himfelf from the face of the
Lord, or where can one hide him that the Lord cannot
finde him?   His words are like fire, and the people
like dry wood, and fhalbe confumed.

Euphues. Although I cannot but reioice to heare
thee acknowledge a God, yet muft I needs lament to
fee thee fo much diftruft him.   The diuel that roaring
Lyon feeing his pray to be taken out of his Iawes
alleadgeth all Scripture, that may condemne the finner,
leauing all out that fhould comfort the forrowfull.   Much
lyke vnto the deceitfull Phifition, which recounteth all
thinges that may endomage his patient, neuer telling
any thing that may recure him.   Let not thy confcience be agrieued, but with a patient heart renounce
all thy former iniquities and thou fhalt receiue eternall
life.   Affure thy felf that as god is a Lord, fo he is a
father, as Chrift is a Iudge fo he is a Sauiour, as there
is a lawe, fo there is a gofpell.   Though God haue
leaden handes which when they ftrike pay home, yet

hath he leaden feete whiche are as flow to ouertake a
finner. Heare therefore the great comfort flowing in
euery leafe and lyne of the Scripture if thou be patient,
I my felfe am euen hee which doth blot out his
tranfgreffions and that for mine own fake, and I wil
not be mindfull of thy fins. Behold the Lords hand
is not fhortned that it cannot faue, neither his eare
heauy yat it cannot heare. If your finnes were as
Crimofin, they fhall bee made whyter then Snowe,
and though they were as redde as Scarlet, they fhall
be made lyke white Wo[o]ll. If wee confeffe our offences he is faithfull and iuft, fo that he will forgiue vs
our finnes. God hath not appointed vs vnto wrath,
but vnto faluation, by the meanes of our Lord Jefus
Chrift, the earth is filled with the mercy of the lord.
It is not ye wil of your father which is in heauen that
any one of the[fe] little ones fhould perifh. God is rich
in mercie, I wil not the death of a finner faith the
Lorde God, returne and liue. The fonne of man
came not to deftroy but to faue. God hath mercy on
al, bicaufe he can do all. God is merciful, long fuffering, and of much mercy. If the wicked man fhall
repent of his wickednes which he hath committed,
and kepe my commaundements doing Iuftice and
Iudgement, he fhall lyue the life, and fhall not dye.
If I fhall fay vnto the finner thou fhalt dye the deathe,
yet if he repent and doe iuftice, he fhal not dye. Call
to thy mind the great goodneffe of God in creating
thee, his finguler loue in giuing his fonne for thee.
So God loued the world that he gaue his only begotten
fonne that whofoeuer beleeued in him might not
perifh but haue euerlafting lyfe. God hath not fent
his fonne to iudge the world, but that the world might
be faued by him. Can the Mother (fayth the Prophet)
forget the childe of hir wombe, and though fhe be fo
vnnaturall, yet will I not be vnmindefull of thee.
There fhall be more ioy in heauen for the repentance
of one finner, then for ninety and nine iuft perfons.
I came not faith Chrift to call the righteous, but

finners to repentance. If any man fin, we haue an
aduocate with the father, Iefus Chrift the righteous,
he is the propitiation for our finnes, and not for our
finnes onely, but for the finnes of the whole world.
I write vnto you lyttle children becaufe your finnes be
forgiuen for his names fake.   Doth not Chrift fay,
that whatfoeuer we fhall afke the father in his name,
we fhall obteyne?   Doth not God fay : This is my
beloued fonne in whom I am well pleafed, heare him.

I haue read of *Themiftocles* which hauing offended
Philip the king of *Macedonia,* and could no way
appeafe his anger, meeting his young fonne *Alexander,*
tooke him in his armes and met *Philip* in the face :
Philip feing the fmiling countenaunce of the childe
was wel pleafed with *Themiftocles.*   Euen fo if through
thy manifolde finnes and haynous offences thou prouoke the heauy difpleafure of thy God, infomuch as
thou fhalt tremble for horror, take his onely begotten
and wel-beloued fonne Iefus in thine armes, and then
hee neither can nor will be angry with thee.   If thou
haue denyed thy God, yet if thou go out with *Peter*
and weepe bitterly, God will not deny thee.   Though
with the prodigall fonne thou wallow in thine owne
wilfulneffe, yet if thou retourne againe forrowfull thou
fhalt bee receyued.   If thou bee a grieuous offender,
yet if thou come vnto Chrift with the woman in *Luke,*
and wafh his feete with thy teares, thou fhalt obteyne
remiffion.   Confider with thy felfe the great loue of
Chrift, and the bitter torments yat he endured for thy
fake, which was enforced through the horror of death
to cry with a loud voyce, *Eloi, Eloi, lama fabacthani.*
My God, my God, why haft thou forfaken me, and
with a groning fpirite to fay, my foule is heauy euen
vnto the death, tary heere and watch : and again,
Father if it be poffible lette this cup paffe from mee.
Remember how hee was crowned with thornes, crucified with theeues, fcourged and hanged for thy faluation,
how he fweat water and bloud for thy remiffion,
how he endured euen the torments of the damned

fpirites for thy redemption, how he ouercame death that thou fhouldeft not dye, howe he conquered the diuel that thou mighteft not be damned.

When thou fhalt record what he hath done to purchafe thy freedome, how canft thou dread bondage? When thou fhalt beholde the agonies and anguifh of minde that he fuffered for thy fake, howe canft thou doubt of the releafe of thy foule? When thy Sauiour fhal be thy Iudge, why fhouldft thou tremble to heare of iudgement? When thou haft a continuall Mediator with God the Father, howe canft thou diftruft of his fauour?

Turne therefore vnto Chrift with a willing heart and a wayling minde for thy offences, who hath promifed that at what time foeuer a finner repenteth him of his finnes, he fhalbe forgiuen, who calleth all thofe that are heauy laden, that they might be refrefhed, who is the doore to them that knock, the way to them that feeke the truth, the rocke, the corner ftone, the fulneffe of time, it is he that can and will poure Oyle into thy wounds.

Who abfolued *Mary Magdalen* from hir finnes but Chrift? Who forgaue the theefe his robbery and manflaughter but Chrift? Who made *Mathew* the Publicane and tollgeatherer an Apoftle and Preacher but Chrift? Who is that good Shephearde that fetcheth home the ftray fheepe fo louingly vppon his fhoulders but Chrift? Who receiued home the loft fonne, was it not Chrift? Who made of *Saul* a perfecutor, *Paul* an Apoftle, was it not Chrift? I paffe ouer diuers other hiftories both of the olde and new Teftament, which do aboundantly declare what great comfort the faithful penitent finners haue alwaies had in hearing the comfortable promifes of Gods mercy. Canft thou then *Atheos* diftruft thy Chrift, who reioyceth at thy repentaunce? Affure thy felfe that through his paffion and bloudfhedding, Death hath loft his ftinge, the Diuell his victory, and that the gates of hell fhall not preuaile againft thee. Lette not therefore the bloude of Chrift be fhedde in vaine by thine obftinate and harde heart. Lette this perfwafion reft in thee, that thou fhalt receiue abfolution freely, and then fhalt thou feele thy foule euen as it were to hunger and thirft after righteoufneffe.

Atheos. Well *Euphues* feeing the holy Ghoft hath made thee the meane to make me a man (for before the taft of the Gofpel I was worfe then a beaft) I hope ye fame fpirite will alfo lighten my confcience with his word and confirme it to the ende in conftancy, that I may not onely confeffe my Chrift faithfully, but alfo preach him freely, that I may not only be a Minifter of his word, but alfo a Martyr for it, if be his pleafure.

O *Euphues*, howe much am I bounde to the goodneffe of almightie God, which hath made me of an Infidell a beleeuer, of a caftaway a Chriftian, of an heathenly Pagan, a heauenly Proteftant. O how comfortable is the feeling and taft of grace, how ioyful are the glad tidings of the Gofpell, the faithfull promifes of faluation, the free redemption of the foule. I will endeauour by all meanes to confute thofe dampnable I know not by what names to terme them, but blafphemers I am fure, which if they be no more, certeinly they can be no leffe. I fee now the ods betwixt light and darkeneffe, faith and frowardeneffe, Chrift and *Belyal.* Be thou *Euphues* a witneffe of my faith, feeing thou haft bene the inftrument of my beliefe, and I will praye that I fhewe it in my lyfe. As for thee, I accompt my felfe fo much in thy debte, as I fhall neuer bee able with the loffe of my lyfe to render thee thy due, but GOD which rewardeth the zeale of all men, will I hope bleffe thee, and I will pray for thee.

Euphues. O *Atheos* lyttle is the debte thou oweft mee, but great is the comfort that I haue receyued by thee. Giue the prayfe to God, whofe goodneffe hath made thee a member of the mifticall body of Chrift, and not onely a brother with his fonne, but alfo coheriter with thy Sauiour.

There is no heart fo hard, no heathen fo obftinate,

no Mifcreaunt or Infidel fo impious, that by grace is not made as fupple as Oyle, as tractable as a Sheepe, as faithfull as any.

The Adamant though it be fo harde that nothing can brufe it, yet if the warme bloud of a Goat be poured vppon it, it burfteth : Euen fo although the heart of the *Atheift* and vnbeleeuer be fo hard that neither reward nor reuenge can mollifie it, fo ftout that no per-fwafion can breake it, yet if the grace of God, purcha-fed by the bloud of Chrift, do but once touch it, it ren-teth in funder, and is enforced to acknowledge an omnipotent and euerlafting *Iehouah?* Let vs therefore both (*Atheos* I will not now call thee but *Theophilus*) fly vnto that Chrift which hath through his mer-cie, not our merits, purchafed for vs the en-heritaunce of euer-lafting life.

(∴)

Certeine Letters writ by

Euphues to his friends.

Euphues to Philautus.

IF the courfe of youth had any refpect to
the ftaffe of age, or the liuing man any
regard to the dying moulde, we would
with greater care when we wer young
fhun thofe things which fhould grieue vs
when we be olde, and with more feueritie direct the
fequele of our life, for the feare of prefent death?
But fuch is either the vnhappines of mans condition,
or the vntowardneffe of his crooked nature, or the
wilfulnes of his minde, or the blindenes of his heart,
that in youth he furfeteth with delights, preuenting
age, or if he liue, continueth in dotage, forgetting
death. It is a world to fee, how in our flourifhing time
when we beft may, we be worft willing to thriue.
And how in the fading of our dayes, when wee moft
fhould, we haue leaft defire to remember our ende.
Thou wilt mufe *Philautus* to heere *Euphues* to preach,
who of late had more minde to ferue his Lady, then
to worfhippe his Lorde. Ah *Philautus*, thou art
now a Courtier in *Italy*, I a Scholler in *Athens*, and as
hard it is for thee to follow good counfaile, as for me
to enforce thee, feeing in thee there is lyttle wil to
amend, and in mee leffe authoritie to commaunde, yet
will I exhort thee as a friende, I woulde I might com-
pell thee as a Father. But I haue heard that it is
peculiar to an *Italian* to ftande in his owne conceipt,
and to a courtier neuer to be controld, which caufeth
me to feare that in thee which I lament in others.
That is, that either thou feeme too wife in thine owne
opinion, thinking fcorn to be taught, or too wilde in
thine attempts in reiecting admonifhment. The one

procedeth of felf loue and fo thy name importeth, the
other of meere folly, and that thy nature fheweth :
thou lokeft I fhold craue pardon for fpeaking fo
boldly. No *Philautus*, I meane not to flatter thee, for
then fhould I incurre the fufpition of frawd. Neither
am I determined to fall out with thee, for then might
the wife conuince me of folly. But thou art in great
credit in the court, and what then ? fhal thy credit
with the Emperour, abate my courage to my God ? or
thy hauty lookes quench my kindeled loue, or thy
gallant fhew aflake my good wil ? hath the courtier any
prerogatiue aboue the clowne, why he fhould not be
reprehended ? Doth his high callyng, not onely giue
him a commiffion to finne, but remiffion alfo if he
offend ? doth his preheminence in the court, warrant
him to oppreffe the poore by might, and acquit him
of punifhment ? No *Philautus*. By how much the
more thou excelleft others in honours, by fo much
the more thou oughteft to exceed them in honeftie,
and the higher thy callyng is, the better ought thy
confcience to be, and as farre it befeemeth a Gentleman to be from pride, as he is from pouertie, and as
neere to gentleneffe in condition, as he is in bloud ?
But I will defcende with thee to perticulars. It is
reported heere for a troth, that *Philautus*, hath giuen
ouer himfelfe to all delicioufneffe, defiring rather to be
dandled in [on] the laps of Ladyes, then bufied in the
ftudie of good letters : And I would this were all,
which is too much, or the reft a lye, which is too
monftrous. It is now in euery mans mouth, that
thou, yea, thou *Philautus*, art fo voyde of curtefie, that
thou haft almoft forgotten common fence and humanitie, hauing neither care of Religion (a thing too common in a courtier) neither regarde of honeftie or any
vertuous behauiour. Oh *Philautus*, doeft thou lyue
as thou fhouldft neuer dye, and laugh as thou fhouldft
neuer mourne, art thou fo fimple as thou doeft not
know from whence thou cameft, or fo finfull that thou
careft not whether thou goeft : what is in thee yat

fhould make thee fo fecure, or what can there be in
any yat may caufe him to glory. *Milo* that great
wraftler beganne to weepe when he fawe his armes
brawnefallen and weake, faying, ftrength, ftrength, is
but vanitie [vaine]. *Helen* in hir new glaffe viewing hir
olde face, with a fmyling countenaunce, cryed : Beauty
where is thy blaze ? *Cræfus* with al his wealth,
Ariftotle with al his wit, all men with all their wifdome, haue and fhall perifh and tourne to duft. But
thou delyghteft to haue the newe fafhion, the *Spanifh*
felte, the *French* ruffe, thy crewe of Ruffians, all thy
attyre miffhapen to make thee a monfter, and all thy
time mifpent to fhewe thee vnhappy : what fhould I
go about to decipher thy life, feeing the beginning
fheweth the ende to bee naught.   Art not you [thou]
one of thofe *Philautus* which fekeft to win credite with
thy fuperiors by flattery, and wring out wealth from
thy inferiors by force, and vndermine thy equals by
frawd : doft thou not make ye court not only a
couer to defend thy felf from wrong, but a coulour
alfo to commit iniury ?   Art not thou one of thofe,
that hauing gotten on their fleeue the cognifance of a
courtier haue fhaken from thy fkirts the regard of curtefie.   I cannot but lament (I would I might remedy)
ye great abufes that raigne in the eyes of the Emperour.   I feare me ye Poet fay to[o] truely.   *Exeat aula
qui vult effe pius, virtus et fumma poteftas non coeunt.*
Is not pietie turned al to pollicy, faith to forefight,
rigor to iuftice : doth not he beft thriue yat worft
deferueth, and he rule al the country, yat hath no
confcience ?   Doth not the emperou[r]s court grow to
this infolent blindnes, that al that fee not their folly
they accompt fooles, and al that fpeak againft it, precife ? laughing at ye fimplicity of the one, and threatning ye boldneffe of the other.   *Philautus*, if thou
wouldeft with due confideration way how farre a
courtiers life is from a found beliefe, thou wouldeft
either frame thy felfe to a new trade, or els amend
thine old manners, yea, thou woldeft with *Crates* leaue

al thy poffeffions, taking thy bookes and trudge to *Athens*, and with *Anaxagoras* difpife wealth to atteyn wifdome, if thou haddeft as great refpect to dye well as thou haft care to liue wantonly, thou wouldeft [fhouldeft] with *Socrates* feeke how thou mighteft yeelde to death, rather then with *Ariftippus* fearch howe to prolong thy lyfe. Doft thou not knowe that where the tree falleth there it lyeth? and euery ones deathes daye is his do[o]mes daye? that the whole courfe of life is but a meditation of death, a pilgrymage, a warfare? Haft thou not read, or doeft thou not regarde what is written, that we fhall all be cyted before the Tribunall feate of God to render a ftraight accompte of our ftewardfhip? if then the reward bee to bee meafured by thy [the] merites, what boote canft thou feeke for, but eternall paine, whiche heere lyueft in continuall pleafure? So fhouldeft thou liue as thou maift dye, and then fhalt thou dye to liue. Wert thou as ftrong as *Sampfon*, as wife as *Solomon*, as holye as *Dauid*, as faithfull as *Abraham*, as zealous as *Mofes*, as good as any that euer lyued, yet fhalt thou dye as they haue done, but not rife againe to lyfe with them, vnleffe thou liue as they dyd. But thou wilt fay that no man ought to iudge thy confcience but thy felfe, feeing thou knoweft it [not] better then any. O *Philautus*, if thou fearch thy felf and fee [finde] not finne, then is thy cafe almoft cureleffe. The patient, if Phifitions are to be credited, and common experience eftemed, is ye neereft death when he thinketh himfelf paft his dif-eafe, and the leffe griefe he fe[e]leth ye greater fits he endureth, ye wound yat is not fearched bicaufe it a little fmarteth, is fulleft of dead flefh, and the fooner it fkinneth, the forer it feftereth. It is faid that Thunder brufeth the tree but breaketh not the barke, and pearceth the blade, and neuer hurteth the fcabberd: Euen fo doth finne wounde the heart, but neuer hurt the eyes, and infect the foule, though outwardly it nothing afflict the body. Defcende therefore into thine own confcience, confeffe thy finnes, reforme thy

manners, contemne the worlde, embrace Chriſt, leaue
the court, follow thy ſtudy, preferre holyneſſe before
honour, honeſtie before promotion, relygion and vp-
rightneſſe of life, before the ouerlaſhinge deſires of the
fleſh.　Reſemble [remember] the Bee, which out of the
dryeſt and bittereſt Time ſucketh moyſt and ſweete
Hunny.　And if thou canſt out of ye court a place of
more pompe then pietie, ſucke out the true iuice of
perfection, but if thou ſee in thy ſelfe a will rather to
goe forwarde, in* thy* loſeneſſe* then* any* meane* to*
goe* backwarde*, if the gliſtering faces of faire Ladyes,
or the glittering ſhew of luſty gallaunts, or courtly ſare,
or any delicate thing ſeeme to entice thee to farther
lewdnes, come from ye court to *Athens*, and ſo in
ſhunning the cauſes of euil, thou ſhalt ſoone eſcape
the effect of thy miſſortune, yat [the] more thoſe things
pleaſe thee, the more thou diſpleaſeſt God, and the
greater pride thou takeſt in ſinne, the greater pain
thou heapeſt to thy ſoule.　Examine thine own conſcience and ſee whether thou haſt done as is required,
if thou haue, thanke the Lorde and pray for encreaſe of
grace, if not, deſire God to giue thee a willing minde to
atteine faith, and conſtancye to continue to the ende.

Euphues and Eubulus.

I Salute thee in the Lord, &c.　Although I was not
　ſo wittie to follow thy graue aduice when I firſt
knew thee : yet doe I not lacke grace to giue thee
thanks ſince I tryed thee.　And if I were as able to
perſwade thee to patience, as thou wert deſirous to
exhort me to pietie, or as wiſe to comfort thee in
thine age, as thou willing to inſtruct me in my youth,
thou ſhouldeſt nowe with leſſe griefe endure thy late
loſſe, and with little care leade thy aged life.　Thou
weepeſt for the death of thy daughter, and I laugh
at the folly of the father, for greater vanitie is there
in the minde of the mourner, then bitterneſſe in the

death of the deceafed. But fhee was amiable, but
yet finful, but fhe was young and might haue liued,
but fhe was mortall and muft haue dyed. I but hir
youth made thee often merry, I but thine age fhold
once make thee wife. I but hir greene yeares wer
vnfit for death, I but thy hoary haires fhould difpyfe
life. Knoweft thou not *Eubulus* that life is the gift
of God, death the due of Nature, as we receiue
the one as a benefite, fo muft we abide the other of
neceffitie. Wife men haue found that by learning
which old men fhould know by experience, that in life
ther is nothing fweete, in death nothing fowre. The
Philofophers accompted it ye chiefeft felicitie neuer to
be borne, the fecond foone to dye. And what hath
death in it fo hard yat we fhould take it fo heauily? is
it ftraunge to fee yat cut off, which by nature is made
to be cut? or that melten, which is fit to be melted?
or that burnt which is apt to be burnt, or man to paffe
that is borne to perifh? But thou graunteft that fhe
fhould haue dyed, and yet art thou gri[ee]ued that fhe
is dead. Is the death the better if ye life be longer?
no truely. For as neither he yat fingeth moft, or
praieth longeft, or ruleth the fterne ofteneft, but he
yat doth it beft deferueth greateft praife, fo he, not yat
hath moft yeares but many vertues, nor he that hath
graieft haires but greateft goodnes, lyueth longeft.
The chiefe beauty of life confifteth not in the num-
bring of many dayes, but in the vfing of vertuous
dooings. Amongft plants thofe be beft eftemed that
in fhorteft time bring foorth much fruite. Be not the
faireft flowers gathered when they be frefheft? the
youngeft beafts killed for facrifice bicaufe they be
fineft? The meafure of life is not length, but honeftie,
neither do we enter into life to the ende we fhould fet
downe ye day of our death, but therfore do we liue,
that we may obey him yat made vs, and be willing to
dye when he fhal cal vs. But I will afke thee this
queftion, whether thou wayle the loffe of thy daughter
for thine owne fake or hirs, if for thine own fake,

bicaufe thou didſt hope in thine age to recouer comfort, then is thy loue to hir but for thy commoditie,
and therin thou art but an vnkinde father, if for hirs,
then doſt thou miſtruſt hir faluation, and therin thou
ſheweſt thy vnconſtant faith. Thou ſhouldſt not weepe
that ſhe hath runne faſt, but that thou haſt gone ſo
ſlow, neither ought it to grieue thee that ſhee is gone
to hir home with a few yeares, but that thou art to go
with many. But why goe I about to vfe a long pro-
ceſſe to a lyttle purpofe ? The bud is blaſted as ſoone
as the blowne Rofe, the winde ſhaketh off the
bloſſome, as well as ye fruit. Death fpareth neither
ye golden locks nor the hoary head. I meane not to
make a treatife in the praife of Death, but to note the
neceſſitie, neither to write what ioyes they receiue that
dye, but to ſhew what paines they endure that liue.
And thou which art euen in the wane of thy life,
whom nature hath nouriſhed ſo long, that now ſhe
beginneth to nod, maiſt wel know what griefes, what
labours, what paines are in age, and yet wouldſt thou
be either young to endure many, or elder to bide
more. But thou thinkeſt it honourable to go to the
graue with a gray head, but I deeme it more glorious
to be buried with an honeſt name. Age faiſt thou is
the bleſſing of God, yet the meſſenger of death.
Deſcend therefore into thine owne conſcience, conſider
the goodneſſe that commeth by the ende, and the
badneſſe which was by the beginning, take the death of
thy daughter patiently, and looke for thine own ſpeedely,
ſo ſhalt thou performe both the office of an honeſt man,
and the honor of an aged father, and ſo farewell.

Euphues to Philautus touching
the death of Lucilla.

I Haue receiued thy letters, and thou haſt deceiued
mine expeĉtation, for thou feemeſt to take more
thought for the loſſe of an harlot, then the life of an
honeſt woman. Thou writeſt that ſhe was ſhamefull

in hir trade, and fhameleffe in hir ende. I beleeue thee, it is no meruaile that fhe which lyuing practifed finne, fhould dying be voyde of fhame, neither coulde there be any great hope of repentaunce at the houre of death, where there was no regard of honeftie in time of life. She was ftriken fodeinely, beeing troubled with no fickeneffe : It may be, for it is commonly feene, that a finfull lyfe, is rewarded with a fodeine death, and a fweet beginning with a fower end. Thou addeft moreouer, that fhe being in great credite with the ftates died in great beggerie in the ftreetes, certes it is an olde faying that who fo liueth in the court, fhall dye in the ftrawe, fhe hoped there by delyghtes to gaine money, and by hir deferts, purchafed mifery: they that feeke to clyme by priuie finne, fhall fall with open fhame, and they that couet to fwim in vice, fhall finke in vanitie, to their owne perills. Thou faift that for beautie fhe was the *Helen* of *Greece*, and I durft fweare that for beaftlyneffe fhe might bee the Monfter of *Italy*. In my minde greater is the fhame to be accompted an harlot, then the praife to be efteemed amiable. But wher thou art in the court there is more regard of beautie then honeftie, and more are they lamented that dye vicioufly, then they loued that liue vertuoufly : for thou giueft as it were a figh, which all thy companions in the Court feeme by thee to founde alfo, that *Lucilla* being one of fo great perfection in all parts of the body, and fo little pietie in the foule, fhould be as it were fnatched out of the iawes of fo many young gentlemen. Wel *Philautus,* thou takeft not fo much care for the loffe of hir as I grieue for thy lewdneffe, neither canft thou forrow more to fee hir dye fodeinely, then I to heare thee liue fhamfully. If thou meane to keepe me as a friend, fhake off thofe vaine toyes and dalyaunces with women, beleeue me *Philautus,* I fpeake it with falt teares tricklyng downe my cheekes, the lyfe thou lyueft in court is no leffe abhorred then the wicked death of *Lucilla* detefted, and more art thou

fcorned for thy folly, then fhe hated for hir filthineffe.

The euil ende of *Lucilla* fhould moue thee to begin
a good [new] lyfe, I haue often warned thee to fhunne
thy wonted trade? and if thou loue me as thou protefteft
in thy letters, then leaue al thy vices, and fhew it in thy
life. If thou meane not to amend thy manners, I
defire thee to write no more to me, for I wil neither
anfwere thee nor read them. The Iennet is broken
as foone with a wand as with the fpurre, a Gentleman
as wel allured with a word, as with a fword. Thou
concludeft in the end that *Liuia* is fick, truly I am
fory, for fhe is a maiden of no leffe comelines then
modeftie, and hard it is to iudge whether fhe deferues
more praife for hir beutie with the amorous, or admiration for hir honeftie of ye vertuous, if thou loue me
embrace hir, for fhe is able both to fatiffie thine eye
for choice, and inftruct thy heart with learning.
Commend me vnto hir, and as I praife hir to thee, fo
wil I pray for hir to god, that either fhee may haue
pacience to endure hir trouble, or deliuerance to fcape
hir perill. Thou defireft me to fende thee the Ser-
mons which were preached of late in *Athens.* I haue
fulfilled thy requeft, but I feare me thou wilt vfe them
as faint *George* doth his horfe, who is euer on his back
but neuer rideth, but if thou wert as willing to read
them as I was to fend them, or as redy to follow them
as defirous to haue them, it fhal not repent thee of thy
labour, nor me of my coft. And thus farewel.

¶ *Euphues to Botonio, to take his exile patiently.*

IF I were as wife to giue thee counfaile, as I am
willing to do thee good, or as able to fet thee at
libertie as defirous to haue thee free, thou fhouldeft
neither want good aduice to guide thee, nor fufficient
help to reftore thee. Thou takeft it heauily that thou
fhouldeft be accufed without colour, and exiled [ban-

iſhed] without cauſe : and I thinke thee happy to be
ſo well rid of the court and bee ſo voyde of crime.
Thou ſayſt baniſhment is bitter to the free born, and I
deeme it the better if thou bee without blame. There
bee manye meates which are ſower in the mouth and
ſharpe in the Mawe, but if thou mingle them with
ſweete ſawces, they yeelde both a pleaſaunt taſt and
wholeſome nouriſhment. Diuers coulours offende the
eyes, yet hauing greene among them, whette the ſight.
I ſpeake this to this ende, that though thy exile ſeeme
grieuous to thee, yet guiding thy ſelfe with the rules of
Philoſophie it ſhal bee more tollerable, hee that is
colde doth not couer himſelfe with care but with
clothes, he that is waſhed in the rayne, dryeth himſelfe
by the fire, not by his fancie, and thou which art
baniſhed oughteſt not with teares to bewayle thy hap,.
but with wiſdome to heale thy hurt.

Nature hath giuen no man a country, no more then
ſhe hath a houſe or lands, or liuings. *Socrates* wold
neither cal himſelf an *Athenian*, neither a *Græcian* but
a citizen of ye world. *Plato* would neuer accompt
him baniſhed yat had ye Sun, Fire, Aire, Water and
Earth, that he had before, where he felt the Winters
blaſt and the Summers blaze, where ye ſame Sun, and
the ſame Moone ſhined, whereby he noted that euery
place was a country to a wiſe man, and al parts a
pallace to a quiet mind. But thou art driuen out of
Naples ? yat is nothing. All the *Athenians* dwel not
in *Colliton*, nor euery *Corinthian* in *Græcia*, nor al the
Lacedemonians in *Pitania*. How can any part of the
world be diſtant farre from the other, when as the
Mathematicians ſet down that the earth is but a point
being compared to ye heauens. Learne of ye Bee as
wel to gather Hunny of ye weede as the flowre, and
out of farre countryes to liue, afwel as in thine own.
He is to be laughed at which thincketh ye Moone
better at *Athens* then at *Corinth*, or the Hunny of the
Bee ſweeter that is gathered in *Hybla*, then that which
is made in *Mantua* ? when it was caſt in *Diogenes*

teeth, yat the *Sinoponetes* had banifhed him *Pontus*,
yea faid he, I them of *Diogenes*.   I may fay to thee as
Straconicus faid to his gueft, who demaunded what
fault was punifhed with exile, and he aunfwering falfe
hoode, why then faid *Straconicus* doft not thou practife
deceit to the ende thou maift auoyd the mifc[h]iefes
that flow in thy country.

And furely if confcience be the caufe thou art
banifhed ye court, I accompt thee wife in being fo
precife yat by the vfing of vertue, thou maift be
exciled the place of vice.   Better it is for thee to
liue with honefty in ye country then with honor in
the court, and greater wil thy praife bee in flying
vanitie, then thy pleafure in followinge traines.   Choofe
that place for thy pallace which is moft quyet, cuftome
will make it thy countrey, and an honeft life will caufe
it a pleafaunt lyuing.   *Philip* falling in the duft, and
feeing the figure of his fhape perfect in fhew.   Good
God faid he, we defire ye whole earth, and fee howe
little ferueth?   *Zeno* hearing that this onely barke
wherin all his wealth was fhipped to haue perifhed,
cryed out, thou haft done wel Fortune to thruft mee
into my gowne againe to embrace Philofophye.   Thou
haft therfore in my minde great caufe to reioyce, that
God by punifhment hath compelled thee to ftrictneffe
of life, which by lybertie might haue ben growen to
lewdneffe.   When thou haft not one place affigned thee
wherein [therein] to liue, but one forbidden thee which
thou muft leaue, then thou being denied but one, that
excepted thou maift choofe any.   Moreouer this difpute with thy felfe, I beare no office wherby I fhould
either for feare pleafe the noble, or for gaine oppreffe
the needy.   I am no arbiterer in doubtful cafes
whereby I fhould either peruerte Iuftice, or incurre
difpleafure.   I am free from the iniuries of the
ftronge, and malice of the weak.   I am out of the
broyles of the feditious, and haue efcaped the threates
of the ambitious.   But as hee that hauing a faire
Orchard, feeing one tree blafted, recomteth the dif-

commoditie of that, and paffeth ouer in filence the fruitefulneffe of the other. So hee that is banyfhed doth alwayes lament the loffe of his houfe, and the fhame of his exile, not reioyfing at the liberty, quietnes and pleafure that he enioyeth by that fweete punifhment. The kings of *Perfia* were deemed happy in that they paffed their Winter in *Babylon* : in *Mediæ* their Summer, and their Spring in *Sufis* : and cer- teinly the Exile in this may be as happy as any king in *Perfia*, for he may at his leafure being at his owne pleafure, lead his Winter in *Athens*, his Summer in *Naples*, his Spring in *Argos*. But if he haue any bufines in hand, he may ftudy without trouble, fleepe without care, and wake at his wil without con- trolment. *Ariftotle* muft dine when it pleafeth *Philip*. *Diogenes* when it lifteth *Diogenes*, the courtier fuppeth when the king is fatiffied, but *Botonio* may now eat when *Botonio* is an hungred. But thou faift that banifhment is fhamefull. No truely, no more then pouertie to the content, or graye haires to the aged. It is the caufe that maketh thee fhame, if thou wert banifhed vpon choler, greater is thy credit in fuf- teining wrong, then thy enuyes in committing iniury, and leffe fhame is it to thee to be oppreffed by might, then theirs that wrought it for malice. But thou feareft thou fhalt not thriue in a ftraunge nation, certeinly thou art more afraide then hurte. The Pine tree groweth as foone in *Pharo* as in *Ida*, ye Nightingale fingeth as fweetly in the defearts, as in ye woods of *Crete*. The wife man liueth as wel in a far country as in his owne home. It is not the nature of the place but the difpofition of the perfon, that maketh the lyfe pleafant. Seing therfore *Botonio*, that al the fea is apt for any fifh, yat it is a bad ground where no flower wil grow, that to a wife man all lands are as fertile as his owne enheritance, I defire thee to tem- per the fharpnes of thy banifhment with the fweetenes of the caufe, and to meafure the cleerenes of thyne owne confcience, with the fpite of thy enimies quarrel, fo fhalt thou reuenge their malyce with patience, and endure thy banifhment with pleafure.

¶ *Euphues to a young gentleman in Naples named Alcius, who leauing his ftudy followed all lightnes and liued both fhamfully and finfully to the griefe of his friends and difcredite of the Vniuerfitie.*

IF I fhould talke in words of thofe things which I haue to conferre with thee in writinges certes thou wouldft blufh for fhame, and I weepe for forrowe: neither could my tongue vtter yat with patience which my hand can fcarce write with modefly, neither could thy ears heare that without glowing which thine eyes can hardly vewe without griefe. Ah *Alcius*, I cannot tel whether I fhould moft lament in thee thy want of learning, or thy wanton lyuinge, in the one thou art inferiour to al men, in the other fuperior to al beafts. Infomuch as who feeth thy dul wit, and marketh thy froward will, may wel fay that he neuer faw fmacke of learning in thy dooings, nor fparke of relygion in thy life. Thou onely vaunteft of thy gentry, truely thou waft made a gentleman before thou kneweft what honefly me[a]nt, and no more haft thou to boaft of thy flocke then he who being left rich by his father, dyeth a begger by his folly. Nobilitie began in thine aunceftors and endeth in thee, and the Generofitie that they gayned by vertue thou haft blotted with vice. If thou claime gentry by pedegree, practife gentleneffe by thine honefly, yat as thou challengeft to be noble in bloud, thou maift alfo proue noble by knowledge, otherwife fhalt thou hang lyke a blaft among the faire bloffomes and lyke a ftaine in a peece of white Lawne.

The Rofe that is eaten with the Canker is not gathered bicaufe it groweth on that ftalke yat the fweet doth, neither was *Helen* made a Starre, bicaufe fhee came of that Egge with *Caftor*, nor thou a gentleman in yat thy aunceftours were of nobilitie. It is not ye defcent of birth but ye confent of conditions that maketh Gentlemen, neither great manors but good manners that expreffe the true Image of dignitie. There is copper coine of the ftampe yat gold is, yet is it not currant, there commeth poyfon of the fifh as wel as good oyle, yet is it not wholfome, and of man may proceede an euill childe and yet no Gentleman. For as the Wine that runneth on the lees, is not there- fore to be accompted neate bicaufe it was drawne of the fame peece. Or as the water that fpringeth from the fountaines head and floweth into the filthy channel is not to be called cleere bicaufe it came of the fame ftreame : fo neither is he that defcendeth of noble parentage, if he defift from noble deedes to be efteemed a Gentleman in yat he iffued from the loyns of a noble fire, for that he obfcureth the parents he came off, and difcrediteth his owne eftate.

There is no Gentleman in *Athens* but forroweth to fee thy behauiour fo far to difagree from thy birthe, for this fay they al (which is the chiefeft note of a gentle- man) that thou fhouldeft as well defire honeftie in thy life, as honor by thy linage : that thy nature fhould not fwerue from thy name, that as thou by dutie woldeft be regarded for thy progenie, fo thou wouldft endea- uour by deferts to be reuerenced for thy pietie.

The pure Coral is chofen as wel by his vertue as his coulour, a king is known better by his courage, then his crowne, a right Gentleman is fooner feene by the tryall of his vertue then blafing of his armes.

But I let paffe thy birth, wifhing thee rather with *Vliffes* to fhew it in workes, then with *Aiax* to boaft of it with words : thy ftocke fhall not be the leffe, but thy modeftie the greater. Thou liueft in *Athens*, as the Wafpe doth among Bees, rather to fting then to gather Hunny, and thou dealeft with moft of thy acquaintaunce as the Dogge doth in the maunger, who neither fuffereth the horfe to eat hay, nor wil himfelfe. For thou being idle, wilt not permit any (as farre as in thee lyeth) to be well employed. Thou art an

heyre to fayre lyuing, that is nothing, if thou be
disherited of learning, for better were it to thee to inherite righteousnesse then riches, and far more seemely
were it for thee to haue thy Studie full of bookes, then
thy pursse full of mony: to get goods is the benefit
of Fortune, to keepe them the gift of Wisedome. As
therfore thou art to possesse them by thy fathers wil, so
art thou to encreafe them by thine owne wit.

But alas, why desirest thou to haue the reuenewes of
thy parent, and nothing regardest to haue his vertues?
seekest thou by succession to enioy thy patrimony, and
by vice to obscure his pietie ? wilt thou haue the title
of his honour, and no touch of his honestie? Ah
Alcius remember yat thou art borne not to liue after
thine own lust, but to learne to dye, wherby thou
maist liue after thy death. I haue often heard thy
father say, and that with a deepe sigh, the teares
trickling downe his gray haires, that thy mother neuer
longed more to haue thee borne when she was in
trauaile, then he to haue thee dead to rid him of
trouble. And not seldome hath thy mother wished,
that either hir wombe had bene thy graue, or the
ground hirs. Yea, all thy friendes with open mouth,
desire either that god will send thee grace to amend
thy life, or griefe to hasten thy death.

Thou wilt demaund of me in what thou dost offend :
and I aske thee in what thou doest not sinne. Thou
sweareft thou art not couetous, but I saye thou arte
prodigall, and as much sinneth he that lauisheth
without meane, as he that hoordeth without measure.
But canst thou excuse thy selfe of vice in that thou
arte not couetous? certeinly no more then the mur-
therer would therefore be guyltlesse, bicause he is no
coyner. But why go I about to debate reason with
thee when thou hast no regard of honestie ? though I
leaue heere to perswade thee, yet will I not ceafe to
pray for thee. In the meane feason I desire thee, yea,
and in gods name commaund thee, yat if neither the
care of thy parents, whom thou shouldest comfort, nor

the counfaile of thy friends which thou fhouldft credite, nor the rigour of the law which thou oughteft to feare, nor the authoritie of the Magiftrate, which thou fhouldft reuerence, can allure thee to grace: yet the law of thy fauiour who hath redeemed thee, and the punifhment of the almightie, who continually threatneth thee, [fhould] draw thee to amendement, otherwife as thou liueft now in finne, fo fhalt thou dye with fhame, and remaine with Sathan. From whom he that made thee, keepe thee.

¶ *Liuia from the Emperours court, to Euphues at Athens.*

IF fickeneffe had not put me to filence, and the weakeneffe of my body hindered the willingneffe of my minde, thou fhouldeft haue had a more fpeedye aunfwere, and I no caufe of excufe. I know it expedient to retourne an aunfwere, but not neceffary to write in poft, for that in things of great importance, we commonly looke before we leape, and where the heart droupeth through faintnes, ye hand is enforced to fhake through feeblenes. Thou faift thou vnderftandeft how men liue in the court, and of me thou defireft to know the eftate of women, certes to diffemble with thee wer to deceiue my felfe, and to cloake the vanities in court, were to clog mine owne confcience with vices.

The Empreffe keepeth hir eftate royall, and hir maidens will not leefe an ynch of their honor, fhe endeauoreth to fet down good lawes, and they to breake them, fhe warneth them of exceffe, and they ftudie to exceed, fhe faith yat decent attire is good, though it be not coftly, and they fweare vnleffe it be deere, it is not comely. She is heere accompted a flutte that commeth not in hir filkes, and fhe that hath not euery fafhion hath no mans fauour. They that be moft wanton are reputed moft wife, and they that be the idleft liuers, are deemed the fineft louers. Ther

N

is great quarrelling for beautie but no queſtion of honeſtie: to conclude, both women and men haue fallen heere in court to ſuch agreement, that they neuer iarre about matters of religion, bicauſe they neuer meane to reaſon of them. I haue wiſhed oftentimes rather in ye country to ſpin, then in the court to daunce, and truly a diſtaffe doth better become a maiden then a Lute, and fitter it is with the needle to praⳍiſe how to lyue, then with the pen to learne how to loue.

The Empreſſe giueth enſample of vertue, and the Ladyes haue no leaſure to follow hir. I haue nothing els to write. Heere is no good newes, as for bad, I haue tolde ſufficient: Yet this I muſt adde that ſome there be, which for their vertue deſerue praiſe, but they are onely commended for their beautie, for this thinke courtiers, that to be honeſt is a certeine kinde of coun-try modeſtie, but to be amiable the courtly courteſie.

I meane ſhortly to ſue to the Empreſſe to be diſ-miſſed of the court, which if I obtaine I ſhall thinke it a good reward for my ſeruice, to be ſo wel ridde from ſuch ſeueritie [ſecuritie], for beleeue me, ther is ſcarce one in court that either feareth God, or meaneth good. I thanke thee for the booke thou diddeſt ſend me, and as occaſion ſhall ſerue I will requite thee.

Philautus beginneth a little to liſten to counſaile, I wiſh him wel, and thee to, of whom to heare ſo much good, it doth me not a lyttle good. Pray for me as I do for thee, and if opportunitie be offered, write to me.

Farewel.

Euphues to his friend Liuia.

DEare *Liuia*, I am as glad to heare of thy welfare, as ſorrowful to vnderſtand thy newes, and it doth me as much good that thou art recouered, as harme to thinke of thoſe which are not to be recured.

Thou haſt ſatiſſied my requeſt and aunſwered my expeⳍation. For I longed to know ye manners of women, and looked to haue them wanton. I like thee wel that thou wilt not conceale their vanities, but I loue thee the better that thou doeſt not follow them : to reproue ſinne is the ſigne of true honour, to renounce it the part of honeſty. Al good men wil accompt thee wiſe for thy truth, and happy for thy tryall, for they ſay, to abſteine from pleaſure is the chiefeſt piety, and I thinke in court to refraine from vice, is no little vertue. Strange it is that the found eye viewing the ſore ſhould not be dimmed, that they [he] that handle[th] pitch ſhoulde not bee defiled, that they yat continue in [the] court ſhold not be infeĉted. And yet it is no great meruaile, for by experience we ſee that the Adamant cannot draw yron, if the Diamond lye by it, nor vice allure the courtyer, if vertue be reteyned.

Thou praiſeſt ye Empreſſe for inſtituting good lawes, and grieueſt to ſee them violated by the Ladyes. I am ſory to thinke it ſhould be ſo, and I ſigh in that it cannot be otherwiſe. Where there is no heede taken of a commaundement, there is ſmall hope to be looked for of amendement. Where duetie can haue no ſhewe, honeſtie can beare no ſway. They that cannot be enforced to obedience by authoritie, wil neuer be won by fauour, for being without feare they commonly are voyd of grace : and as farre be they careleſſe from honour as they be from awe, and as ready to diſpiſe the good counſaile of their Peeres, as to contemne the good lawes of their prince. But the breaking of lawes doth not accuſe the Empreſſe of vice, neither ſhall hir making of them, excuſe the Ladyes of vanities. The Empreſſe is no more to be ſuſpeĉted of erring, then the Carpenter that buildeth the houſe be accuſed bicauſe theeues haue broken it, or the Mintmaſter condemned for his coyne bicauſe the traitor hath clipped it. Certeinely God will both reward the godly zeale of the Prince, and reuenge the godleſſe doings of the people. Moreouer thou ſaiſt that in the court all be ſluts that ſwim not in ſilkes, and that the ideleſt liuers are accompted ye braueſt louers. I can not tell whether I ſhould rather laugh at their folly, or lament their phrenfie, neither do I know whether ye fin be greater in apparell which moueth to pride, or in affection which entifeth to peeuifhnes. The one caufeth them to forget them-felues, the other to forgo their fences, each doe deceiue their foule, they that thinke one cannot be cleanly without pride, will quickly iudge none to be honeft without pleafure, which is as hard to confeffe as to fay no meane to bee without exceffe : thou wifheft to be in the Country with thy diftaffe, rather then to continue in the court with thy delyghts. I cannot blame thee. For *Greece* is as much to be commended for learning as the court for brauery, and here maift thou liue with as good report for thine honefty, as they with renowne for their beauty. It is better to fpinne with *Penelope* all night, then to finge with *Helen* all daye. Hufwifery in the Country is as much praifed as honour in the court. Wee thinke it as great mirth to fing Pfalmes, as you melody to chaunt Sonets, and we accompt them as wife that keepe their owne lands with credite, as you thofe that get others liuinges by craft. There-fore if thou wilt follow my aduice, and profecute thine owne determination, thou fhalt come out of a warme Sunne into Gods bleffing. Thou addeft (I feare me alfo thou erreft) that in the court ther be fome of great vertue, wifedome and fobrietie : if it be fo, I like it, and in that thou faift it is fo, I beleeue it. It may bee, and no doubt it is in the courte as in all ryuers, fome Fifh fome Frogges, and as in all gardeins, fome flowers, fome weedes, and as in al trees, fome bloffoms fome blafts. *Nylus* breedeth the precious ftone and the poyfoned ferpent. The court may as wel nourifh vertuous Matrones, as the lewd minion. Yet this maketh me mufe that they fhoulde rather be com-mended for their beautie then for their vertue, which is an infallible argument that the delights of ye flefh are preferred before the holyneffe of the fpirite. Thou faift thou wilt fue to leaue thy feruice, and I wil pray for thy good fucceffe, when thou art come into the

country, I would haue thee firſt learne to forget all
thoſe things which thou haſt ſeene in the court. I
would *Philautus* wer of thy minde, to forſake his
youthfull courſe, but I am glad thou writeſt yat he
beginneth to amend his conditions, he runneth far that
neuer returneth, and he ſinneth deadly that neuer repenteth. I would haue him end as ˙*Lucilla* began
without vice, and not begin as ſhe ended without
honeſtie. I loue the man well, but I cannot brooke
his manners. Yet I conceiue a good hope, that in his
age he will be wiſe, for that in his youth I perceiued
him wittie. He hath promiſed to come to *Athens*,
which if he do, I will ſo handle the matter, that either
he ſhal abiure the court for euer, or abſent himſelf for
a yeare. If I bring the one to paſſe he ſhall forgoe
his olde courſe, if the other forget his il conditions.
He that in court wil thriue to reape wealth, and liue
wary to get worſhip, muſt gaine by good conſcience,
and clime by wiſdome, otherwiſe his thrift is but theft,
wher ther is no regard of gathering, and his honour
but ambition, wher ther is no care but of promotion.
Philautus is too ſimple to vnderſtand the wiles in
court, and too young to vndermine any by craft. Yet
hath he ſhown himſelf as far from honeſtie as he is
from age, and as full of craft as he is of courage.
If it wer for thy preferment, and his amendment,
I wiſh you were both maryed, but if hee ſhould
continue his folly whereby thou ſhouldeſt fall from
thy dutie, I rather wiſh you both buryed. Salute
him in my name, and haſten his iourney, but forget
not thine owne. I haue occaſion to go to *Naples*,
that I may with more ſpeede ariue in *England*,
where I haue heard of a woman yat in al qualities
excelleth any man. Which if it be ſo, I ſhal thinke
my labour as well beſtowed as *Saba* did hirs, when
ſhe trauailed to ſee *Salomon*. At my going if thou be
in *Naples* I will viſite thee, and* at my retourn I wil
tell thee my iudgement. If *Philautus* come this
winter, he ſhall in this my pilgrimage be a partner, a

pleafaunt companion is a bait in a iourny. We fhal ther as I heare, fee a court both brauer in fhew, and better in fubftaunce, more gallant courtiers, more godly confciues, as faire ladies and fairer conditions. But I will not vaunt, before the victorie, nor fweare it is fo, vntil I fee it be fo. Farewel, vnto whom aboue all I wifh well.

I Haue finifhed the firft part of *Euphues*, whom now I left readye to croffe the Seas to *England*, if the winde fende him a fhort cut you fhall in the fecond part heare what nevves he bringeth, and I hope to haue him retourned within one Summer. In the meane fefon, I wil ftay for him in the country, and as foone as he ariueth you shall know of his comming.

FINIS.

¶ Imprinted at London, by Thomas Eaft, for *Gabriel Cawood, dwelling in Paules Church-yard.* *1579.*

[Colophon of Edition, 1581.

¶ Imprinted at London by

Thomas Eaft, for Gabriel Cawood, dwelling in Paules Church- yard. *1581.*]

Edition. 1581.

Title-page, The Epistle Dedicatorie, &c.

from

the copy in the Grenville collection, in the

Britiſh Muſeum.

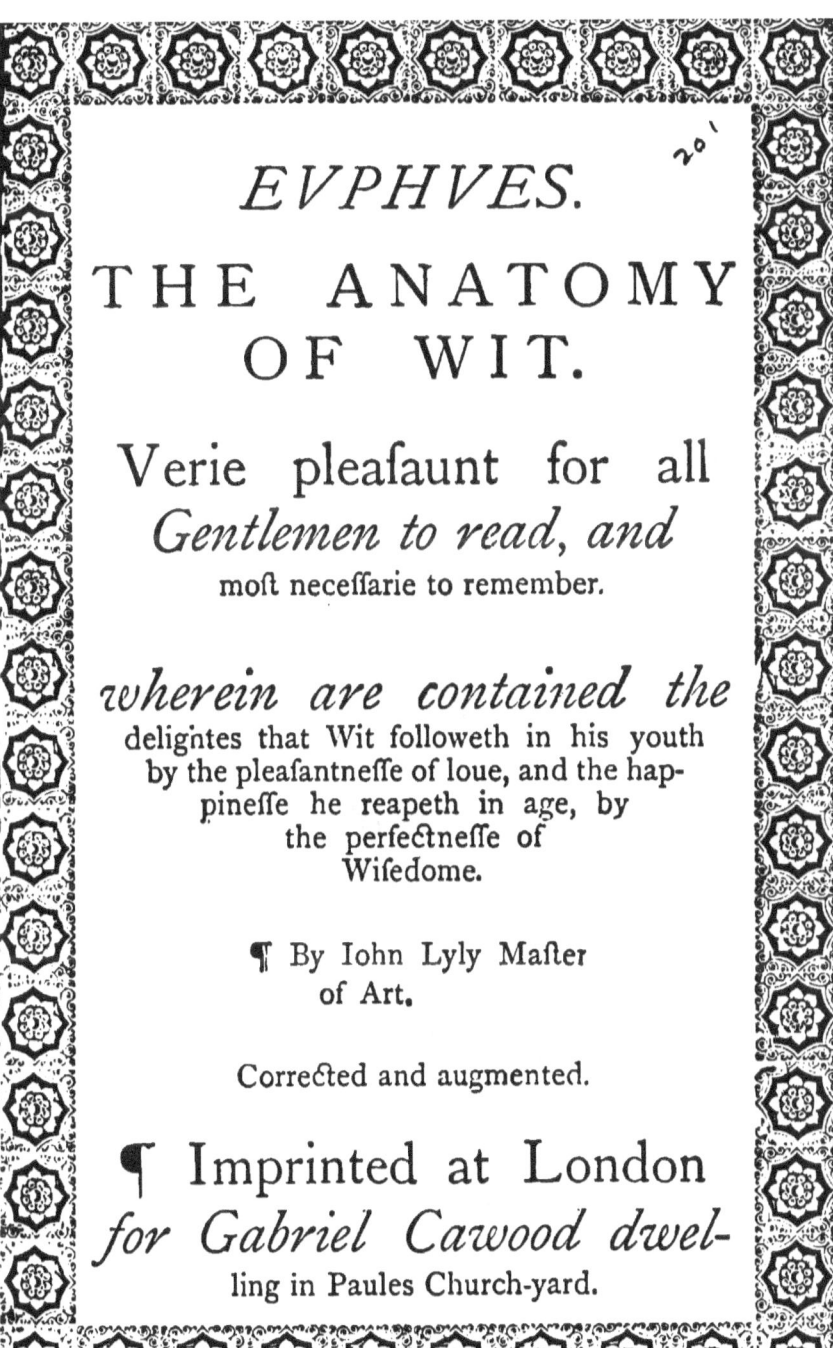

EVPHVES.

20'

THE ANATOMY OF WIT.

Verie pleafaunt for all *Gentlemen to read, and* moft neceffarie to remember.

wherein are contained the delightes that Wit followeth in his youth by the pleafantneffe of loue, and the happineffe he reapeth in age, by the perfectneffe of Wifedome.

¶ By Iohn Lyly Mafter of Art.

Corrected and augmented.

¶ Imprinted at London *for Gabriel Cawood dwel-* ling in Paules Church-yard.

To the right Honorable my verie

good Lord and Mafter, Sir William Weft, Knight,
Lord De la Warre : Iohn Lyly wifheth long
life, with increafe of
honour.

ARRHASIVS drawing the counterfaite of Helen (Right Honourable)
made the attire of her head loofe,
who being demaunded why he did
fo, aunfwered fhe was loofe. Vulcan
was painted curioufly, yet with a
polt foote, Læda cunningly, yet
with her blacke haire. Alexander
hauing a fkarre in his cheeke, held his finger vppon
it, that Appelles might not paint it, Appelles painted
him, with his finger cleauing to his face, vvhy
quod Alexander, I laid my finger on my skarre,
becaufe I would not haue thee fee it, (yea faid
Appelles) and I drew it there, becaufe none els fhould
perceiue it, for if thy finger had bene avvaie, either
thy fkarre wold haue bene feene or my arte mifliked :
whereby I gather, that in all perfect works, as vvell
the fault as the face is to be fhovven. The faireft
Leopard is made vvith his fpots, the fineft cloth with
his lift, the fmootheft fhooe hath his lafte. Seeing
then that in euerie counterfaite, as vvell the blemifh
as the beautie is coloured : I hope I fhall not incurre
the difplefure of the vvife, in that in the difcourfe of
Euphues, I haue as vvell touched the vanities of his
loue, as the vertues of his life. The Perfians, vvho
aboue all their kings moft honored Cyrus, caufed
him to be ingrauen as vvell vvith his hooked nofe as
his high forhead. He that loued Homere beft, concealed not his flattering; and he that praifed Alexander moft, bevvraied his quaffing. Demonides muft haue
a crooked fhooe for his vvrie foote, Damocles a
fmooth gloue for his ftraight hand.

For as euery Painter that fhadoweth a man in all

partes, giueth euerie peece his iuſt proportion, ſo he that difciphereth the qualities of the minde, ought as well to ſhew euerie humor in his kinde, as the other doth euery parte in his colour. The Surgion that maketh the Anatomie, ſheweth as well the muſcles in the heele, as the vaines of the heart. If then the firſt ſight of Euphues ſhall ſeeme too light to be read of the wiſe, or too fooliſh to be regarded of the learned, they ought not to impute it to the iniquitie of the Author, but to the neceſſitie of the Hiſtorie. Euphues beginneth with loue, as allured by wit, but endeth not with luſt, as bereft of wiſedome. He wooeth women, prouoked by youth, but weddeth not himſelfe to wantonneſſe, as pricked by pleaſure. I haue ſet downe the follies of his wit without breach of modeſtie, and the ſparkes of his wiſedome without fuſpition of diſhoneſtie. And certes I thinke there be mo ſpeaches, which for grauitie will miſlike the fooliſh : then vnſeemly termes, which for vanitie may offende the wiſe. Which difcourſe (right Honorable) I hope you will the rather pardon for the rudenes, in that is the firſt, and proteℭt it the more willingly if it offend, in that it maye be the laſt. It may be that fine wits will defcant vpon him that hauing no wit, goeth about to make the Anatomye of wit: and certainly their ieſting in my minde is tollerable. For if the Butcher ſhould take vppon him to cut the Anatomie of a man, becauſe he hath ſkill in opening an Oxe, he would proue himſelfe a Calfe, or if the horſeleach would aduenture to miniſter a potion to a ſick patient, in that hee hath knowledge to giue a drench to a diſeaſed horſe, he wold make himſelfe an Aſſe. The ſhomaker muſt not goe aboue his latchet, nor the hedger meddle with any thing but his bil. It is vnſeemly for the Painter to feather a ſhafte, or the Fletcher to handle the penℭill. All which things make moſt againſt me, in that a foole hath intruded himſelf to difcourſe of wit : but as I was vvilling to commit the fault, ſo am I content to make amendes. Hovvloeuer the caſe ſtandeth, I look for no praiſe for my labour, but pardon for my good will : it is the greateft revvarde I dare aske, and the leaft that they can offer, I defire no more, I deferue no leffe. Though the ftile nothing delight the daintie eare of the curious fifter, yet vvill the matter recreate the minde of the curteous Reader: the varietie of the one will abate the harfhneffe of the other. Things of greateft profit, are fet forth with leaft price, where the vvine is neat, ther needeth no Iuie-bufh, the right Corall needeth no colouring, vvhere the matter it felfe bringeth credit, the man with his glofe winneth fmall commendation. It is therefore me thinketh a greater fhevve of a pregnaunt vvit, then perfecte wifdome, in a thing of fufficient excellencie to vfe furperfluous eloquence. We commonly fee that a blacke ground doth beft befeeme a white counterfaite, and Venus according to the iudgement of Mars, vvas then moft amiable when fhe fate clofe by Vulcan. If thefe thinges be true, which experience trieth, that a naked tale doeth moft truelye fet foorth the naked trueth, that where the countenaunce is faire, there neede no colours, that painting is meeter for ragged walls than fine marble, that veritie then fhineth moft bright, when fhe is in leaft brauerie, I fhall fatiffie mine ovvne minde, though I cannot feed their humors, which greatly feeke after thofe that fift the fineft meale, and beare the whiteft mouthes. It is a world to fee hovv Englifhmen defire to heare finer fpeech then the language will allovve, to eate finer bread then is made of wheat, to vveare finer cloth then is vvrought of vvoll: but I let paffe their finenes, vvhich can no vvay excufe my folly. If your Lordfhip fhall accept my good vvil vvhich I haue alvvaies defired, I vvill patiently beare the ill vvill of the malitious, vvhich I neuer deferued.

Thus committing this fimple Pamphlet to your Lordfhippes patronage, and your Honour to the Almighties protection: For the preferuation of the which, as moft bounden, I will piaie continuallie, I ende.

Your Lordfhips feruant to commaund. I. Lily.

To the Gentlemen Readers.

Was driuen into a quandarie Gentlemen, whether I might fende this my Pamphlet to the Printer or to the pedler, I thought it too bad for the preffe, and too good for the packe, but feeing my folly in writing to be as great as others, I was willing my fortune fhould be as ill as anies. We commonly fee the booke that at Eafter lyeth bounde on the Stacioners ftall, at Chriftmaffe to be broken in the Haberdafhers fhop, which fith it is the order of proceeding, I am content this Summer to haue my dooinges read for a toye, that in Winter they may be readye for trafh. It is not ftrange when as the greateft wonder lafteth but nine daies, that a new worke fhuld not endure but three months. Gentlemen vfe bookes as Gentlewomen handle their flowers, who in the morning ftick them in their heads, and at night ftrawe them at their heeles. Cherries be fulfom when they be through ripe, becaufe they be plentie, and bookes be ftale when they be printed in that they be common. In my minde Printers and Tailers are chiefely bound to pray for Gentlemen, the one hath fo many fantafies to print, the other fuch diuers fafhions to make, that the preffing yron of the one is neuer out of the fire, nor the printing preffe of the other at any time lieth ftill. But a fafhion is but a daies wearing and a booke but an houres reading : which feeing it is fo, I am of the fhoomakers minde, who careth not fo the fhooe hold the plucking on, nor I, fo my labours laft the running ouer. He that commeth in print becaufe he woulde be knowen, is like the foole that commeth into the Market becaufe he would be feene. I am not he that feeketh praife for his labour, but pardon for his offence, neyther doe I fet this forth for anie deuotion in Print, but for duetie which I owe to

my Patron. If one write neuer fo well, he cannot
pleafe all, and write he neuer fo ill, hee fhall pleafe
fome. Fine heads will picke a quarrell with me, if all
be not curious, and flatterers a thanke if anye thing be
currant: but this is my minde, let him that findeth
fault amend it, and him that liketh it, vfe it. Enuye brag-
geth, but draweth no bloud: ye malitious haue more
minde to quip, then might to cut. I fubmit myfelfe to the
iudgement of the wife, and little efteeme the cenfure of
fooles: the one will be fatiffied with reafon: the other are
to be aunfwered with filence. I know Gentlemen will
finde no fault without caufe, and beare with thofe
that deferue blame, as for others I care
not for their ieftes, for I neuer
meant to make them
my Iudges.

Farewell.

To my verie good friends the
Gentlemen Schollers of Oxford.

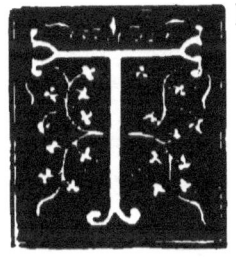HERE is no priuiledge that needeth a pardon, neither is there any remiffion to bee afked where a commiffion is graunted. I fpeake this Gentlemen, not to excufe the offence which is taken, but to offer a defence where I was miftaken. A cleere confcience is a fure carde, truth hath the pre-rogatiue to fpeake with plainenefle, and the modeftie to beare with patience. It was reported of fome, and beleeued of many, that in the Education of *Ephœbus*, where mention is made of Vniuerfities, that *Oxford* was too much either defaced or defamed. I knowe not what the enuious haue picked out by mallice, or the curious by wit, or the guilty by their owne galled confciences, but this I fay, yat I was as far from think-ing ill, as I finde them from iudging well. But if I fhould now goe about to make amends, I were then faultie in fomewhat amiffe, and fhould fhew my felfe like *Apelles* Prentice, who coueting to mend the nofe, marred the cheeke, and not vnlike the foolifh Dyar, who neuer thought his cloth blacke vntill it was burned. If anie fault be committed, impute it to *Euphues* who knew you not, not to *Lyly* who hate you not.

Yet may I of all the reft moft condempne *Oxfora* of vnkindneffe, of vice I cannot, who feemnd to weane mee before fhe brought mee forth, and to giue mee boanes to gnawe, before I could get the teate to fucke. Wherein fhe played the nice mother in fending me into the Countrie to nurfe, where I tyred at a drie breaft three yeares, and was at the laft inforced to weane my felfe. But it was deftinie, for if I had not ben gathered from the tree in the bud, I fhould being blowen haue prooued a blaft, and as good it is to be an addle egge, as an idle bird.

· *Euphues* at his arriuall I am affured will viewe *Oxford*, where he will either recant his fayinges, or renewe his complaints, he is now on the feas, and how he hath bene tooffed I know not, but whereas I thought to receiue him at *Douer*, I mufte meete him at *Hampton*.

Nothing can hinder his comming but death, neither anie thing haften his departure but vnkindneffe.

Concerning my felfe, I haue alwayes thought fo reuerently of *Oxford*, of the Schollers, of the manners, that I feemed to be rather an Idolater then a blafphemer.

They that inuented this toie were vnwife, and they that reported it vnkinde, and yet none of them can proue mee vnhoneft.

But fuppofe I glaunced at fome abufes : did not *Iupiters* egge bring forth as well Helen a light hufwife in earth, as *Caftor* a light Starre in heauen? The Eftritch that taketh the greateft pride in her feathers, picketh fome of the worft out, and burneth them : there is no tree but hath fome blaft, no countenaunce but hath fome blemifh, and fhall *Oxford* then be blameleffe? I wifh it were fo, but I cannot thinke it is fo. But as it is it may be better, and were it badder, it is not the worft.

I thinke there are fewe Vniuerfities that haue leffe faultes then *Oxford*, many that haue more, none but haue fome.

But I commit my caufe to the confciences of thofe that either know what I am, or can geffe what I fhould bee, the one will anfwere themfelues in conftruing friendly, the other if I knew them, I would fatiffie reafonably.

Thus loth to incur the fufpition of vnkindneffe in not telling my minde, and not willing to make anie excufe where there neede no amends, I can neither craue pardon, leaft I fhould confeffe a fault, nor conceale my meaning, leaft I fhould be thought a foole. And fo I end, yours affured to vfe.

Iohn Lyly.

Euphues and his England.

Text. *Editio princeps*, 1580.

Profeffor Morley's copy.

Completed, after 'Or a Foxe,' p. 475, to the end, from
the Bodleian copy, of the fame year.

Collation. Edition, 1582.

excepting
eight leaves wanting (=pp. 362-3, and 463-478)
which have been compared with
the edition of 1586.

Both copies in the collection of
H. Pyne, Efq., Affiftant Tithe Commiffioner.

EUPHUES AND HIS ENGLAND.

THE PRINCIPAL CHARACTERS PRESENT IN THE ACTION.

EUPHUES.

PHILAUTUS.

FIDUS, *an old Englijh courtier, now a keeper of bees.*

SURIUS, *a young Englijh gentleman, ' of great birth and noble blood.'*

PSELLUS, *an Italian gentleman, reputed 'great in Magick.'*

MARTIUS, *an Englijhman, ' not very young.'*

CAMILLA, *a young Englijhwoman of eighteen years. 'Of no great birth,' but ' of greater beauty than birth.' ' Such a one jhe was, as almojt they all are that ferve fo noble a Prince, fuch virgins carry lights before fuch a* VESTA, *fuch nymphes, arrows with fuch a* DIANA.' p. 311.

The Lady FLAVIA, *an Englijhwoman. ' One of the Ladies who delighted much in mirth.'*

Mijtrejs FRANCES, *niece to the Lady* FLAVIA.   PHILAUTUS' *Violet.*

SCENE AND TIME.

.·. Yet Philautus' last letter—received by Euphues in ATHENS not passing one quarter of a yeare after he left England, p. 464— is dated 1. February 1579[80].

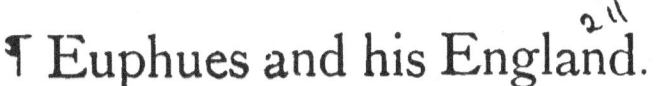

¶ Euphues and his England. ²¹¹

CONTAINING

his voyage and aduentures, myxed with
fundry pretie difcourfes of honeft
Loue, the difcription of the
countrey, the Court, and
the manners of that
Ifle.

DELIGHTFVL TO

be read, and nothing hurtfull to be regarded : wher-in there is fmall offence
by lightneffe giuen to the wife,
and leffe occafion of loofe-
nes proffered to the
wanton.

¶ By Iohn Lyly, Maifter
of Arte.

Commend it, or amend it.

Imprinted at London for

Gabriell Cawood, dwelling in
Paules Church-yard.
1580.

To the Right Honourable my

very good Lorde and Maifter, Edward de Vere,
Earle of Oxenforde, Vicount Bulbeck, Lorde of
Efcales and Badlefmere, and Lorde great
Chamberlaine of England, *Iohn Lyly*
wifheth long lyfe, with en-
creafe of Honour.

HE firft picture that Phydias the
firft Paynter fhadowed, was the pro-
traiture of his owne perfon, faying
thus : if it be well, I will paint many
befides Phydias, if ill, it fhall offend
none but Phydias.

In the like manner fareth it with
me (Right Honourable) who neuer
before handling the penfill, did for my fyrft counterfaite,
coulour mine owne Euphues, being of this minde, that
if it wer[e] lyked, I would draw more befides Euphues,
if loathed, grieue none but Euphues.

Since that, fome there haue bene, that either diffembling the faultes they faw, for feare to difcourage me,
or not examining them, for the loue they bore me,
that praifed mine olde worke, and vrged me to make a
new, whofe words I thus anfwered. If I fhould coyne
a worfe, it would be thought that the former was framed
by chaunce, as Protogenes did the foame of his dogge,
if a better, for flatterie, as Narciffus did, who only was
in loue with his own face, if none at all, as froward as
the Mufition, who being entreated, will fcarfe fing fol
fa, but not defired, ftraine aboue Ela.

But their importunitie admitted no excufe, in-fomuch that I was enforced to preferre their friendfhip
before mine owne fame, being more carefull to fatiffie
their requeftes, then fearefull of others reportes : fo that

at the laſt I was content to ſet an other face to Euphues, but yet iuſt behind the other, like the Image of
Ianus, not running together, lik[e] the Hopplitides of
Parrhaſius leaſt they ſhould ſeeme ſo vnlike Brothers,
that they might be both thought baſtardes, the piᶜture
wherof I yeeld as common all to view, but the patronage onely to your Lordſhippe, as able to defend, knowing that the face of Alexander ſtamped in copper doth
make it currant, that the name of Cæſar, wrought in
Canuas, is eſteemed as Cambricke, that the very feather of an Eagle, is of force to conſume the Beetle.
I haue brought into the worlde two children, of the
firſt I was deliuered, before my friendes thought mee
conceiued, of the ſecond I went a whole yeare big, and
yet when euerye one thought me ready to lye downe,
I did then quicken : But good hufwiues ſhall make my
excuſe, who know that Hens do not lay egges when
they clucke, but when they cackle, nor men ſet forth
bookes when they promiſe, but when they performe.
And in this I reſemble the Lappwing, who fearing hir
young ones to be deſtroyed by paſſengers, flyeth with
a falſe cry farre from their [the] neſtes, making thoſe
that looke for them ſeeke where they are not : So I ſuſ-
peᶜting that Euphues would be carped of ſome curious
Reader, thought by ſome falſe ſhewe to bringe them
in hope of that which then I meant not, leading them
with a longing of a ſecond part, that they might ſpeake
well of the firſt, being neuer farther fiom my ſtudie,
then when they thought mee houering ouer it.

My firſt burthen comming before his time, muſt
needes be a blind whelp, the ſecond brought forth after
his time muſt needes be a monſter, the one I ſent to
a noble man to nurſe, who with great loue brought him
vp, for a yeare : ſo that where-ſoeuer he wander, he
hath his Nurſes name in his forhead, wher ſucking his
firſt milke, he can-not forget his firſt Maſter.

The other (right Honourable) being but yet in his
ſwathe cloutes, I commit moſt humbly to your Lordſhips proteᶜtion, that in his infancie he may be kepte

by your good care from fal[l]s, and in his youth by your great countenaunce ſhielded from blowes, and in his age by your gracious continuaunce, defended from contempt. He is my youngeſt and my laſt, and the paine that I fuſtained for him in trauell, hath made me paſt teeming, yet doe I thinke my ſelfe very fertile, in that I was not altogether barren. Glad I was to ſende them both abroad, leaſt making a wanton of my firſt, with a blinde conceipt, I ſhould reſemble the Ape, and kill it by cullyng it, and not able to rule the ſecond, I ſhould with the Viper, looſe my bloud with mine own brood. Twinnes they are not, but yet Brothers, the one nothing reſemblyng the other, and yet (as all children are now a dayes) both like the father. Wherin I am not vnlike vnto the vnskilfull Painter, who hauing drawen the Twinnes of Hippocrates, (who wer as lyke as one peaſe is to an other) and being told of his friends that they wer[e] no more lyke than Saturne and Appollo, he had no other ſhift to manifeſt what his worke was, then ouer their heads to write: The Twinnes of Hippocrates. So may it be, that had I not named Euphues, fewe woulde haue thought it had bene Euphues, not that in goodnes the one ſo farre excelleth the other, but that both beeing ſo bad, it is hard to iudge which is the worſt.

This vnskilfulneſſe is no wayes to be couered, but as Accius did his ſhortneſſe, who beeing a lyttle Poet, framed for himſelfe a great picture, and I being a naughtie Painter, haue gotten a moſt noble Patron: being of Vlyſſes minde, who thought himſelfe ſafe vnder the Shield of Aiax.

I haue now finiſhed both my labours, the one being hatched in the hard winter with the Alcyon, the other not daring to bud till the colde were paſt, like the Mulbery, in either of the which or in both, if I ſeeme to gleane after an others Cart, for a few eares of corne, or of the Taylors ſhreds to make me a lyuery, I will not deny, but that I am one of thoſe Poets, which the painters faine to come vnto Homers baſon, there to lap vp, that he doth caſt vp.

In that I haue written, I deſire no praiſe of others
but patience, altogether vnwillyng, bicauſe euery way
vnworthy, to be accompted a workeman.

It ſufficeth me to be a water bough, no bud, ſo I may
be of the ſame roote, to be the yron, not ſteele, ſo I be
in the ſame blade, to be vineger, not wine, ſo I be in
the ſame caſke, to grinde colours for Appelles, though
I cannot garniſh, ſo I be of the ſame ſhop. What
I haue done, was onely to keepe my ſelfe from ſleepe,
as the Crane doth the ſtone in hir foote, and I would
alſo with the ſame Crane, I had bene ſilent holding a
ſtone in my mouth.

But it falleth out with me, as with the young wraſt-
ler, that came to the games of Olympia, who hauing
taken a ſoyle, thought ſcorne to leaue, till he had receiued a fall, or him that being pricked in the finger
with a Bramble, thruſteth his whole arme among the
thornes, for anger. For I ſeeing my ſelfe not able to
ſtande on the yce, did neuertheleſſe aduenture to runne,
and being with my firſt booke ſtriken into diſgrace,
could not ceaſe vntil I was brought into contempt by
the ſecond : wherein I reſemble thoſe that hauing
once wet their feete, care not how deepe they wade.

In the which my wading (right Honourable) if the
enuious ſhal clap lead to my heeles to make me ſinke,
yet if your Lordſhip with your lyttle finger doe but
holde me vp by the chinne, I ſhall ſwimme, and be ſo
farre from being drowned, that I ſhall ſcarce be duckt.

When Bucephalus was painted, Appelles craued the
iudgement of none but Zeuxis : when Iuppiter was carued, Priſius aſked the cenſure of none but Lyſippus :
now Euphues is ſhadowed, only I appealé to your
honour, not meaning thereby to be careleſſe what others
thinke, but knowing that if your Lordſhip allowe it,
there is none but wil lyke it, and if ther be any ſo nice,
whom nothing can pleaſe, if he will not commend it,
let him amend it.

And heere right Honourable, although the Hiſtorie
ſeeme vnperfect, I hope your Lordſhip will pardon it.

Appelles dyed not before he could finiſh Venus, but before he durſt, Nichomachus left Tindarides rawly, for feare of anger, not for want of Art, Timomachus broke off Medea ſcarce halfe coloured, not that he was not willing to end it, but that he was threatned : I haue not made Euphues to ſtand without legges, for that I want matter to make them, but might to maintein them : ſo that I am enforced with the olde painters, to colour my picture but to the middle, or as he that drew Ciclops, who in a little table made him to lye behinde an Oke, wher[e] one might perceiue but a peece, yet conceiue that al the reſt lay behinde the tree, or as he that painted an horſe in the riuer with halfe legges, leauing the paſternes for the viewer, to imagine as in the water. For he that vieweth Euphues, wil ſay that he is drawen but to the waſt, that he peepeth, as it were behinde ſome ſcreene, that his feet are yet in the water : which maketh me preſent your Lordſhip, with the mangled body of Hector, as it appeared to Andromache, and with half a face as the painter did him that had but one eye, for I am compelled to draw a hoſe on, before I can finiſh the legge, and in ſteed of a foot to ſet downe a ſhoe. So that whereas I had thought to ſhew the cunning of a Chirurgian by mine Anatomy with a knife, I muſt play the Tayler on the ſhoppe boorde with a paire of ſheeres. But whether Euphues lympe with Vulcan, as borne lame, or go on ſtilts with Amphionax, for lack of legs, I truſt I may ſay, that his feet ſhold haue ben, olde Helena : for the poore Fiſher-man that was warned he ſhould not fiſh, did yet at his dore make nets, and the olde Vintener of Venice, that was forbidden to ſell wine, did notwithſtanding hang out an Iuie buſh. This Pamphlet right honorable, conteining the eſtate of England, I know none more fit to defend it, then one of the Nobilitie of England, nor any of the Nobilitie, more auntient or more honorable then your Lordſhip, beſides that, deſcribing the condition of the Engliſh court, and the maieſtie of our dread Souereigne, I could not finde one more noble in court, then your

Honor, who is or ſhould be vnder hir Maieſtie chiefeſt
in court, by birth borne to the greateſt Office, and therfore me thought by right to be placed in great authoritie:
for who ſo compareth the honor of your L. noble house,
with the fidelitie of your aunceſtours, may wel ſay,
which no other can truly gainſay, *Vero nihil ve-
rius.* So that I commit the ende of al my pains vnto
your moſt honorable protection, aſſuring my ſelf that
the little Cock boat is ſafe, when it is hoiſed into a tall
ſhip, that the Cat dare not fetch the mouſe out of the
Lions den, that Euphues ſhal be without daunger by
L[ordſhips] Patronage, otherwiſe, I cannot ſee, wher[e]
I might finde ſuccour in any noble perſonage. Thus
praying continually for the encreaſe of your Lordſhips
honour, with all other things that either you woulde
wiſh, or God will graunt, I ende.

Your Lordſhips moſt dutifully to commaund,

JOHN LYLY.

¶ *TO THE LADIES*

and Gentlewoemen of England,
Iohn Lyly wiſheth what
they would.

*Rachne* hauing wouen in cloth of Arras, a Raine-bow of ſundry ſilkes, it was obieꞔted vnto hir by a Ladie more captious then cunning, that in hir worke there wanted ſome cou-lours : for that in a Raine-bow there ſhould bee all : Vnto whom ſhe re-plyed, if the coulours lacke thou lookeſt for, thou muſt imagine that they are on the other ſide of the cloth : For in the Skie wee canne diſcerne but one ſide of the Raine-bowe, and what couloures are in the other, ſee wee can-not, geſſe wee may.

In the like manner (Ladies and Gentlewoemen) am I to ſhape an aunſwere in the behalfe of *Euphues,* who framing diuers queſtions and quirkes of loue, if by ſome more curious then needeth, it ſhall be tolde him that ſome ſleightes are wanting, I muſt ſaye they are noted on the back ſide of the booke. When *Venus* is paynted, we can-not ſee hir back, but hir face, ſo that all other thinges that are to be recounted in loue, *Euphues* thinketh them to hang at *Venus* back in a budget, which bicauſe hee can-not ſee, hee will not ſet downe.

Theſe diſcourſes I haue not clapt in a cluſter, think-

ing with my felfe, that Ladies had rather be fprinckled
with fweete water, then wafhed, fo that I haue fowed
them heere and there, lyke Strawberies, not in heapes,
lyke Hoppes : knowing that you take more delyght, to
gather flowers one by one in a garden, then to fnatche
them by handfulles from a Garland.

It refteth Ladies, that you take the paines to read
it, but at fuch times, as you fpend in playing with your
little Dogges, and yet will I not pinch you of that
paftime, for I am content that your Dogges lye in
your laps : fo *Euphues* may be in your hands, that when
you fhall be wearie in reading of the one, you may be
ready to fport with the other : or handle him as you
doe your Iunckets, that when you can eate no more,
you tye fome in your napkin for children, for if you be
filled with the firft part, put the fecond in your pocket
for your wayting Maydes : *Euphues* had rather lye fhut
in a Ladyes cafket, then open in a Schollers ftudie.

Yet after dinner, you may ouerlooke him to keepe
you from fleepe, or if you be heauie, to bring you a
fleepe, for to worke vpon a full ftomacke is againft
Phificke, and therefore better it were to holde *Euphues* in your hands, though you let him fal[l], when
you be willing to winke, then to fowe in a clout, and
pricke your fingers, when you begin to nod.

What-foeuer he hath written, it is not to flatter, for
he neuer reaped anye rewarde by your fex, but repentaunce, neyther canne it be to mocke you, for hee
neuer knewe anye thing by your fexe, but righteoufneffe.

But I feare no anger for faying well, when there is
none, but thinketh fhe deferueth better.

She that hath no glaffe to dreffe hir head, will vfe
a bole of water, fhee that wanteth a fleeke-ftone to
fmooth hir linnen, wil take a pebble, the country dame
girdeth hir felfe as ftraight in the waft with a courfe
caddis, as the Madame of the court with a filke
riband, fo that feeing euerye one fo willing to be
pranked, I could not thinke any one vnwilling to be
praifed.

One hand wafheth an other, but they both wafh the face, one foote goeth by an other, but they both carrye the body, *Euphues* and *Philautus* prayfe one an other, but they both extoll woemen : Therfore in my minde you are more beholding to Gentlemen that make the coulours, then to the Painters, that drawe your counterfaites : for that *Apelles* cunning is nothing if hee paint with water, and the beautie of women not much · if they go vnpraifed.

If your thinke this Loue dreamed not done, yet mee thinketh you may as well like that loue which is penned and not practifed, as that flower that is wrought with the needle, and groweth not by nature, the one you weare in your heades, for the faire fight, though it haue no fauour, the other you may reade for to paffe the time, though it bring fmall paftime. You chufe cloth that will weare whiteft, not that will laft longeft, coulours that looke frefheft, not that endure foundeft, and I would you woulde read bookes that haue more fhewe of pleafure, then ground of profit, then fhould *Euphues* be as often in your hands, being but a toy, as Lawne on your heads, being but trafh, the one will be fcarfe liked after once reading, and the other is worne out after the firft wafhing.

There is nothing lyghter then a feather, yet is it fette a loft in a woemans hatte, nothing flighter then haire, yet is it moft frifled in a Ladies head, fo that I am in good hope, though their [there] be nothing of leffe accounte then *Euphues*, yet he fhall be marked with Ladies eyes, and lyked fomtimes in their eares : For this I haue diligently obferued, that there fhall be nothing found, that may offend the chaft minde with vnfeemely tearmes, or vncleanly talke.

Then Ladies I commit my felfe to your curtefies, crauing this only, that hauing read, you conceale your cenfure, writing your iudgments as you do the pofies in your rings, which are alwayes next to the finger, not to be feene of him that holdeth you by the hands, and yet known to you that wear them on your hands :

If you be wronge [wroong] (which cannot be done without wrong) it were better to cut the fhooe, then burne
the laft.

If a Tailour make your gowne too little, you couer
his fault with a broad ftomacher, if too great, with a
number of plights, if too fhort, with a faire garde, if
too long, with a falfe gathering, my truft is you will deale
in the like manner with *Euphues,* that if he haue nut
fead [fedde] your humor, yet you will excufe him more
then the Tailour: for could *Euphues* take the meafure
of a womans minde, as the Tailour doth of hir bodie,
hee would go as neere to fit them for a fancie, as the
other doth for a fafhion.

Hee that weighes wind, muft haue a fteadie hand to
holde the ballaunce, and he that fe[a]rcheth a woemans
thoughts muft haue his own ftayed.   But leaft I make
my Epiftle as you do your new found bracelets, end-
leffe, I will frame it like a bullet, which is no fooner
in the mould but it is made.   Committing your Ladifhips to the Almightie, who graunt you al[l] you would
haue, and fhould haue: fo your wifhes ftand with his
will   And fo humbly I bid you farewell.

Your Ladifhips to commaund

IOHN LYLY.

¶ To the Gentlemen Readers.

Entlemen, Euphues is come at the length though too late, for whofe abfence, I hope three badde excufes, fhall ftande in fteede of one good reafon.

Firft in his trauaile, you muft think he loytered, tarying many a month in Italy viewing the Ladyes in a Painters fhop, when he fhould haue bene on the Seas in a Merchaunts fhip, not vnlike vnto an idle Hufwife, who is catch-ing of flyes, when fhe fhould fweepe downe copwebs.

Secondly, being a great ftart from Athens to Eng-land, he thought to ftay for the aduantage of a Leape yeare, and had not this yeare leapt with him, I think he had not yet leapt hether.

Thirdly, being arriued, he was as long in viewing of London, as he was in comming to it, not farre differing from Gentlewomen, who are longer a drefsing their heads then their whole bodyes.

But now he is come Gentlemen, my requeft is onely to bid him welcome, for diuers ther[e] are, not that they miflike the matter, but that they hate the man, that wil[l] not ftick to teare Euphues, bicaufe they do enuie Lyly : Where-in they refemble angry Dogges, which byte the ftone, not him that throweth it, or the cho-laricke Horfe-rider, who being caft from a young Colt, and not daring to kill the Horfe went into the ftable to cutte the faddle.

Thefe be they, that thought Euphues to be drowned

and yet were neuer troubled with drying of his clothes, but they geffed as they wifhed, and I woulde it had happened as they defired.

They that loath the Fountaines heade, will neuer drinke of the lyttle Brookes: they that feeke to poyfon the Fifh, will neuer eate the fpawme : they that lyke not mee, will not allowe anye thing, that is mine.

But as the Serpent Porphirius, though he bee full of poyfon yet hauing no teeth, hurteth none but himfelfe, fo the enuious, though they fwell with malyce till they burft, yet hauing no teeth to bite, I haue no caufe to feare.

Onely my fute is to you Gentlemen, that if anye thing bee amiffe, you pardon it : if well, you defende it : and how-foeuer it bee, you accepte it.

Faultes efcaped in the Printing, correcte with your pennes : omitted by my neglygence, ouerflippe with patience: committed by ignoraunce, remit with fauour.

If in euery part it feeme not alyke, you know that it is not for him that fafhioneth the fhoe, to make the graine of the leather.

The olde Hermit will haue his talke fauour of his Cell: the olde Courtier, his loue tafte of Saturne : yet the laft Louer, may happely come fomwhat neere Iuppiter.

Louers when they come into a Gardeine, fome gather Nettles, fome Rofes, one Tyme, an other Sage, and euerye one, that, for his Ladyes fauour, that fhe fauoureth : infomuch as there is no Weede almofte, but it is worne. If you Gentlemen, doe the lyke in reading, I fhall bee fure all my difcourfes fhall be regarded, fome for the fmell, fome for the fmart, all
for a kinde of a louing fmacke : Lette
euerye one followe his fancie, and fay
that is beft, which he lyketh beft.
And fo I commit euerye mans
delight to his own choice, &
my felfe to all your
courtefies.

Yours to vfe,
Iohn Lyly.

¶ Euphues and his England.

Vphues hauing gotten all things ne-
ceffary for his voyage into *England*,
accompanied onelye with *Philautus*,
tooke fhipping the firft of December,
1579, by our Englifh Computation :
Who as one refolued to fee that with
his eies, which he had oftentimes
heard with his eares, began to vfe this
perfwafion to his friend *Philautus*, afwell to counfell him
how he fhould behaue him-felfe in *England*, as to comfort him beeing nowe on the Seas.

As I haue found thee willing to be a fellow in my
trauell, fo would I haue thee ready to be a follower of
my counfell : in the one fhalt thou fhew thy good will,
in the other manifeft thy wifdome.   Wee are now fayling into an Iland of fmal compaffe as I geffe by their
Maps, but of great ciuility as I hear by their man[n]ers,
which if it be fo, it behooueth vs to be more inquifitiue of their conditions, then of their countrey : and
more carefull to marke the natures of their men, then
curious to note the fituation of the place.   And furely
me thinketh we cannot better beftow our time on the
Sea, then in aduife how to behaue our felues when we
come to ye fhore : for greater daunger is ther to ariue
in a ftraunge countrey where the inhabitants be polli-
tique, then to be toffed with the troublefome waues,

where the Mariners be vnfkilfull. Fortune guideth
men in the rough Sea, but Wifdome ruleth them in a
ftraunge land.

If Trauailers in this our age were as warye of their
conditions, as they be venterous of their bodyes, or as
willing to reape profit by their paines, as they are to
endure perill for their pleafure, they would either prefer their own foyle before a ftraunge Land, or good
counfell before their owne conceyte. But as the young
fcholler in *Athens* went to heare *Demofthenes* eloquence
at *Corinth*, and was entangled with *Lais* beautie, fo
moft of our trauailers which pretend to get a fmacke
of ftraunge language to fharpen their wits, are infected with vanity by [in] following their wils.   Daunger
and delight growe both vppon one ftalke, the Rofe
and the Canker in one bud, white and blacke are
commonly in one border.   Seeing then my good
Philautus, that we are not to conquer wilde beafts by
fight, but to confer with wife men by pollicie: We
ought to take greater heede that we be not intrapped
in follye, then feare to bee fubdued by force.   And
heere by the way it fhall not be amiffe, afwell to driue
away the tedioufneffe of time, as to delight our felues
with talke, to rehearfe an olde treatife of an auncient
Hermitte, who meeting with a pylgrime at his Cell,
vttered a ftraunge and delightfull tale, which if thou
Philautus art difpofed to heare, and thefe prefent attentiue to haue, I will fpende fome time about it, knowing
it both fit for vs that be trauailers to learne wit, and
not vnfit for thefe that be Merchaunts to get wealth.

Philautus although the ftumpes of loue fo fticked in
his mind, that he rather wifhed to heare an Eelegie in
Ouid, then a tale of an Hermit: yet was hee willing
to lend his eare to his friende, who had left his heart
with his Lady, for you fhal vnderftand that *Philautus*
hauing read the Cooling Carde which *Euphues* fent
him, fought rather to aunfwere it, then allowe it.   And
T doubt not but if *Philautus* fall into his olde vaine in
England, you fhall heare of his new deuice in *Italy*.

And although fome fhall thinke it impertinent to the hiſtorie, they fhall not finde it repugnant, no more then in one nofegay to fet two flowers, or in one counterfaite two coulours, which bringeth more delight, then difliking.

Philautus aunfwered *Euphues* in this manner.

MY good *Euphues*, I am as willing to heare thy tale, as I am to be pertaker of thy trauaile, yet I knowe not howe it commeth to paffe, that my eyes are eyther heauy againſt foule weather, or my head fo drowfie againſt fome ill newes, that this tale fhall come in good time to bring me a fleepe, and then fhall I get no harme by the Hermit, though I get no good : the other that wer then in the fhippe flocked about *Euphues*, who began in this manner.

THere dwelt fome-tymes in the Iland *Scyrum*, an auncient gentleman called *Caffander*, who afwell by his being a long gatherer, as his trad[e] being a lowd [lewde] vfurer, waxed fo wealthy, that he was thought to haue almoſt all the money in that countrey, in his owne coffers, being both aged and fickly, found fuch weakneffe in him-felfe, that he thought nature would yeeld to death, and phificke to his difeafes. This Gentleman had one onely fonne, who nothing refem-bled the father either in fancie or fauour, which the olde manne perceiuing, diffembled with him both in nature and honeſtie, whom he caufed to be called vnto his bedfide, and the chamber beeing voyded, he brake with him in thefe tearmes.

Callimachus (for fo was hee called) thou art too young to dye, and I too old to lyue : yet as nature muſt of neceffitie pay hir debt to death, fo muſt fhe alfo fhew hir deuotion to thee, whome I aliue had to be the comfort of myne age, and whome alone I muſt leaue behynde mee, for to bee the onely maynteiner of all myne honour. If thou couldeſt afwell conceiue the care of a father, as I can leuel at the nature of a childe, or wer I as able to vtter my affeċtion towards a fonne as thou oughteſt to fhew thy duety to thy fire, then wouldeſt thou defire my life to enioy my counſell, and I ſhould correct [corrupt] thy life to amend thy conditions: yet ſo tempered, as neyther rigor might detract any thing from affection in me, or feare any whit from thee, in duety. But feeing my felfe fo feeble that I cannot liue to bee thy guyde, I am refolued to giue thee fuch counſell as may do thee good, wher-in I ſhal ſhew my care, and difcharge my duetie.

My good ſonne, thou art to receiue by my death wealth, and by my counfel wifdom, and I would thou wert as willing to imprint the one in thy hart, as thou wilt be ready to beare the other in thy purfe: to bee rich is the gift of Fortune, to bee wife the grace of God. Haue more minde on thy bookes then my [thy] bags, more defire of godlineſſe then gold, greater affection to dye well, then to liue wantonly.

But as the Cypreſſe tree, the more it is watered, the more it withereth, and the oftner it is lopped, the fooner it dyeth, ſo vnbrideled youth, the more it is alfo by graue aduife counfelled, or due correction controlled, the fooner it falleth to confuſion, hating all reaſons that would bring it from folly, as that tree doth all remedies, that ſhould make it fertile.

Alas *Callimachus*, when wealth commeth into the handes of youth before they can vfe it, then fall they to al diforder that may be, tedding that with a forke in one yeare, which was not gathered together with a rake, in twentie.

But why difcourfe I with thee of worldly affaires, being my felf going to heauen, heere *Callimachus* take the key of yonder great barred Cheſt, wher thou ſhalt finde fuch ſtore of wealth, that if thou vfe it with difcretion, thou ſhalt become the onely rich man of the world. Thus turning him on his [the] left fide, with a deepe figh and pitifull grone, gaue vp the ghoaſt.

Callimachus, hauing more minde to looke to the locke, then for a ſhrowding ſheete, the breath beeing ſcarce out of his fathers mouth, and his body yet panting with heate, opened the Cheſt, where he found nothing, but a letter written very faire, fealed vp with his Signet of armes, with this fuperfcription:

¶ *In finding nothing, thou fhalt gaine all things.*

Callimachus, although hee were abaffhed at [the] fight of the emptie Cheft, yet hoping this letter would direct him to the golden Myne, he boldly opened it, the contents whereoff, follow[ed] in thefe termes.

Wifedome, is great wealth. Sparing, is good getting. Thrift confifteth not in golde, but grace. It is better to dye with-out mony, then to liue with out modeftie. Put no more clothes on thy back, then will expell colde : neither any more meat in thy belly, then may quench hunger. Vfe not chaunge in attire, nor varietie in thy dyet : the one bringeth pride, the other furfets. Each vaine, voyd of pietie : both coftly, wide of profit.

Goe to bed with the Lambe, and rife with the Larke : Late watching in the night, breedeth vnquyet: and long fleeping in the day, vngodlineffe : Flye both : this, as vnwholfome : that, as vnhoneft.

Enter not into bands, no not for thy beft friends : he that payeth an other mans debt feeketh his own decay, it is as rare to fee a rich Surety, as a black Swan, and he that lendeth to all that will borowe, fheweth great good will, but lyttle witte. Lende not a penny without a pawne, for that will be a good gage to borowe. Be not haftie to marry, it is better to haue one plough going, then two cradells : and more profit to haue a barne filled then a bedde. But if thou canft not liue chaftly, chufe fuch an one, as maye be more commended for humilitie, then beautie. A good hufwife, is a great patrimony : and fhe is moft honourable, that is moft honeft. If thou defire to be olde, beware of too much wine : If to be healthy, take heede of many women : If too be rich, fhunne playing at al games. Long quaffing, maketh a fhort lyfe : Fonde luft, caufeth drye bones: and lewd paftimes, naked purffes. Let the Cooke be thy Phifition, and the fhambles thy Apothecaries fhop: He that for euery qualme wil take a Receipt, and can-not make two meales, vnleffe *Galen* be his Gods good: fhall be fure to make the Phifition rich, and himfelfe a begger: his bodye will neuer be with-out difeafes, and his purffe euer with-out money.

Be not too lauifh in giuing almes, the charitie of this Countrey, is, God helpe thee: and the courtefie, I haue the beft wine in towne for you.

Liue in the Countrey, not in the Court: where neither Graffe will growe, nor Moffe cleaue to thy heeles.

Thus haft thou if thou canft vfe it, the whole wealth of the world: and he that can-not follow good counfel, neuer can get commoditie. I leaue thee more, then my father left me: For he dying, gaue me great wealth, without care how I might keepe it: and I giue thee good counfell, with all meanes how to get riches. And no doubt, what fo is gotten with witte, will bee kept with warineffe, and encreafed with Wifedome.

God bleffe thee, and I bleffe thee: and as I tender thy fafetie, fo God deale with my foule.

Callimachus was ftroken into fuch a maze, at this his fathers laft Will, that he had almoft loft his former wit: And being in an extreame rage, renting his clothes and tearing his haire, began* to* [he] vtter[ed] thefe words.

IS this the nature of a Father to deceiue his fonne, or the part of crabbed age, to delude credulous youth? Is the death bedde which ought to bee the ende of deuotion, become the beginning of deceipt? Ah *Caffander*, friend I can-not terme thee, feeing thee fo vnkinde: and father I will not call thee, whome I finde fo vnnaturall.

Who fo fhall heare of this vngratefulneffe, will rather lament thy dealyng, then thy death: and maruel yat a man affected outwardly with fuch great grauitie, fhould inwardly be infected with fo great guile. Shall I then fhew the duetie of a childe, when thou haft forgotten the Nature of a Father? No, no, for as the Torch tourned downewarde, is extinguifhed with the felfe fame waxe which was the caufe of his lyght : fo Nature tourned to vnkindeneffe, is quenched by thofe meanes it fhoulde be kindeled, leauing no braunch of loue, where it founde no roote of humanitie.

Thou haft caryed to thy graue more graye haires, then yeares : and yet more yeares, then vertues. Couldeft thou vnder the Image of fo precife holyneffe, harbour the expreffe patterne of barbarous crueltie ? I fee now, that as the Canker fooneft entreth into the white Rofe, fo corruption doth eaflieft creepe into the white head.

Would *Callimachus* could afwell difgeft thy malyce with patience, as thou diddeft difguife it with craft : or would I might either burie my care with thy carcaffe, or that thou hadft ended thy defame with thy death. But as ye hearb *Moly* hath a floure as white as fnow, and a roote as blacke as incke : fo age hath a white head, fhowing pietie, but a black hart, fwelling with mifchiefe.

Wher-by I fee, that olde men are not vnlyke vnto olde Trees, whofe barkes feemeth to be found, when their bodies are rotten.

I will mourne, not that thou art now dead, but bicaufe thou haft liued fo long : neither doe I weepe to fee thee without breath, but to finde thee without mony.

In fteede of coyne, thou haft left me counfaile : O polytique olde man. Didft thou learne by experience, that an edge can be any thing worth, if it haue nothing to cut, or yat Myners could worke without mettals, or Wifedome thriue, with-out where-with.

What auayleth it to be a cunning Lapidarie, and haue no ftones ? or a fkilfull Pilot, and haue no fhip ? or a thriftie man, and haue no money. Wifdome hath no Mint, Counfell is no Coyner. He that in thefe dayes feeketh to get wealth by wit, with-out friends, is lyke vnto him, that thinketh to buye meate in the

market for honeftie with-out money : which thriueth
on either fide fo well, that the one hath a wittie head
and an emptie purffe : the other a godly minde, and an
emptie belly.

Yea, fuch a world it is, that Gods can do nothing
with-out golde, and who of more might? nor Princes
any thing with-out gifts, and who of more Maieftie? nor
Philofophers any thing with-out guylt [gylte], and who of
more wifedome? For as among the *Aegyptians*, there
was no man efteemed happie, that had not a beaft full
of fpots, fo amongft vs ther is none accompted wife
that hath not a purfe full of golde. And haddeft thou
not loued money fo well, thou wouldeft neuer haue
liued fo warily and died fo wickedly, who either bury-
ing thy treafure, doeft hope to meete it in hell, or
borowing it of the Diuel haft rendred him the whole,
the intereft where-of, I feare me, commeth to no leffe
then the price of thy foule.

But whether art thou caried, *Callimachus*, rage can
neither reduce thy fathers life, nor recouer his treafure.
Let it fuffice thee, that he was vnkinde, and thou vnfortunate, that he is dead and heareth thee not, that
thou art aliue and profiteft nothing.

But what, did my father think, that too much wealth
would make me proud, and feared not too great mifery
would make me defperate? Whileft he was beginning
a frefh to renew his complaints and reuile his parents,
his kinffolke affembled, who caufed him to bridle his
lauifh tongue, although they meruailed at his pitious
tale : For it was well knowne to them all, that
Caffander had more mony then halfe the countrey,
and loued *Callimachus* better then his own felfe.

Callimachus by the importunitie of his allies, repreffed his rage, fetting order for all thinges requifite
for his fathers funeralles, who being brought with due
reuerence vnto the graue, hee returned home, making
a fhort Inuentorie to his fathers long Wil. And
hauing made ready money of fuch mouables as were
in his houfe, putte both them and his houfe into his

purfe, refoluing now with him-felfe in this extremitie, eyther with the hazarde of his labour to gayne wealth, or by myffortune to feeke death, accompting it [as] great fhame to liue with-out trauell, as griefe to bee left with-out treafure, and although hee were earneftly entreated, as well by good proffers of gentle per-fwafions to weane him-felfe from fo defolate, or rather defperate lyfe, hee would not hearken eyther to his owne commodities or their counfelles : For feeing (fayd hee) I am left heyre to all the worlde, I meane to execute my authoritie, and clayme my lands in all places of the world. Who now fo rich as *Callimachus?* Who had as many reuenues euery where as in his owne countrey? Thus beeyng in a readines to departe, apparrelled in all coulours, as one fitte for all companies, and willing to fee all countries, iournyed three or foure dayes verye deuoutlye lyke a pilgrime, who ftraying out of his pathway, and fomwhat weary, not vfed to fuch day-labours, refted him-felf vppon the fide of a filuer ftreame, euen almoft in the grifping of the euening, where thinking to fteale a nappe, beganne to clofe his eyes. As he was thus between flumbring and waking, he heard one cough pitioufly, which caufed him to ftart: and feeing no creature, hee fearched diligently in euery bufhe and vnder euery fhrubbe, at the laft he lyghted on a little caue, where thrufting in his head more bolde then wife, hee efpyed an olde man cladde all in gray, with a head as white as Alablafter, his hoarie beard hanging downe well neere to his knees, with him no earthly creature, fauing onelye a Moufe fleeping in a Cattes eare. Ouer the fyre this good olde man fatte, leaning his head to looke into a little earthen veffell which floode by him.

Callimachus delyghted more then abafhed at this ftraunge fight, thought to fee the manner of his hofte, before he would be his gueft.

This olde manne immediatelye tooke out of his potte certayne rootes, on the which hee fedde hunger-lye, hauing no other drinke then fayre water. But that which was mofte of all to bee confidered and noted, the Moufe and the Catte fell to their victualles, beeing fuch reliques as the olde manne had left, yea and that fo louinglye, as one woulde haue thought them both married, iudging the Moufe to be verye wilde, or the Cat very tame.

Callimachus coulde not refrayne laughter to beholde the folempne feafte, at the voyce where-of the olde manne arofe, and demaunded who was there : vnto whome *Callimachus* aunfwered : Father, one that wifheth thee both greater cheere and better feruaunts : vnto whome hee replyed fhoaring vp his eyes, by Iis fonne, I accompt the cheere good, which maintayneth health, and the feruauntes honeft, whome I finde fayth-full. And if thou neyther thinke fcorne of my com-pany nor my Cell, enter and welcome : the which offer *Callimachus* accepted with great thankes, who thought his lodging would be better then his fupper.

The next morning the olde manne being very inquifitiue of *Callimachus* what he was, wher he dwelt, and whether he would, *Callimachus* difcourfed with him in perticulers, as before, touching his Fathers death and defpite, againft whome hee vttered fo many bytter and burning wordes, as the olde Hermittes eares gloed to heare them, and my tonge would blyfter if I fhould vtter them. More-ouer he added that he was determined to feeke aduentures in ftraunge lands, and either to fetch the golden fleece by trauaile, or fufteine the force of Fortune by his owne wilfull follye.

Now *Philautus*, thou fhalt vnderftand that this olde Hermitte, whiche was named alfo *Caffander*, was Brother to *Callimachus* Father, and Vncle to *Calli-machus*, vnto whom *Caffander* had before his death conueyed the fumme of tenne thoufand poundes, to the vfe of his fonne in his moft extremitie and neceffitie, knowing or at the leaft forefeeing that his young colt will neuer beare a white mouth with-out a harde bridle. Alfo hee affured him-felfe that his brother fo little ten-dred money being a profeffed Hermitte, and fo much tendred and efteemed *Callimachus*, beeing his neere kinfman, as he put no doubt to ftand to his deuotion.

Caffander this olde Hermitte hearing it to bee *Callimachus* his Nephewe, and vnderftanding of the death of his brother, diffembled his griefe although he were glad to fee thinges happen out fo well, and determined with him-felfe to make a Cofinne of his young Neuew [Nephew], vntyll hee had bought witte with the price of [his] woe, wherefore he affayed firft to ftaye him from trauell, and to take fome other courfe, more fitte for a Gentleman. And to the intent fayde hee, that I may perfwade thee, giue eare vnto my tale, and this is the tale *Philautus* that I promifed thee, which the Hermitte fitting nowe in the Sunne, began to vtter to *Callimachus.*

WHen I was younge as thou nowe art, I neuer thought to bee olde, as nowe I am, which caufed luftye bloud to attempte thofe thinges in youth, which akyng boanes haue repented in age. I hadde one onely Brother, which alfo bore my name, being both borne at one tyme as twinnes, but fo farre dyfagreeing in nature, as hadde not as well the refpecte of the iuft tyme, as alfo the certeyntie and affuraunce of our Mothers fidelitie, perfwaded the worlde wee hadde one Father: It would verye hardelye haue beene thought, that fuch contrarye difpofitions coulde well haue beene bredde in one wombe, or iffued from ones loynes. Yet as out of one and the felfe-fame roote, commeth as well the wilde Olyue, as the fweete, and as the Palme *Perfian* Fig tree, beareth as well Apples, as Figs: fo our mother thruft into the world at one time, the bloffome of grauitie and lyghtneffe.

We were nurffed both with one teate, where my brother fucked a defire of thirft [thrift], and I of theft: which euidently fheweth that as the breath of the Lyon, engendreth afwell the Serpent, as the Ant : and as the felfe fame deaw forceth the Earth to yeelde both the Darnell and Wheat : or as the Eafterly winde maketh the bloffomes to blaft, and the buddes to blowe : fo one wombe nourifhed contrary wits, and one milke diuers manners, which argueth fomething in Nature I know not what, to be meruaylous, I dare not faye monftrous.

As we grew olde in yeares, fo began we to be more oppofit in opinions : He graue, I gamefome : he ftudious, I carelefte : he without mirth, and I without modeftie.

And verely, had we refembled each other, as little in fauour, as we did in fancie, or difagreed as much in fhape as we did in fence : I know not what *Dedalus* would haue made a *Laborynth* for fuch Monfters, or what *Appelles* could haue couloured fuch Miffhapes.

But as the Painter *Tamantes* could no way expreffe the griefe of *Agamemnon* who faw his onely daughter facraficed, and therefore drew him with a vale ouer his face, whereby one might better conceiue his anguifh, then he colour it : fo fome *Tamantes* feeing vs, would be conftrained with a Curtaine to fhadow that deformitie, which no counterfait could portraie lyuely. But nature recompenfed ye diffimilitude [fimilitude] of mindes, with a *Sympathy* of bodies, for we were in all parts one fo like the other, that it was hard to diftinguifh either in fpeach, countenaunce, or height, one from the other : fauing that either car[r]ied the motion of his mind, in his manners, and that the affects of the hart were bewrayed by the eyes, which made vs knowen manifeftly. For as two Rubies be they neuer fo lyke, yet if they be brought together one ftaineth the other, fo we beeing clofe one to the other, it was eafely to imagine by the face whofe vertue deferued moft fauour, for I could neuer fee my brother, but his grauitie would make me blufh, which caufed me to refemble the Thrufhe, who neuer fingeth in the companye of the Nightingale. For whileft my Brother was in prefence, I durft not prefume to talke, leaft his wifedome might haue checked my wildneffe : Much lyke to *Rofcius*, who was alwayes dumbe, when he dined with *Cato*. Our Father being on his death bed, knew not whom to

ordein his heire, being both of one age : to make both,
woulde breede as he thought, vnquiet: to appoint but
one, were as he knew iniury: to deuide equally, were to
haue no heire : to impart more to one then to ye other,
were partiality: to difherite me of his wealth, whom
Nature had difherited of wifedome, were againſt reafon:
to barre my brother from golde, whome God feemed
to endue with grace, were flatte impietie: yet calling
vs before him, he vttered with watrie eyes, thefe words.

WEre it not my fonnes, that Nature worketh
more in me, then Iuftice, I fhould difherite
the one of you, who promifeth by his folly to fpende
all, and leaue the other nothing, whofe wifedome
feemeth to purchafe all things.   But I well know, that
a bitter roote is amended with a fweete graft, and
crooked trees proue good Cammocks, and wilde
Grapes, make pleafaunt Wine.   Which perfwadeth me,
that thou (poynting to me) wilt in age repent thy
youthly affeċtions, and learne to dye as well, as thou
haſt lyued wantonly.   As for thee (laying his hande on
my brothers head) although I fee more then commonly
in any of thy yeares, yet knowing that thofe that giue
themfelues to be bookifh, are oftentimes fo blockifh,
that they forget thrift: Where-by the olde Saw is verified, that the greateſt Clearkes are not the wifeſt men,
who digge ſtill at the roote, while others gather the fruite,
I am determined to helpe thee forward, leaſt hauing
nothing thou defire nothing, and fo be accompted as
no body.   He hauing thus faid, called for two bags,
the one ful of gold, the other ſtuft with writings, and
cafting them both vnto us, fayd this : There my fonnes
deuide all as betweene you it fhal be beſt agreed,
and fo rendred vp his ghoaſt, with a pitifull grone.

My brother as one that knew his owne good, and my
humour, gaue me leaue to chufe which bag I lyked, at
the choice I made no great curiofitie, but fnatching the
gold, let go ye writings, which wer as I knew Euidences
for land, oblygations for debt, too heauy for me to

cary, who determined (as now thou doeſt *Callima-chus*) to feeke aduentures. My purſſe now ſwelling with a timpany, I thought to ferch al countries for a remedy, and ſent many golden Angels into euery quarter of ye world, which neuer brought newes again to their maſter, being either ſoared into heauen, wher I cannot fetch them, or ſunke into Hell for pride, wher I meane not to follow them. This life I continued ye ſpace of. xiiij. yeares, vntil I had viſited and viewed euery country, and was a ſtranger in mine owne : but finding no treaſure to be wrapped in trauell, I returned with more vices, then I went forth with pence, yet with ſo good a grace, as I was able to ſinne both by experience and authoritie, vſe framing me to the one, and the Countryes to the other. There was no cryme ſo barbarous, no murther ſo bloudy, no oath ſo blaf-phemous, no vice ſo execrable, but yat I could readely recite where I learned it, and by roate repeate the peculiar crime, of euerye perticular Country, Citie, Towne, Village, Houſe, or Chamber.

If I met with one of *Creete*, I was ready to lye with him for the whetſtone. If with a *Grecian*, I could diſſemble with *Synon*. I could court it with the *Italian*, carous it with the *Dutch-man*. I learned al kinde[s] of poyſons, yea, and ſuch as were fit for the Popes holy-neſſe. In *Aegypt* I worſhipped their ſpotted God, at *Memphis*. In *Turkey*, their *Mahomet*. In *Rome*, their Maſſe : which gaue me not onely a remiſſion for my ſinnes paſt without penaunce, but alſo a commiſſion to ſinne euer after with-out preiudice.

There was no faſhion but fitted my backe, no fanci? but ſerued my tourne : But now my Barrell of golde, which Pride ſet a broche, Loue began to ſet a tilte, which in ſhort time ranne ſo on the lees, that the Diuell daunced in the bottome, where he found neuer a croſſe. It were too tedious to vtter my whole lyfe in this my Pilgrimage, the remembraunce where-off, doth nothing but double my repentaunce.

Then to grow to an ende, I ſeeing my money waſted,

my apparell worne, my minde infected with as many
vices, as my body with difeafes, and my bodye with
more maladyes, then the Leopard hath markes, hauing
nothing for amends but a few broken languages, which
ferued me in no more fteede, then to fee one meat
ferued in diuers difhes: I thought it beft to retourne
into my natiue foyle, where finding my brother as farre
now to exceede others in wealth, as hee did me in wit,
and that he had gayned more by thrift, then I could
fpende by pride, I* neither enuyed his eftate, nor
pityed mine owne: but opened the whole courfe of
my youth, not thinking there-by to recouer that of him
by requeft, which I had loft my felfe by riot, for cafting in my minde the miferie[s] of the world with the
mifchiefes of my life, I determined from that vnto my
liues end, to lead a folitary life in this caue, which I
haue don[e] the tearm of ful forty winters, from whence,
neither the earneft entreatie of my Brother, nor the
vaine pleafures of the world could draw me, neyther
fhall any thing but death.

Then my good *Callimachus*, recorde with thy felfe
the inconueniences that come by trauailing, when on
the Seas euery ftorme fhall threaten death, and euery
calme a daunger, when eyther thou fhalt be compelled
to boord others as a pyrate, or feare to be boorded of
others as a Marchaunt: when at all times thou muft
haue the back of an Affe to beare all, and the fnowt of
a fwine to fay nothing, thy hand on thy cap to fhew
reuerence to euery rafcall, thy purfe open to be prodigall to euery Boore, thy fworde in thy fheath, not
once daring either to ftrick or ward, which maketh me
think that trauailers are not onely framed not to commit iniuries, but alfo to take them. Learne *Callimachus*, of the Byrde *Acanthis*, who being bredde in
the thiftles will liue in the thiftles, and of the Grafhopper, who being fproung of the graffe, will rather
dye then depart from the graffe. I am of this minde
with *Homer*, that as the Snayle that crept out of hir
fhell was turned eftfoones into a Toad, and therby

was forced to make a ftoole to fit on, difdaining hir
own houfe: fo the Trauailer that ftragleth from his
own countrey, is in fhort tyme tranfformed into fo
monftrous a fhape, that hee is faine to alter his man-
fion with his manners, and to liue where he canne, not
where he would.   What did *Vlyffes* wifh in the middeft
of his trauailing, but onely to fee the fmoake of his
owne Chymnie?   Did not all the *Romaines* faye that
he that wandered did nothing els but heap forowes to
his friends, and fhame to himfelf, and refembled thofe
that feeking to light a Lynke, quenched a Lamp, imitating the barbarous *Gothes*, who thought the rootes in
Alexandria, fweeter then ye refons [Raifons] in *Barbary*:
But*he*that*leaueth*his*own*home,*is*worthy*no*
home.*   In my opinion it is a homely kinde of dealing
to preferre the curtefie of thofe he neuer knew, before
the honefty of thofe among whom he was born : he that
cannot liue with a gro[a]t[e] in his own country, fhal neuer
enioy a penny in an other nation.   Litle doft thou know
Callimachus with what wood trauailers are warmed,
who muft fleepe with their eies open, leaft they be
flain in their beds, and wake with their eyes fhut, leaft
they be fufpected by their lookes, and eat with their
mouths clofe, leaft they be poyfoned with theyr meates.
Where if they wax wealthy, thou fhalt be enuied, not
loued : If poore punifhed, not pittied : If wife, accounted efpials : If foolifh, made drudges.   Euery
Gentle-man will be thy peere though they be noble, and
euery pefaunt their Lord if they [he] be gentle.   Hee
therefore that leaueth his own houfe to feeke aduentures, is like the Quaile that forfaketh the Malowes to
eat Hemlock, or the Fly that fhunneth the Rofe, to
light in a cowfhard.   No *Callimachus*, there wil no
Moffe fticke to the ftone of *Sifiphus*, no graffe hang on
[the] heeles of *Mercury*, no butter cleaue on ye bread of
a trauailer.   For as the E[a]gle at euery flight loofeth
a fether, which maketh hir bald in hir age : fo the
trauailer in euery country loofeth fome fleece, which
maketh him a begger in his youth, buying that with a

pound, which he cannot fell againe for a penny, repentaunce. But why go I about to diffwade thee from
that, which I my felf followed, or to perfwade thee to
that which thou thy felfe flyeft? My gray haires are
like vnto a white froft, thy read [redde] bloud not vnlike
vnto a hot fyre: fo that it cannot be yat either thou
fhouldeft follow my counfell, or I allow thy conditions:
fuch a quarrel hath ther alwaies bin betwene the graue
and the cradle, that he yat is young thinketh the olde
man fond, and the olde knoweth the young man to be
a foole. But *Callimachus,* for the towardnes I fee in
thee, I muft needs loue thee, and for thy frowardnes,
of force counfel thee: and do in ye fame fort, as
Phœbus did yat [ye] daring boy *Phœton.* Thou goeft
about a great matter, neither fit for thy yeares being very
young, nor thy profit being left fo poore, thou defireft
yat which thou knoweft not, neither can any performe
yat which thou feemeft to promife. If thou couet to
trauaile ftraunge countries, fearch the Maps, there
fhalt thou fee much, with great pleafure and fmal
paines, if to be conuerfant in al courts, read hiftories,
where thou fhalt vnderftand both what the men haue
ben, and what their maners are, and me thinketh ther
muft be much delight, when ther is no daunger. And
if thou haue any care either of ye greene bud which
fpringeth out of the tender ftalke, or the timely fruite
which is to grow of fo good a roote, feeke not to kill the
one, or haften ye other: but let time fo work that grafts
may be gathered off [on] the tree, rather then fticks to
burn. And fo I leaue thee, not to thy felf, but to him yat
made thee, who guid[e] thee with his grace, whether thou
go as thou wouldeft, or tarry at home as thou fhouldeft.

Callimachus obftinate in his fond conceit, was fo far
from being perfwaded by this old Hermit, yat he rather
made it a greater occafion of his pilgrimage, and with an
anfwer betwen fcorning and re[a]foning, he replied thus.

Father or friend (I know not verye well howe to
tearme you) I haue beene as attentiue to heare your
good difcourfe, as you were willing to vtter it: yet mce

thinketh you deale maruailouflye with youth, in feeking by fage counfell to put graye hayres on their chins, before nature hath giuen them almoft any hayres on their heades: where-in you haue gone fo farre, that in my opinion your labour had bene better fpent in trauailing where you haue not lyued, then in talking wher you cannot be beleeued. You haue bene a Trauailer and tafted nothing but fowre, therefore who-foeuer trauaileth, fhall eate of the fame fauce: an Argument it is, that your fortune was ill, not that others fhould be as bad, and a warning to make you wife, not a warning to proue others vnfortunate. Shal a fouldier that hath receiued a fkar in the bat-taile, giue out that all warriours fhall be maymed? Or the Marchaunt that hath loft by the Seas, be a caufe that no other fhould venture, or a trauailer that hath fuftained harm by finifter fortune, or bene infected by his own folly, diffwade al Gentlemen to reft at their own home till they come to their long home? Why then let al men abftaine from wine, bicaufe it made *Alexander* tipfie, let no man loue a woman for yat *Tarquine* was banifhed, let not a wife man play at al, for yat a foole hath loft al: which in my minde would make fuch medly, that wee fhould bee enforced to leaue things that were beft, for feare they may bee badde, and that were as fond as not to cut ones meate with that knife yat an other hath cut his finger. Things are not to be iudged by the euent, but by the ende, nor trauailing to be condemned by yours or manies vnluckie fucceffe, but by the common and moft approued wifdome of thofe that canne better fhew what it is then I, and will better fpeake of it then you doe.

Where you alledge *Vliffes* that he defired nothing fo much, as to fee the fmoake of *Ithaca*, it was not bicaufe he loued not to trauaile, but yat he longed to fee his wife after his trauaile: and greater commendation brought his trauail to him, then his wit: the one taught but to fpeake, the other what he fhould fpeake. And in this you tourne the poynt of your owne bodkin into your owne bofome. *Vliſſes* was no leſſe eſteemed for knowledge he had of other countryes, then for ye reuenewes he had in his own, and wher in ye ende, you feeme to refer me to yat [the] viewing of Maps, I was neuer of that minde to make my fhip in a Painters fhop, which is lyke thofe, who haue great fkill in a wooden Globe, but neuer behold the Skie. And he that feeketh to bee a cunning trauailer by feeing the Mappes, and an expert Aftronomer, by turning the Globe may be an Apprentice for *Appelles*, but no Page for *Vliſſes*.

Another reafon you bring, that trauailing is coftly, I fpeake for my felfe : He that hath lyttle to fpende, hath not much to lofe, and he that hath nothing in his owne countrey, can-not haue leffe in any.

Would you haue me fpend the floure of my youth, as you doe the withered rafe of your age ? can ye faire bloud of youth creepe into the ground as it were froft bitten ? No Father Hermit, I am of *Alexanders* minde, if there were as many worlds, as there be cities in the world, I would neuer leaue vntill I had feene all the worlds, and each citie in euerie world. Therefore to be fhort, nothing fhall alter my minde, neither penny nor *Pater noſter*.

This olde man feeing him fo refolute, refolued to let him depart, and gaue him this Fare-well.

MY good fonne though thou wilt not fuffer mee to perfwade thee, yet fhalt thou not let mee to pittie thee, yea and to pray for thee : but the tyme will come when comming home by weeping croffe, thou fhalt confeffe, that it is better to be at home in the caue of an Hermit then abroad in the court of an Emperour, and that a cruft with quietneffe, fhall be better then Quayles with vnreft. And to the ende thou maift proue my fayings as true, as I know thy felfe to bee wilfull, take the paines to retourne by [to] this poore Cel[l], where thy fare fhall be amended, if thou amende thy fault, and fo farewell.

Callimachus courteoufly tooke his leaue, and went

his waye : but we will not leaue him till we haue him
againe, at the Cell, where we found him.

NOw *Philautus* and Gentlemen all, fuppofe that
    Callimachus had as il fortune, as euer had any,
his minde infected with his body, his time confumed
with his treafure : nothing won, but what he cannot
loofe though he would, Miferie.   You muft imagine
(bicaufe it were too long to tell all his iourney) that he
was Sea ficke, (as thou beginneft to be *Philautus*) that
he hardly efcaped death, that he endured hunger and
colde, heate with-out drinke, that he was entangled with
women, entrapped, deceiued, that euery ftoole he fate
on, was penniles bench, that his robes were rags, that
he had as much neede of a Chirurgian as a Phifition,
and that thus he came home to the Cell, and with
fhame and forrow, began to fay as followeth.

I Finde too late yet at length that in age there is a
   certeine forefight, which youth can-not fearch, and
of a kinde of experience, vnto which vnripened yeares
cannot come : fo that I muft of neceffitie confeffe, that
youth neuer raineth wel, but when age holdeth the
bridell, you fee (my good father) what I would fay by
outward fhew, and I neede not tell what I haue tryed,
bicaufe before you tolde me I fhould finde it : this I
fay, that whatfoeuer miferie happened either to you or
any, the fame hath chaunced to me alone.   I can fay
no more, I haue tryed no leffe.

The olde Hermit glad to fee this ragged Colte retourned, yet grieued to fee him fo tormented, thought
not to adde fower words to augment his fharp woes,.
but taking him by the hande, and fitting down, began
after a folempn manner, from the beginning to ye ende,
to difcourfe with him of his fathers affaires, euen after
the fort that before I rehearfed, and delyuered vnto
him his money, thinking now that miferie woulde
make him thriftie, defiring alfo, that afwell for the
honour of his Fathers houfe, as his owne credite, hee

would retourne againe to the Iflande, and there be a comfort to his friends, and a reliefe to his poore neighbours, which woulde be more worth then his wealth, and the fulfilling of his Fathers laft Will.

Callimachus not a little pleafed with this tale, and I thinke not much difpleafed with the golde, gaue fuch thankes, as to fuch a friend appertained, and following the counfel of his vnckle, which euer after he obeyed as a commaundement, he came to his owne houfe, liued long with great wealth, and as much worfhip as any one in *Scyrum,* and whether he be now lyuing, I know not, but whether he be or no, it fkilleth not.

Now *Philautus,* I haue tolde this tale, to this ende, not that I thinke trauailing to be ill if it be vfed wel, but that fuch aduice be taken, yat the horfe carry not his own bridle, nor youth rule himfelf in his own conceits. Befides yat, fuch places are to be chofen, wher-in to inhabit as are as commendable for vertue, as buildings: where the manners are more to be marked, then ye men feene. And this was my whole drift, either neuer to trauaile, or fo to trauaile, as although ye purffe be weakened, ye minde may be ftrengthened. For not he yat hath feene moft countries is moft to be efteemed, but he that learned beft conditions: for not fo much are ye fcituation of the places to be noted, as the vertues of the perfons. Which is contrarie to the common practife of our trauailers, who goe either for gaine, and returne without knowledge, or for fafhion fake, and come home with-out pietie: Whofe eftates are as much to be lamented, as their follyes are to be laughed at.

This caufeth youth, to fpende their golden time, with-out either praife or profit, pretending a defire of learning, when they onely followe loytering. But I hope our trauell fhal be better employed, feeing vertue is the white we fhoote at, not vanitie: neither the Englifh tongue (which as I haue heard is almoft barbarous) but the Englifh manners, which as I thinke are moft precife. And to thee *Philautus* I begin to addreſſe my ſpeach, hauing made an end of mine [my] hermits tale, and if theſe few precepts I giue thee be obſerued, then doubt not but we both ſhall learne that we beſt lyke.   And theſe they are.

A T thy comming into *England* be not too in-quiſitiue of newes, neither curious in matters of State, in aſſemblies aſke no queſtions, either concerning manners or men.   Be not lauiſh of thy tongue, either in cauſes of weight, leaſt thou ſhew thy ſelfe an eſpyall, or in wanton talke, leaſt thou proue thy ſelfe a foole.

It is the Nature of that country to ſift ſtraungers: euery one that ſhaketh thee by the hand, is not ioyned to thee in heart.   They thinke *Italians* wanton, and *Grecians* ſubtill, they will truſt neither they are ſo incredulous: but vndermine both, they are ſo wiſe. Be not quarrellous for euery lyght occaſion : they are impatient in their anger of any equal, readie to reuenge an iniury, but neuer wont to profer any: they neuer fight without prouoking, and once pro-uoked they neuer ceaſe.   Beware thou fal not into ye ſnares of loue, ye women there are wiſe, the men craftie : they will gather loue by thy lookes, and picke thy minde out of thy hands.   It ſhal be there better to heare what they ſay, then to ſpeak what thou thinkeſt : They haue long ears and ſhort tongues, quicke to heare, and ſlow to vtter, broad eyes and light fingers, ready to eſpy and apt to ſtricke.   Euery ſtraunger is a marke for them to ſhoote at : yet this muſt I ſay which in no country I can tell the iike, that it is as ſeldome to ſee a ſtraunger abuſed there, as it is rare to ſee anye well vſed els where : yet preſume not too much of the curteſies of thoſe, for they differ in natures, ſome are hot, ſome cold, one ſimple, an other wilie, yet if thou vſe few words and fayre ſpeaches, thou ſhalt commaund any thing thou ſtandeſt in neede of.

Touching the ſituation of the ſoile I haue read in my ſtudie, which I partly beleeue (hauing no worſe Author then *Cæſar*) yet at my comming, when I ſhal

conferre the thinges I fee, with thofe I haue read, I
will iudge accordingly. And this haue I heard, that
the inner parte of *Br:'taine* is inhabited by fuch as
were born and bred in the Ifle, and the Sea-choaft by
fuch as haue paffed thether out of *Belgick* to fearch
booties and to make war. The country is meruailouf-
lye replenifhed with people, and there be many
buildings almoft like in fafhion to the buildings of ·
Gallia, there is great ftore of cattell, ye coyn they
vfe is either of braffe or els rings of Iron, fifed at a
certein weight in fteede of money. In the inner parts
of the Realme groweth tinne, and in the fea coaft
groweth yron. The braffe yat they occupy is brought
in from beyond-fea. The ayre is more temperate in
thofe places then in *Fraunce*, and the colde leffer. The
Ifland is in fafhion three cornered, wher-of one fide is
toward *Fraunce*, the one corner of this fide which is
in* Kent, where for the moft part Shippes ariue out of
Fraunce, is in the Eaft, and the other nethermore, is
towardes the South. This fide containeth about fiue
hundred miles, an other fide lyeth toward *Spain* and
the Sunne going down, on the which fide is *Ireland*,
leffe then *Brittain* as is fuppofed by the one halfe :
but the cut betweene them, is like the diftaunce
that is betweene *Fraunce* and *Brittaine*. In the
middeft of this courfe is an Ifland called *Man*, the
length of this fide is (according to the opinion
of the Inhabiters) feuen hundred miles. The third
fide is northward, and againft it lyeth no land, but
the poynt of that fide butteth moft vppon *Germany*.
This they efteeme to be eight hundred miles long,
and fo the circuit of the whole Ifland is two thouf-
and miles. Of al the Inhabitants of this Ifle, the
Kentifh men are moft ciuileft, the which country
marcheth altogether vpon the fea, and differeth not
greatly from the man[n]er of *France*. They that dwell
more in the hart of the Realme fow corne, but liue by
milk and flefh, and cloth themfelues in lether. All
the *Brittaines* doe die them-felues with woad, which

fetteth a blewifh coulour vpon them, and it maketh
them more terrible to beholde in battaile.   They weare
their hayre long and fhaue all partes of their bodyes,
fauing the head and the vpper lippe.   Diuers other
vfes and cuftomes among them, as I haue read
Philautus : But whether thefe be true or no, I wil not
fay : for me thinketh an Ifland fo well gouerned in
peace then, and fo famous in victories, fo fertile in all
refpects, fo wholfome and populous, muft needes in
the terme of a thoufand yeares be much better, and I
beleeue we fhall finde it fuch, as we neuer read the like
of any, and* vntil we ariue there, we wil fufpend our
iudgementes: Yet do I meane at my returne from thence
to draw the whole difcription of the Land, the cuftomes,
ye nature of ye people, ye ftate, ye gouernment, and
whatfoeuer deferueth either meruaile or commendation.

Philautus not accuftomed to thefe narrow Seas,
was more redy to tell what wood the fhip was made
of, then to aunfwer to *Euphues* difcourfe : yet between
waking and winking, as one halfe ficke and fome-what
fleepy, it came in his braynes, aunfwered thus.

In fayth *Euphues* thou haft told a long tale, the beginning I haue forgotten, ye middle I vnderftand not,
and the end hangeth not together: therfore I cannot
repeat it as I would, nor delight in it as I ought : yet
if at our arriuall thou wilt renew thy tale, I will rub my
memorie : in the meane feafon, would I wer either
again in *Italy*, or now in *England*.   I cannot brook
thefe Seas, which prouoke my ftomack fore.   I haue
an appetite, it wer beft for me to take a nap, for euery
word is brought forth with a nod.

Euphues replied.   I cannot tell *Philautus* whether
the Sea make thee ficke, or fhe that was borne of the
Sea : if the firft, thou haft a que[a]fie ftomacke : if the
latter, a wanton defire.   I wel beleue thou remembreft
nothing yat may doe thee good, nor forgetteft any
thing, which can do thee harme, making more of a
foare then a plaifter, and wifhing rather to be curffed
then cured, where-in thou agreeft with thofe which

hauing taken a furfet, feeke the meanes rather to fleepe then purge, or thofe that hauing ye greene ficknes, and are brought to deaths dore follow their own humour, and refufe the Phifitions remedy. And fuch *Philautus* is thy defeafe, who pining in thine owne follies, chufeft rather to perifh in loue, then to liue in wifdome, but what-foeuer be the caufe, I wifh the effect may anfwer my friendly care : then doubtlefs you [thou] fhalt neither die being feafick, or doat being loue fick. I would ye Sea could afwel purge thy mind of fond conceits, as thy body of grofe humours. Thus ending, *Philautus* againe began to vrge.

Without dou[b]t *Euphues* you [thou] doft me great wrong, in feeking a fkar in a fmo[o]th fkin, thinking to ftop a vain wher none [is] opened, and to caft loue in my teeth, which I haue already fpit out of my mouth, which I muft needes thinke proceedeth rather for lacke of matter, then any good meaning, els wo[u]ldeft thou neuer harp on yat ftring which is burft in my hart, and yet euer founding in thy eares. Thou art like thofe that procure one to take phifick before he be fick, and to apply a fearcloth to his bodye, when he feeleth no ach, or a vomit for a furfet, when his ftomacke is empty. If euer I fall to mine old Byas, I muft put thee in the fault that talkes of it, feeing thou didft put me in the minde to think of it, wher-by thou feemeft to blow ye co[a]le which thou woldeft quench, fetting a teene edge, wher thou defireft to haue a fharp poynt, ymping a fether to make me flye, when thou oughteft rather to cut my wing for feare of foaring.

Lucilla is dead, and fhe vpon whome I geffe thou harpeft is forgotten : the one not to be redeemed, the other not to be thought on : Then good *Euphues* wring not a horfe on the withers, with a falfe faddle, neither imagin what I am by thy thoughts, but by mine own doings : fo fhalt thou haue me both willing to followe good counfell, and able hereafter to giue thee comfort. And fo I reft halfe fleepy with the Seas.

With this aunfwere *Euphues* held him-felf content,

but as much wearyed with talke as the other was with
trauaile, made a pyllow of his hand, and there let them
both fleepe their fill and dreame with their fancies
[fantafies], vntill either a ftorme caufe them to wake,
or their hard beds, or their iournies ende.

Thus for the fpace of an eight weekes *Euphues* and
Philautus failed on ye feas, from their firft fhipping,
betwen whome diuers fpeaches were vttered, which to
refite were nothing neceffary in this place, and weighing the circumftances, fcarfe expedient, what tempefts
they endured, what ftraung[e] fights in ye element, what
monftrous fifhes were feene, how often they were in
daunger of drowning, in feare of boording, how wearie,
how fick, how angrie, it were tedious to write, for that
whofoeuer hath either* read of trauailing, or [hath] himfelfe vfed it, can fufficiently geffe what is to be fayd.
And this I leaue to the iudgement of thofe that in the like
iourney haue fpent their time from *Naples* to *England*,
for if I fhould faine more then others haue tryed, I
might be thought too Poeticall : if leffe, partiall : therefore I omit the wonders, the Rockes, the markes, the
goulfes, and whatfoeuer they paffed or faw, leaft I
fhould trouble diuers with things they know, or may
fhame my felfe, with things I know not.   Lette this
fuffice, that they are fafely come within a ken of *Douer*,
which the Mafter efpying, with a cheerefull voyce
waking them, began to vtter thefe words vnto them.

G Entlemen and friends, the longeft Summers day
    hath his euening, *Vliffes* arriueth at laft, and
rough windes in time bring the fhip to fafe Road.
We are now with-in foure houres fayling of our Hauen,
and as you wil thinke of an earthly heauen.   Yonder
white Cliffes which eafely you may perceiue, are *Douer*
hils, where-vnto is adioyning a ftrong and famous
Caftle, into the which *Iulius Cæfar* did enter, where
you fhall view many goodly monuments, both ftraunge
and auncient.   Therefore pull vp your harts, this
merry winde will immediately bring vs to an eafie bayte.

Philautus was glad he flept fo long, and was awaked in fo good time, beeing as weary of the feas, as he that neuer vfed them. *Euphues* not forrowfull of this good newes, began to fhake his eares, and was foone apparailed. To make fhort, the windes were fo fauorable, the Mariners fo fkilfull, the waye fo fhort, that I feare me they will lande before I can defcribe the manner how, and therefore fuppofe them now in *Douer* Towne in the noble Ifle of *England*, fomwhat benighted, and more apt to fleepe then fuppe. Yet for manners fake they enterteined their Mafter and the reft of the Merchants and Marriners, wher hauing in due time both recorded their trauailes paft, and ended their repaft, euery one went to his lodging, where I wil leaue them foundly fleeping vntill the next day.

The next day they fpent in viewing the Caftle of *Douer*, the Pyre, the Cliffes, the Road, and Towne, receiuing as much pleafure by the fight of auncient monuments, as by their curteous enterteinment, no leffe praifing ye perfons for their good mindes, then the place for ye [their] goodly buildings : and in this fort they refrefhed themfelues 3.or.4. daies, vntil they had digefted ye feas, and recouered again their healths, yet fo warely [warilye] they behaued themfelues, as they wer neuer heard, either to enquire of any newes, or point to any fortres, beholding the bulwarkes with a flight and careles regard, but ye other places of peace, with admiration. Folly it wer to fhew what they faw, feing heereafter in ye defcription of *England*, it fhall moft manifeftly appeare. But I will fet them forwarde in their iourney, where now with-in this two houres, we fhall finde them in *Caunterbury*.

Trauailing thus like two Pilgrimes, they thought it moft neceffary to direct their fteppes toward *London*, which they h[e]ard[e] was the moft royall feat of the Queene of *England*. But firft they came to *Caunterbury*, an olde Citie, fomewhat decayed, yet beautiful to behold, moft famous for a Cathedrall Church, the very Maieftie whereoff, ftroke them into a maze, where they faw many monuments, and heard tell of greater, than either they euer faw, or eafely would beleeue.

After they had gone long, feeing them-felues almoſt benighted, determined to make the nexte houfe their Inne, and efpying in their way euen at hande a very pleafaunt garden, drew neere : where they fawe a comely olde man as bufie as a Bee among his Bees, whofe countenaunce bewrayed his conditions, this auncient Father, *Euphues* greeted in this manner.

FAther, if the courtefie of *Englande* be aunfwerable to the cuſtome of Pilgrimes, then will the na-ture of the Countrey, excufe the boldneffe of ſtraungers: our requeſt is to haue fuch enterteinment, beeing al-moſt tyred with trauaile, not as diuers haue for ac-quaintaunce, but as all men haue for their money, which curtefie if you graunt, we will euer remaine in your debt, although euery way difcharge our due : and rather we are importunate, for that we are no leffe delighted with the pleafures of your garden, then the fight of your grauitie. Vnto whom the olde man fayd.

GEntlemen you are no leffe I perceiue by your manners, and you can be no more beeing but men. I am neither fo vncourteous to miſlyke your requeſt nor fo fufpicious to miſtruſt your truthes, al-though it bee no leffe perillous to be fecure, then peeuifh to be curious. I keepe no victualling, yet is my houfe an Inne, and I an Hoſte to euery honeſt man, fo far as they with courtefie wil, and I may with abilytie. Your enterteinment fhal be as fmal for cheere, as your acquaintaunce is for time, yet in my houfe ye may happely finde fome one thing cleanly, nothing courtly : for that wifedome prouideth things neceffarie, not ſuperfluous, and age feeketh rather a Modicum for fuſtenaunce, then feaſtes for furfets. But vntil fome thing may be made ready, might I be fo bold as enquire your names, countreys, and ye caufe of your pilgrimage, where-in if I fhalbe more inqufitiue then I ought, let my rude birth excufe [fatiffie] my bolde requeft, which I will not vrge as one importunate (I might fay) impudent.

Euphues, feeing this fatherly and friendlye Sire, (whom we will name *Fidus*) to haue no leffe inwarde courtefie, then outward comelyneffe, conie&ured (as well he might) that the profer of his bountie, noted the nobleneffe of his birth, beeing wel affured that as no *Therfites* could be tranfformed into *Vliffes*, fo no *Alexander* could be couched in *Damocles.*

Thinking therefore now with more care and ad-uifedneffe to temper his talke, leaft either he might feeme foolyfh or curious, he aunfwered him, in thefe termes.

Good fir, you haue bound vs vnto you with a double chaine, the one in pardoning our pre-fumption, the other in graunting our peticion. Which great and vndeferued kindeneffe, though we can-not requit[e] with the lyke, yet if occafion fhall ferue, you fhall finde vs heereafter as willing to make amends, as we are now ready to giue thankes.

Touching your demaunds, we are not fo vnwife to miflyke them, or fo vngratefull to deny them, leaft in concealing our names, it might be thought for fome trefpaffe, and couering our pretence, we might be fuf-pe&ed of treafon. Know you then fir, that this Gentleman my fellow, is called *Philautus*, I *Euphues :* he an *Italian*, I a *Grecian :* both fworne friendes by iuft tryall, both Pilgrimes by free will. Concerninge the caufe of our comming into this Iflande, it was onely to glue our eyes to our eares, that we might iuftifie thofe things by fight, which we haue oftentimes with incredible admiration vnderftoode by hearing : to wit, the rare qualyties as well of the body as the minde, of your moft dreade Souereigne and Queene, the brute of the which hath filled euery corner of the worlde, infomuch as there is nothing that moueth either more matter or more meruaile then hir excellent maieftie,

with [which] fame when we faw, with-out comparifon
and almoft aboue credit, we determined to fpend fome
parte of our time and treafure in the Englifh court,
where if we could finde the reporte but to be true in
halfe, wee fhoulde not onelye thinke our money and
trauayle well employed, but returned with intereft
more then infinite. This is the onely ende of our
comming, which we are nothing fearefull to vtter,
trufting as well to the curtefie of your countrey, as the
equitie of our caufe.

Touching the court, if you can giue vs any inftruc-
tions, we fhal think the euening wel fpent, which pro-
curing our delight, can no way worke your difliking.

Entle-men (aunfwered this olde man) if bicaufe I
G entertaine you, you feeke to vndermin[e] me,
you offer me great difc[o]urtefie: you muft needes thinke
me verye fimple, or your felues very fubtill, if vpon fo
fmall acquaintaunce I fhould anfwer to fuch demaunds,
as are neither for me to vtter being a fubiect, nor for
you to know being ftraungers. I keepe hiues for
Bees, not houfes for bufibodies (pardon me Gentlemen, you haue moued my patience) and more welcome fhal a wafp be to my honny, then a priuy enimy
to my houfe. If the rare reporte of my moft gracious
Ladye haue brought you hether, mee thinketh you
haue done very ill to chufe fuch a houfe to confirme
your mindes, as feemeth more like a prifon then a
pallace, where-by in my opinion, you meane to dero-
gate from the worthines of the perfon by ye vilnes of
the place, which argueth your pretences to fauor of
malice more then honeft meaning. They vfe to confult of *Ioue* in ye Capitol, of *Cæfar*, in the fenat, of
our noble Queene, in hir owne court. Befides that,
Alexander muft be painted of none but *Appelles*, nor
engrauen of any but *Lifippus*, nor our *Elizabeth* fet
forth of euery one that would in duety, which are all,
but of thofe that can in fkyll, which are fewe, fo furre
[farre] hath nature ouercome arte, and grace eloquerce,

that the paynter draweth a vale ouer that he cannot
fhaddow, and the Orator holdeth a paper in his hand,
for that he cannot vtter. But whether am I wandring,
rapt farther by deuotion then I can wade through with
difcretion. Ceafe then Gentle-men, and know this,
that an Englifh-man learneth to fpeake of menne, and
to holde his peace of the Gods. Enquire no farther
then befeemeth you, leaft you heare that which can-not
like you. But if you thinke the time long before your
repaft, I wil finde fome talk which fhall breede your
delight touching my Bees.

And here *Euphues* brake him off, and replyed :
though not as bitterly as he would, yet as roundlye as
he durft, in this manner.

We are not a little fory fyr, not that we haue opened
our mindes, but that we are taken amiffe, and where
[when] we meant fo well, to be entreated fo ill, hauing
talked of no one thing, vnleffe it be of good wil towar[d]s
you, whome [we] reuerenced [reuerence] for age, and of
dutye towarde your Souereigne, whom we meruailed at
for vertue : which good meaning of ours mifconftrued by
you, hath bread fuch a diftemperature in our heads, that
we are fearfull to praife hir, whom al the world extolleth,
and fufpitious to truft you, whom aboue any in the
worlde we loued. And wheras your greateft argument
is, the bafenes of your houfe, me thinketh that maketh
moft againft you. *Cæfar* neuer reioyced more, then
when hee heard that they talked of his valyant exploits
in fimple cotages, alledging this, that a bright Sunne
fhineth in euery corner, which maketh not the beames
worfe, but the place better, when (as I remember)
Agefilaus fonne was fet at the lower end of the table,
and one caft it in his teeth as a fhame, he anfwered :
this is the vpper end where I fit, for it is not the place
that maketh the perfon, but the perfon that maketh
the place honorable. When it was told *Alexander*
that he was much prayfed of a Myller, I am glad quoth
he, that there is not fo much as a Miller but loueth
Alexander. Among other fables, I call to my remem-

brance one, not long, but apt, and as fimple as it is,
fo fit it is, that I cannot omit it for ye opportunitie of
the time, though I might ouer-leap it for the bafeneffe
of the matter. When all the Birds wer appointed to
meete to talke of ye Eagle, there was great contention,
at whofe neft they fhould affemble, euery one willing
to haue it at his own home, one preferring the nobilitie
of his birth, an other the ftatelynes of his building :
fome would haue it for one qualitie, fome for an other :
at the laft the Swalow, faid they fhould come to his
neft (being commonly of filth) which all the Birds
difdaining, fayd : why thy houfe is nothing els but
durt, and therfore aunfwered ye Swalow, would I
haue talke there of the Eagle : for being the bafeft, the
name of an Eagle wil make it ye braueft. And fo
good father may I fay of thy cotage, which thou
feemeft to account of fo hom[e]ly, that mouing but
fpe[e]ch of thy Souereigne, it will be more like a court
then a cabin, and of a prifon the name of *Elizabeth*
wil make it a pallace. The Image of a Prince ftampt
in copper goeth as currant, and a Crow may cry *Aue
Cæfar* with-out any rebuke.

The name of a Prince is like the fweete deaw, which
falleth as well vppon lowe fhrubbes, as hygh trees,
and refembleth a true glaffe, where-in the poore maye
fee theyr faces with the rych, or a cleare ftreame
where-in all maye drincke that are drye : not they
onelye that are wealthy. Where you adde, that wee
fhoulde feare to moue anye occafion touching talke of
fo noble a Prince, truly our reuerence taketh away the
feare of fufpition. The Lambe feareth not the Lion,
but the Wolfe : the Partridge dreadeth not the Eagle,
but the Hawke : a true and faythfull heart ftandeth
more in awe of his fuperior whom he loueth for feare,
then of his Prince whom he feareth for loue. A cleere
confcience needeth no excufe, nor feareth any accufa-
tion. Laftly you conclude, that neither arte nor heart
can fo fet forth your noble Queene, as fhe deferueth. I
graunt it, and reioyce at it, and that is the caufe of

our comming to fee hir, whom none can fufficiently commend : and yet doth it not follow, that bicaufe wee cannot giue hir as much as fhe is worthy off, therefore wee fhould not owe hir any. But in this we will imitate the olde paynters in *Greece*, who drawing in theyr Tables the portra[i]ture of *Iupiter*, were euery houre mending it, but durft neuer finifh it : And being demaunded why they beganne that, which they could not ende, they aunfwered, in that we fhew him to bee *Iupiter*, whome euery one may beginne to paynt, but none can perfeĉt. In the lyke manner meane we to drawe in parte the prayfes of hir, whome we cannot throughly portraye, and in that we fignifie hir to be *Elyzabeth.* Who enforceth euery man to do as much as he can, when in refpeĉt of hir perfeĉtion, it is nothing. For as he that beholdeth the Sunne fted-faftly, thinking ther-by to defcribe it more perfeĉtly, hath his eies fo dafeled, that he can difcerne nothing, fo fareth it with thofe that feeke marueiloufly to praife thofe, yat are without ye compaffe of their iudgements, and al comparifon, yat the more they defire, the leffe they difcern, and the neerer they think them [f]elues in good wil, the farther they finde themfelues of[f] in wifdom, thinking to me[a]fure yat by the ynch, which they cannot reach with ye ell. And yet father, it can be neither hurtful to you, nor hateful to your Prince, to here the com-mendation of a ftraunger, or to aunfwere his honeft requeft, who will wifh in heart no leffe glorye to hir, then you doe : although they can wifh no more. And therfore me thinketh you haue offered a little difcourtefie, not to aunfwere vs, and to fufpeĉt vs, great iniury : hauing neither might to attempt any thing which may do you harme, nor malice to reuenge, wher we finde helpe. For mine owne part this I fay, and for my friend prefent the lyke I dare fweare, how boldly I can-not tell, how truely I know : that there is not any one, whether he be bound by benefit or duetie, or both : whether linked by zeale, or time, or bloud,

R

or al : that more humbly reuerenceth hir Maieftie, or
meruaileth at hir wifedome, or prayeth for hir long
profperous and glorious Reigne, then we : then whom
we acknowledge none more fimple, and yet dare
auowe, none more faithfull. Which we fpeake not to
get feruice by flatterie, but to acquite our felues of
fufpition, by faith : which is al that either a Prince can
require [defire] of his fubiect, or a vaffal yeeld to his
Souereign, and that which we owe to your Queene, and
all others fhould offer, that either for feare of punifhment dare not offend, or for loue of vertue, will not.

Heere olde *Fidus* interrupted [interrupting] young
Euphues, being almoft induced by his talke, to aunfwere
his requeft, yet as one neither too credulous, nor altogether miftruftful, he replyed as a friend, and fo wifely
as he glaunced from the marke *Euphues* fhot at, and
hit at [the] laft the white which *Philautus* fet vp, as fhall
appeare heereafter. And thus he began.

MY fonnes (mine age giueth me the priuiledge
of that terme, and your honefties can-not
refufe it) you are too young to vnderftand matters of
ftate, and were you elder to knowe them it were not
for your eftates. And therfore me thinketh, the time
were but loft, in pullyng *Hercules* fhooe vppon an
Infants foot, or in fetting *Atlas* burthen on a childes
fhoulder, or to brufe your backes, with the burthen of
a whole kingdome, which I fpeake not, that either I
miftruft you (for your reply hath fully refolued yat
feare) or yat I malice you (for my good will maye
cleare me of yat fault) or that I dread your might
(for your fmal power cannot bring me into fuch a folly)
but that I haue learned by experience, yat to reafon
of Kings or Princes, hath euer bene much miflyked of
ye wife, though much defired of fooles, efpecially wher
old men, which fhould be at their beads, be too bufie
with the court, and young men which fhold follow their
bookes, be to[o] inquifitiue in ye affaires of princes.
We fhold not looke at yat we cannot reach, nor long

for yat we fhold not haue : things aboue vs, are not
for vs, and therfore are princes placed vnder ye gods,
yat they fhould not fee what they do, and we vnder
princes, that we might not enquire what they doe.
But as ye foolifh Eagle yat feing ye fun coueteth to
build hir neft in ye fun, fo fond youth, which viewing
ye glory and gorgeoufneffe of ye court, longeth to know
the fecrets in [of] ye court. But as ye Eagle, burneth
out hir eyes with that proud luft : fo doth youth break
his hart with yat peeuifh conceit. And as *Satirus* not
knowing what fire was, wold needs embrace it, and
was burned, fo thefe fonde *Satiri* not vnderftanding
what a Prince is, runne boldly to meddle in thofe matters
which they know not, and fo feele worthely ye heat
they wo[u]ld not. And therfore good *Euphues* and *Philautus* content your felues with this, yat to be curious in
things you fhould not enquire off, if you know them,
they appertein not vnto you : if you knew them
not, they cannot hinder you. And let *Appelles* anfwere to *Alexander* be an excufe for me. When *Alexander* would needes come to *Appelles* fhop and paint,
Appelles placed him at his backe, who going to his
owne worke, did not fo much as caft an eye back, to
fee *Alexanders* deuifes, which being wel marked, *Alexander* faid thus vnto him : Art not thou a cunning
Painter, and wilt thou not ouer-looke my picture, and
tel me wherin I haue done wel, and wherin ill, whom
he anfwered wifely, yet merily : In faith O king it is
not for *Appelles* to enquire what *Alexander* hath done,
neither if he fhew it me, to iudge how it is done, and
therefore did I fet your Maieftie at my back, yat I
might not glaunce towards a kings work, and that
you looking ouer my head might fee mine, for *Appelles*
fhadowes are to be feene of *Alexander*, but not *Alexanders* of *Appelles*. So ought we *Euphues* to frame
our felues in all our actions and deuifes, as though
the King ftood ouer vs to behold vs, and not to
locke what the King doth behinde vs. For whatfoeuer he painteth it is for his pleafure, and wee muft

think for our profit, for *Appelles* had his reward though
he faw not the worke.

I haue heard of a *Magnifico* in *Millaine* (and I
thinke *Philautus* you being an *Italian* do remember
it,) who hearing his fonne inquifitiue of the Emperours
lyfe and demeanour, reprehended him fharply, faying:
that it befeemed not one of his houfe, to enquire how
an Emperour liued, vnleffe he himfelf were an Emperour: for yat the behauiour and vfage of fo honourable perfonages are not to be called in queftion of
euery one that doubteth, but of fuch as are their
equalls.

Alexander being commaunded of *Philip* his Father
to wraftle in the games of *Olympia*, aunfwered he
woulde, if there were a King to ftriue with him, where-
by I haue noted (that others feeme to inforce) that as
Kings paftimes, are no playes for euery one: fo their
fecretes, their counfells, their dealings, are not to be
either fcanned or enquired off any way, vnleffe of
thofe that are in the lyke place, or ferue the lyke
perfon. I can-not tell whether it bee a *Caunterbury*
tale, or a Fable in *Aefope*, (but pretie it is, and true in
my minde) That the Foxe and the Wolfe, gooing
both a filching for foode, thought it beft to fee whether
the Lyon were a fleepe or awake, leeft beeing too bolde,
they fhould fpeede too bad. The Foxe entring into
the Kings denne, (a King I call the Lyon) brought word
to the Wolfe, that he was a fleepe, and went him-felfe
to his owne kenell, the Wolfe defirous to fearche in the
Lyons denne, that hee might efpye fome fault, or fteale
fome praye, entered boldly, whom the Lyon caught in
his pawes and afked what he would? The fillye Wolfe
(an vnapte tearme, for a Wolfe, yet fit, being in a Lyons
handes) aunfwered, that vnderftanding by the Foxe
he was a fleepe, hee thought he might be at lybertie
to furuey his lodging: vnto whome the princelye Lyon
with great difdaine though little defpite (for that there
can be no enuy in a King) fayde thus: Doeft thou
thinke that a Lyon, thy Prince and gouernour can

fleepe though he winke, or dareft thou enquire, whether he winke or wake? The Foxe had more craft then thou, and thou more courage (courage I wil not fay, but boldnes : and boldnes is too good, I may fay defperateneffe) but you fhal both wel know, and to your griefs feele, yat neither ye wilines of the Fox, nor ye wildnes of ye Wolf, ought either to fee, or to afke, whether ye Lyon either fleepe or wake, bee at home or abroad, dead or alyue. For this is fufficient for you to know, that there is a Lyon, not where he is, or what he doth. In lyke manner *Euphues*, is the gouernment of a Monarchie (though homely bee the comparifon, yet apte it is) that it is neither the wife Fox, nor the malitious Wolfe, fhould venture fo farre, as to learne whether the Lyon fleepe or wake in his denne, whether the Prince faft or feafte in his court: but this fhoulde bee their order, to vnderftand there is a king, but what he doth is for the Goddes to examine, whofe ordinaunce he is, not for men, whofe ouer-feer he is. Then how vaine is it *Euphues* (too mylde a worde for fo madde a minde) that the foote fhould negleét his office to correét the face, or that fubieétes fhoulde feeke more to knowe what their Princes doe, then what they are : where-in they fhewe them-felues as badde as beafts, and much worfe then my Bees, who in my conceite though I maye feeme partiall, obferue more order then they, (and if I myght faye fo of my good Bees,) more honeftie : honeftie my olde Graund-father called that, when menne lyued by law, not lyft : obferuing in all thinges the meane, which wee name vertue, and vertue we account nothing els but to deale iuftly and temperately.

And if I myght craue pardon, I would a little ac-quaint you with the common wealth of my Bees, which is neyther impertinent to the matter we haue now in hand, nor tedious to make you weary.

Euphues delighted with the difcourfes of old *Fidus*, was content to heare any thing, fo he myght heare him fpeake fome thing, and confenting willingly, hee

defired *Fidus* to go forward: who nowe remouing
him-felfe neerer to the Hyues, beganne as followeth.

Entlemen, I haue for ye fpace of this twenty
G yeares dwelt in this place, taking no delight in
any thing but only in keeping my Bees, and marking
them, and this I finde, which had I not feene, I fhold
hardly haue beleeued. That they vfe as great wit by
indution, and arte by workmanfhip, as euer man hath,
or can, vfing be[t]weene themefelues no leffe iuftice then
wifdome, and yet not fo much wifdome as maieftie: infomuch as thou wouldeft thinke, that they were a kinde
of people, a common wealth for *Plato*, where they all
labour, all gather honny, flye all together in a fwarme,
eate in a fwarm, and fleepe in a fwarm, fo neate and
finely, that they abhorre nothing fo much as vncleane-
nes, drinking pure and cleere water, delighting in
fweete and found Mufick, which if they heare but once
out of tune, they flye out of fight : and therefore are
they called the *Mufes* byrds, bicaufe they folow not
the found fo much as the confent. They lyue vnder
a lawe, vfing great reuerence to their elder, as to the
wifer. They chufe a King, whofe pallace they frame
both brauer in fhow, and ftronger in fubftaunce: whome
if they finde to fall, they eftablifh again in his thron[e],
with no leffe duty then deuotion, garding him continually, as it were for feare he fhould mifcarry,
and for loue he fhould not : whom they tender with
fuch fayth and fauour, that whether-foeuer he
flyeth, they follow him, and if hee can-not flye, they
carry him : whofe lyfe they fo loue, that they will
not for his fafety ftick to die, fuch care haue they for
his health, on whome they build all their hope. If
their Prince dye, they know not how to liue, they lan-
guifh, weepe, figh, neither intending their work, nor
keeping their olde focietie.

And that which is moft meruailous, and almofte incredible : if ther be any that hath difobeyed his commaundements, eyther of purpofe, or vnwittingly, hee

kylleth him-felfe with his owne fting, as executioner of
his own ftubborneffe. The King him-felfe hath his
fting, which hee vfeth rather for honour then punifh
ment: And yet *Euphues*, al-beit they lyue vnder a
Prince, they haue their priueledge, and as great liber-
ties as ftraight lawes.

They call a Parliament, wher-in they confult, for
lawes, ftatutes, penalties, chufing officers, and creating
their king, not by affection but reafon, not by the
greater part, but ye better. And if fuch a one by
chaunce be chofen (for among men fom-times the
worft fpeede beft) as is bad, then is there fuch ciuill
war and diffention, that vntill he be pluckt downe,
there can be no friendfhip, and ouer-throwne, there is
no enmitie, not fighting for quarrelles, but quietneffe.

Euery one hath his office, fome trimming the honny,
fome working the wax, one framing hiues, an other
the combes, and that fo artificially, that *Dedalus* could
not with greater arte or excellencie, better difpofe the
orders, meafures, proportions, diftinctions, ioynts and
circles. Diuers hew, others polifh, all are carefull to
doe their worke fo ftrongly, as they may refift the craft
of fuch drones, as feek to liue by their labours, which
maketh them to keepe[,to] watch and warde, as
lyuing in a campe to others, and as in a court to
them-felues. Such a care of chaftitie, that they neuer
ingender, fuch a defire of cleanneffe, that there is not
fo much as meate in all their hiues. When they go
forth to work, they marke the wind, the clouds, and
whatfoeuer doth threaten either their ruine, or raign
[reigne], and hauing gathered out of euery flower honny
they return loden in their mouthes, thighs, wings, and
all the bodye, whome they that tarried at home receyue
readily, as eafing their backes of fo great burthens.

The Kyng him-felfe not idle, goeth vp and downe,
entreating, threatning, commaunding, vfing the counfell of a fequel[l], but not loofing the dignitie of a Prince,
preferring thofe yat labour to greater authoritie, and
punifhing thofe that loyter, with due feueritie. All

which thinges being much admirable, yet this is moſt, that they are ſo profitable, bringing vnto man both honnye and wax, each ſo wholſome that wee all deſire it, both ſo neceſſary that we cannot miſſe them. Here *Euphues* is a common wealth, which oftentimes calling to my minde, I cannot chuſe but commend aboue any that either I haue heard or read of. Where the king is not for euery one to talke of, where there is ſuch homage, ſuch loue, ſuch labour, that I haue wiſhed oftentimes, rather be a Bee, then not be as I ſhould be.

In this little garden with theſe hiues, in this houſe haue I ſpent the better parte of my lyfe, yea and the beſt : I was neuer buſie in matters of ſtate, but referring al my cares vnto the wiſdom of graue Counſellors, and my confidence in the noble minde of my dread Souereigne and Queene, neuer aſking what ſhe did, but alwayes praying ſhe may do well, not enquiring whether ſhe might do what ſhe would, but thinking ſhe would do nothing but what ſhe might.

Thus contented with a meane eſtate, and neuer curious of the high eſtate, I found ſuch quiet, that mee thinketh, he which knoweth leaſt, lyueth longeſt : inſomuch that I chuſe rather to be an Hermitte in a caue, then a Counſellor in in the court,

Euphues perceyuing olde *Fidus*, to ſpeake what hee thought, aunſwered him in theſe ſhorte wordes.

He is very obſtinate, whome neither reaſon nor experiynce can perſwade : and truly ſeeing you haue alledged both, I muſt needes allow both. And if my former requeſt haue bred any offence, let my latter repentaunce make amends. And yet this I knowe, that I enquyred nothing that might bring you into daunger, or me into trouble : for as young as I am, this haue I learned, that one maye poynt at a Starre, but not pull at it, and ſee a Prince but not ſearch him : And for mine own part, I neuer mean to put my hand betweene the barke and the tree, or in matters which are not for me to be ouer curious.

The common wealth of your Bees, did fo delight me, that I was not a lyttle fory yat either their eftate haue not ben longer, or your leafure more, for in my fimple iudgement, there was fuch an orderlye gouern-ment, that men may not be afhamed to imitate them, nor you wearie to keepe them.

They hauing fpent much time in thefe difcourfes, were called in to Supper, *Philautus* more willing to eate, then heare their tales, was not the laft yat went in : where being all fet downe, they were ferued al in earthen difhes, al things fo neat and cleanly, that they perceiued a kinde of courtly Maieftie in the minde of their hoft, though he wanted matter to fhew it in his houfe. *Philautus* I know not whether of nature melancholy, or feeling loue in his bofome, fpake fcarce ten words fince his comming into the houfe of *Fidus*, which the olde man well noting, began merily thus to *parle* with him.

I Meruaile Gentleman that all this time, you haue bene tongue tyed, either thinking not your felfe welcome, or difdayning fo homely enterteinment : in the one you doe me wrong, for I thinke I haue not fhewed my felfe ftraunge : for the other you muft pardon me, for that I haue not to do as I would, but as I may : And though *England* be no graunge, but yeeldeth euery thing, yet is it heere as in euery place, al for money. And if you will but accept a willing minde in fteede of a coftly repaft, I fhall thinke my felfe beholding vnto you : and if time ferue, or [and] my Bees profper, I wil make you part of amends, with a better breakfaft.

Philautus thus replyed : I know good Father, my welcome greater then any wayes I can requite, and my cheere more bountifull then euer I fhall deferue, and though I feeme filent for matters that trouble me, yet I would not haue you thinke me fo foolifh, that I fhould either difdaine your company, or miflyke your cheere, of both the which I thinke fo well, that if time

might aunfwere my true meaning, I would exceede
in coft, though in courtefie I know not how to compare with you, for (without flatterie be it fpoken) if
the common courtefie of *Englande* be no worfe then
this towarde ftraungers, I muft needes thinke them
happy that trauaile into thefe coafts, and the inhabi-
taunts the moft courteous, of all countreyes.

Heere began *Euphues* to take the tale out of *Philautus* mouth, and to play with him in his melan-
cholicke moode, beginning thus.

N O Father I durft fweare for my friend, that both
he thinketh himfelfe welcome, and his fare good,
but you muft pardon a young courtier, who in the
abfence of his Lady thinketh himfelfe forlorne : And
this vile Dog Loue will fo ranckle where he biteth, that
I feare my friends fore, will breed to a *Fiftula* : for you
may perceiue that he is not where he liues, but wher
he loues, and more thoughts hath he in his head, then
you Bees in your Hiues : and better it were for him
to be naked among your Wafpes, though his bodye
were al bliftered, then to haue his heart ftong fo with
affection, where-by he is fo blinded. But beleeue mee
Fidus, he taketh as great delight to courfe a cogitacion
of loue, as you doe to vfe your time with Honny. In
this plight hath he bene euer fince his comming out
of *Naples*, and fo hath it wrought with him (which I
had thought impoffible) that pure loue did make
him Seaficke, infomuch as in all my trauaile with him,
I feemed to euery one to beare with me the picture
of a proper man, but no liuing perfon, the more pitie,
and yet no force. *Philautus* taking *Euphues* tale by
the ende, and the olde man by the arme, betweene
griefe and game, ieft and earneft, aunfwered him thus.

E *Vphues* would dye if he fhould not talke of loue
once in a day, and therfore you muft giue him
leaue after euery meale to cloafe his ftomacke with
Loue, as with Marmalade, and I haue heard, not thofe

that fay nothing, but they that kicke oftenest against loue, are euer in loue: yet doth he vfe me as the meane to moue the matter, and as the man to make his Myrrour, he himfelfe knowing best the price of Corne, not by the Market folkes, but his owne foote-fteppes. But if he vfe this fpeach either to make you merrye, or to put me out of conceipt, he doth well, you must thanke him for the one, and I wil thinke on him for the other. I haue oftentimes fworne that I am as farre from loue as he, yet will he not beleeue me, as incredulous as thofe, who thinke none balde, till they fee his braynes.

As *Euphues* was making aunfwere, *Fidus* preuented him in this manner.

THere is no harme done *Philautus*, for whether you loue, or *Euphues* iest, this fhall breed no iarre. It may be when I was as young as you, I was as idle as you (though in my opinion, there is none leffe idle then a louer.) For to tell the truth, I my felf was once a Courtier, in the dayes of that most noble King of famous memorie *Henry* the eight, Father to our most gratious Lady *Elizabeth*.

Where, and with that, he paufed, as though the remembraunce of his olde lyfe, had ftopped his newe fpeach, but *Philautus* eytching [itching] to hear what he would fay, defired him to goe forward, vnto whome *Fidus* fetching a great figh fayd, I will. And there agayne made a full poynt. *Philautus* burning as it were, in defire of this difcourfe, vrged him againe with great entreatie: then the olde man commaunded the boorde to be vncouered, grace being fayd, called for stooles, and fitting al by the fire, vttered the whole difcourfe of his loue, which brought *Philautus* a bedde, and *Euphues* a fleepe.

And now Gentlemen, if you will giue eare to the tale of *Fidus*, it may be fome will be as watchfull as *Philautus*, though many as droufie as *Euphues*. And thus he began with a heauie countenaunce (as tho ıgh his paines were prefent, not paſt) to frame his tale.

I Was borne in the wylde of *Kent*, of honeſt Parents, and worſhipfull, whoſe tender cares, (if the fondneſſe of parents may be ſo termed) prouided all things euen from my very cradell, vntil their graues, that might either bring me vp in good letters, or make me heire to great lyuings. I (with-out arrogancie be it ſpoken) was not inferiour in wit to manye, which finding in my ſelfe, I flattered my ſelfe, but in ye ende, deceiued my ſelfe : For being of the age of. xx. yeares, there was no trade or kinde of lyfe that either fitted my humour or ſerued my tourne, but the Court : thinking that place the onely meanes to clymbe high, and ſit ſure : Wherin I followed the vaine of young Souldiours, who iudge nothing ſweeter then warre til they feele the weight, I was there enterteined as well by ᵗhe great friends my father made, as by mine own forwardneſſe, where it being now but Honnie Moone, I endeauoured to courte it with a grace, (almoſt paſt grace,) laying more on my backe then my friendes could wel beare, hauing many times a braue cloke and a thredbare purſe.

Who ſo conuerſant with the Ladyes as I ? who ſo pleaſaunt ? who more prodigall ? In-ſomuch as I thought the time loſt, which was not ſpent either in their company with delight, or for their company in letters. Among all the troupe of gallant Gentle-men, I ſingled out one (in whome I myſliked nothing but his grauitie) that aboue all I meant to truſt : who aſwell for ye good qualities he ſaw in me, as the little gouernment he feared in mee, beganne one night to vtter theſe fewe wordes.

Friend *Fidus* (if Fortune allow a tearm ſo familiar) I would I might liue to ſee thee as wife, as I percieue thee wittie, then ſhould thy life be ſo ſeaſoned, as neyther too much witte might make thee proude, noɪ too great ryot poore. My acquaintaunce is not great

with thy perfon, but fuch infight haue I into tny conditions, that I feare nothing fo much, as that, there
thou catch thyfall, where thou thinkeft to take thy rifing.
Ther belongeth more to a courtier then brauery,
which ye wife laugh at, or perfonage, which ye
chaft mark not, or wit, which the moft part fee not.
It is fober and difcret behauiour, ciuil and gentle demeanor, that in court winneth both credit and commoditie : which counfel thy vnripened yeares, thinke
to proceede rather of the malice of age, then the good
meaning. To ryde well is laudable, and I like it, to
runne at the tilt not amiffe, and I defire it, to reuell
much to be praifed, and I haue vfed it : which thinges
as I know them all to be courtly, fo for my part I
accompt them neceffary, for where greateft affemblies
are of noble Gentle-men, there fhould be the greateft
exercife of true nobilitie. And I am not fo prefife,
[precife] but that I efteeme it as expedient in feates of
armes and actiuitie to employ the body, as in ftudy
to waft the minde : yet fo fhould the one be tempered
with the other, as it myght feeme as great a fhame to
be valiaunt and courtly with-out learning, as to bee
ftudious and bookifh with-out valure.

But there is an other thing *Fidus*, which I am to
warn thee of, and if I might to wreaft thee from : not
that I enuy thy eftate, but that I would not haue thee
forget it. Thou vfeft too much (a little I thinke to bee
too much) to dallye with woemen, which is the next
way to doate on them : For as they that angle for the
Tortois, hauing once caught him, are dryuen into fuch
a lythemeffe, that they loofe all their fprightes [fpirites],
being beenummed, fo they that feeke to obtayne the
good-will of Ladyes, hauing once a little holde of their
loue, they are driuen into fuch a traunce, that they let
go the holde of their libertie, bewitched like thofe
that viewe the head of *Medufa*, or the Viper tyed to
the bough of the Beech tree, which keepeth him in a
dead fleepe, though it beginne with a fweete flumber.
I my felfe haue tafted new wine, and finde it to bee

more pleafaunt then wholfome, and Grapes gathered before they bee rype, maye fet the eyes on luft, but they make the teeth an edge, and loue defired in the budde, not knowing what the bloffome were, may delight the conceiptes of the head, but it will deftroye the contemplature of the heart. What I fpeake now is of meere good-will, and yet vpon fmall prefumption, but in things which come on the fodaine, one cannot be too warye to preuent, or too curious to myftruft : for thou art in a place, eyther to make thee hated for vice, or loued for vertue, and as thou reuerenceft the one before the other, fo in vprightneffe ot lyfe fhewe it. Thou haft good friendes, which by thy lewde delights, thou mayft make great enimies, and heauy foes, which by thy well doing thou mayft caufe to be earneft abettors of thee, in matters that nowe they canuaffe agaynft thee.

And fo I leaue thee, meaning herafter to beare the reign of thy brydell in myne hands : if I fee thee head ftronge : And fo he departed.

I gaue him great thanks, and glad I was we were parted : for his putting loue into my minde, was like the throwing of Bugloffe into wine, which encreafeth in him that drinketh it a defire of luft, though it mitti-gate the force of drunkenneffe.

I now fetching a windleffe, that I myght better haue a fhoote, was preuented with ready game, which faued me fome labour, but gained me no quiet. And I would gentlemen yat you could feel the like impref-fions in your myndes at the reherfall of my mifhappe, as I did paffions at the entring into it. If euer you loued, you haue found the like, if euer you fhall loue, you fhall tafte no leffe. But he† fo e[a]ger of an end, as one leaping ouer a ftile before hee come to it, defired few parenthefes or digreffions or glofes, but the text, wher he him-felf, was co[a]ting in the margant [mar-gent]. Then faid *Fidus*, thus it fell out.

It was my chaunce (I know not whether chaunce or

† *I.e.*, Philautus.

deftime) that being inuited to a banket where many
Ladyes were and too many by one, as the end tryed,
though then to[o] many by al fauing yat one, as I thought,
I caft mine eies fo earneftly vpon hir, yat my hart
vowd hir the miftris of my loue, and fo fully was I refolued to profecut[e] my determination, as I was earneft
to begin it. Now Gentlemen, I commit my cafe to
your confiderations, being wifer then I was then, and
fomwhat as I geffe elder : I was but in court a nouice,
hauing no friende, but him before rehearfed, (whome
in fuch a matter I was lyklier to finde a brydell,
then a fpurre) I neuer before that tyme could imagin
what loue fhould meane, but vfed the tearm as a flout
to others, which I found now as a feuer in my felfe,
neither know[ing] from whence the occafion fhould arife,
nor where I might feeke the remedy. This diftreffe I
thought youth would haue worne out, or [by] reafon, or
time, or abfence, or if not euery one of them, yet all.
But as fire getting hould in the bottome of a tree,
neuer leaueth till it come to the toppe, or as ftronge
poyfon *Antidotum* being but chafed in the hand, pearceth at the laft the hart, fo loue which I kept but low,
thinking at my will to leaue, entred at the laft fo farre
that it held me conquered. And then difputing with
my felfe, I played this on the bit.

Fidus, it ftandeth thee vppon eyther to winne thy
loue, or to weane thy affections, which choyce is fo
hard, that thou canft not tel whether the victory wil
be the greater in fubduing thy felfe, or conquering hir.

To loue and to lyue well is wifhed of myne [manye],
but incident to fewe. To liue and to loue well is
incident to fewe, but indifferent to all. To loue without reafon is an argument of luft, to lyue with-out
loue, a token of folly. The meafure of loue is to haue
no meane, the end to be euerlafting.

Thefius had no neede of *Ariadnes* threed to finde
the way into the *Laborinth*, but to come out, nor thou of
any help how to fal into thefe bracks [brakes], but to fall
from them. If thou be [be]witched with eyes, weare the

eie [eyes] of a wefill in a ring, which is an enchauntment
againſt ſuch charmes, and reaſon with thy ſelf whether
ther be more pleaſure to be accounted amorous, or
wife.   Thou art in the view of the whole court, wher
the ielous wil ſuſpecteth vppon euery light occaſion,
where of the wife thou ſhalt be accounted fond, and
of* the fooliſh amorous : the Ladies themſelues, how-
ſoeuer they looke, wil thus imagine, that if thou take
thought for loue, thou art but a foole, if take it lyghtly,
no true feruaunt.   Beſides this thou art to be bounde
as it were an Apprentice feruing feauen yeares for that,
which if thou winne, is loſt in feauen houres, if thou
loue thine equall, it is no conqueſt : if thy ſuperiour,
thou ſhalt be enuyed : if thine inferiour, laughed at.
If one that is beautifull, hir colour will chaunge before
thou get thy defire : if one that is wife, ſhe will ouer-
reache thee ſo farre, that thou ſhalt neuer touch hir :
if vertuous, ſhe will efchue ſuch fonde affection : if
one deformed, ſhe is not worthy of any affection : if
ſhe be rich, ſhe needeth thee not : if poore, thou
needeſt not hir : if olde, why ſhouldeſt thou loue hir,
if young, why ſhould ſhe loue thee.

Thus Gentlemen, I fed my ſelfe with mine owne
deuices, thinking by peecemeale to cut off that which
I could not diminiſh : for the more I ſtriued with
reaſon to conquere mine appetite, the more againſt
reaſon, I was fubdued of mine affections.

At the laſt calling to my remembrance, an olde rule
of loue, which a courtier then tolde me, of whom when
I demaunded what was the firſt thing to winne my
Lady, he aunſwered, Opportunitie, aſking what was
the ſecond, he ſayd Opportunitie : deſirous to know
what might be the thirde, he replyed Opportunitie.

Which aunſweres I marking, as one that thought to
take mine ayme of ſo cunning an Archer, coniectured
that to the beginning, continuing an[d] ending of loue,
nothing could be more conuenient then Opportunitie,
to the getting of the which I applyed my whole ſtudie,
and wore my wits to the hard* ſtumpes, aſſuring my

felfe, that as there is a time, when the Hare will lycke
the Houndes eare, and the fierce Tigreffe play with
the gentle Lambe : fo ther was a certein feafon, when
women were [are] to be won, in the which moment
they haue neither will to deny, nor wit to miftruft.

Such a time I haue read a young Gentleman found
to obtaine the loue of the Ducheffe of *Millayne*:
fuch a time I haue heard that a poore yeoman chofe
to get the faireft Lady in *Mantua*.

Vnto the which time, I trufted fo much, that I folde
the fkinne before the Beafte was taken, rec[k]oning
with-out mine hoaft, and fetting downe that in my
bookes as ready money, which afterwards I found to
be a defperate debt.

I T chaunced that this my Lady (whome although
I might name for the loue I bore hir, yet I will
not for the reuerence I owe hir, but in this ftorye call
hir *Iffida*) for to recreate hir minde, as alfo to folace
hir body, went into the countrey, where fhe determined
to make hir abode for the fpace of three moneths,
hauing gotten leaue of thofe that might beft giue it. And
in this iourney I founde good Fortune fo fauourable,
yat hir abiding was within two miles of my Fathers
mantion houfe, my parents being of great familiaritie
with the Gentleman, where my *Iffida* lay. Who now
fo fortunate as *Fidus*? who fo fralicke? She being in
ye countrey, it was no being for me in ye court?
wher euery paftime was a plague to the minde yat
lyued in melancholy. For as the Turtle hauing loft
hir mate, wandreth alone, ioying in nothing but in
folitarineffe, fo poore *Fidus* in the abfence of *Iffida*,
walked in his chamber as one not defolate for lacke of
company, but defperate. To make fhort of ye circum-
ftaunces, which holde you too long from that you would
heare, and I faine vtter, I came home to my father
[Fathers], wher at mine entraunce, fupper being fet on
the table, I efpyed *Iffida*, *Iffida* Gentlemen, whom I
found before I fought, and loft before I wonne. Yet

S

leaſt the alteration of my face, might argue ſome ſuſ‑
pition of my follyes, I, as courtly as I could, though
god knowes but courſly at that time behaued my ſelfe,
as though nothing payned me, when in truth nothing
pleaſed me.　In the middle of ſupper, *Iffida* as well
for the acquaintance, we had in court as alſo the
courteſie ſhe vſed in generall to all, taking a glaſſe in
hir hand filled with wine, dranke to me in this wiſe.
Gentleman, I am not learned, yet haue I heard, that
the Vine beareth three grapes, the firſt altereth, the
ſecond troubleth, the third dulleth.　Of what Grape this
Wine is made I cannot tell, and therefore I muſt craue
pardon, if either this draught chaunge you, vnleſſe it
be to the better or grieue you, except it be ſor greater
gaine, or dull you, vnleſſe it be your deſire, which long
preamble I vſe to no other purpoſe, then to warne you
from wine heere-after, being ſo well counſelled before.
And with that ſhe drinking, deliuered me the glaſſe.
I now taking heart at graſſe, to ſee hir ſo gameſome, as
merely [merrily] as I could, pledged hir in this manner.

I T is pitie Lady you want a pulpit, hauing preached
ſo well ouer the pot, wherin you both ſhewe
the learning, which you profeſſe you haue not, and a
kinde of loue, which would you had : the one appear‑
eth by your long ſermon, the other by the deſire you
haue to keepe me ſober, but I wil refer mine anſwere
till after ſupper, and in the meane ſeaſon, be ſo tempe‑
rate, as you ſhall not thinke my wit to ſmell of the
wine, although in my opinion, ſuch grapes ſet rather
an edge vpon wit, then abate the point.　If I may
ſpeak in your caſt, quoth *Iffida* (the glaſſe being at my
noſe) I thinke, wine is ſuch a whetſtone for wit, that
if it be often ſet in that manner, it will quickly grinde
all the ſteele out, and ſcarce leaue a back wher it found
an edge.

　　With many like ſpeaches we continued our ſupper,
which I will not repeat, leaſt you ſhould thinke vs
Epicures to ſit ſo long at our meate : but all being

ended, we arofe, where as the manner is, thankes and
curfie made to each other, we went to the fire, wher
I boldened now without blufhing, tooke hir by the
hand, and thus began to kindle the flame which I
fhoulde rather haue quenched, feeking to blow a cole,
when I fhould haue blowne out the candle.

Entlewoman either thou thoughts my wits verye
short, yat a fippe of wine could alter me, or els
yours very fharpe, to cut me off so roundly, when as I
(without offence be it fpoken) haue heard, that as
deepe drinketh the Goofe as the Gander.

Gentleman (quoth fhe) in arguing of wittes, you
miftake mine, and call your owne into queftion. For
what I fayd proceeded rather of a defire to haue you
in health, then of malyce to wifh you harme. For you
well know, that wine to a young blood, is in the fpring
time, Flaxe to fire, and at all times either vnwholfome,
or fuperfluous, and fo daungerous, that more perifh by
a furfet then the fword.

I haue heard wife Clearkes fay, that *Galen* being
afked what dyet he vfed that he lyued fo long, aunfwered: I haue dronke no wine, I haue touched no
woman, I haue kept my felfe warme.

Now fir if you will lycence me to proceede, this I
thought, yat if one of your yeares fhould take a dram
of *Magis*, wherby confequently you fhold fal to an ounce
of loue, and then vpon fo great heat take a little colde,
it were inough to caft you away, or turne you out of
the way. And although I be no Phifition, yet haue I
bene vfed to attend ficke perfons, where I founde nothing
to hurt them fo much as Wine, which alwayes drew
with it, as the Adamant doth the yron, a* defire of women: how hurtfull both haue bene, though you be too
young to haue tryed it, yet you are olde enough to beleeue it. Wine fhould be taken as the Dogs of *Egypt*
drinke water, by fnatches, and fo quench their thirft,
and not hynder theyr running, or as the Daughters
of *Lyfander* vfed it, who with a droppe of wine tooke

a fpoonefull of water, or as the Virgins in *Rome*, whoe dryncke but theyr eye full, contenting them-felues as much with the fight, as the tafte.

Thus to excufe my felfe of vnkindneffe, you haue made me almoft impudent, and I you (I feare mee) impatient, in feeming to prefcribe a diette wher there is no daunger, [in] giuing a preparatiue when the body is purged : But feeing all this talke came of drinkeing, let it ende with drinking.

I feeing my felfe thus rydden, thought eyther fhee fhould fit faft, or els I would caft hir. And thus I replyed.

Lady, you thinke to wade deepe, where the Foorde is but fhallow, and to enter into the fecretes of my minde, when it lyeth open already, wher-in you vfe no leffe art to bring me in doubt of your good wil, then craft to put me out of doubt, hauing bayted your hooke both with poyfon and pleafure, in that, vfing the meanes of phyficke (where-of you fo talke) myngling fweete firroppes with bitter dragges [dregs]. You ftand in feare that wine fhould inflame my lyuer and con-uert me to a louer : truely I am framed of that met-tall, that I canne mortifye anye affeētions, whether it bee in dryncke or defire, fo that I haue no neede of your playfters, though I muft needes giue thankes for your paynes.

And nowe *Philautus*, for I fee *Euphues* begynne to nodde, thou fhalt vnderftand, that in the myddeft of my replye, my Father with the refte of the com-panye, interrupted mee, fayinge they woulde all fall to fome paftyme, whiche bycaufe it groweth late *Philautus*, wee wyll deferre tyll the morning, for age muft keepe a ftraight dyot [dyette], or els a fickly life.

Philautus tyck[e]led in euerye vaine [veyne] with delyght, was loath to leaue fo, although not wylling the good olde manne fhould breake his accuftomed houre, vnto whome fleepe was the chiefeft fuftenaunce. And fo waking *Euphues*, who hadde taken a nappe, they all went to their lodging, where I thinke *Phi-*

lautus was muſing vppon the euent of *Fidus* his loue :
But there I will leaue them in their beddes, till the
next morning.

Entle-menne and Gentle-woemenne, in the diſcourſe of this loue, it maye ſeeme I haue taken
a newe courſe : but ſuch was the tyme then, that
it was ſtraunge to loue, as it is nowe common, and
then leſſe vſed in the Courte, then it is now in the
countrey : But hauing reſpecte to the tyme paſt, I
truſt you will not condempne my preſent tyme, who
am enforced to ſinge after their plaine-ſonge, that was
then vſed, and will followe heare-after the Crotchetts
that are in theſe dayes cunninglye handled.

For the mindes of Louers alter with the madde
moodes of the Muſitions : and ſo much are they within
fewe yeares chaunged, that we accompt their olde
wooing and ſinging to haue ſo little cunning, that we
eſteeme it barbarous, and were they liuing to heare
our newe quoyings, they woulde iudge it to haue ſo
much curioſitie, that they would tearme it fooliſh.

In the time of *Romulus* all heades were rounded
of his faſhion, in the time of *Cæſar* curled of his
manner. When *Cyrus* lyued, euerye one prayſed the
hooked noſe, and when hee dyed, they allowed the
ſtraight noſe.

And ſo it fareth with loue, in tymes paſt they vſed
to wooe in playne tearmes, now in piked [picked] ſentences, and hee ſpeedeth beſt, that ſpeaketh wiſeſt :
euery one following the neweſt waye, which is not euer
the neereſt way : ſome going ouer the ſtile when the
gate is open, and other [another] keeping the right
beaten path, when hee maye croſſe ouer better by the
fieldes. Euery one followeth his owne fancie, which
maketh diuers leape ſhorte for want of good ryſinge,
and many ſhoote ouer for lacke of true ayme.

And to that paſſe it is come, that they make an
arte of that, which was woont to be thought naturall :
And thus it ſtandeth, that it is not yet determyned,

whether in loue *Vlyffes* more preuailed with his wit, or *Paris* with his perfonage, or *Achilles* with his proweffe.

For euerye of them haue *Venus* by the hand, and they are all affured and certaine to winne hir heart.

But I hadde almoft forgotten the olde manne, who vfeth not to fleepe compaffe, whom I fee with *Euphues* and *Philautus* now alreadye in the garden, readye to proceede with his tale : which if it feeme tedious, wee will breake of[f] againe when they go to dynner.

F*Idus* calling thefe Gentle-men vppe, brought them into his garden, where vnder a fweete Arbour of Eglentine, be [the] byrdes recording theyr fweete notes, hee alfo ftrayned his olde pype, and thus beganne.

G Entle-menne, yefter-nyght I left of[f] abruptlye, and therefore I muft nowe begynne in the* like manner.

My Father placed vs all in good order, requefting eyther by queftions to whette our wittes, or by ftories to trye our memoryes, and *Iffiyda* that might beft there bee bolde, beeing the beft in the companye, and at all affayes too good for me, began againe to preach in this manner.

Thou art a courtier *Fidus*, and therefore beft able to refolue any queftion : for I knowe thy witte good to vnderftand, and ready to aunfwere : to thee therfore I addreffe my talke.

T Here was fom-time in *Sienna* a *Magnifico*, whom God bleffed with three Daughters, but by three wiues, and of three fundrye qualities : the eldeft was verye fayre, but a very foole : the fecond meruailous wittie, but yet meruailous wanton : the third as ver-tuous as any liuing, but more deformed then any that euer lyued.

The noble Gentle-man their father difputed for the beftowing of them with him-felfe thus.

I thank the Gods, that haue giuen me three Daugh-ters, who in their bofomes carry their dowries, in-fomuch as I fhall not neede to difburfe one myte for all theyr marryages. Maydens be they neuer fo foolyfhe, yet beeynge fayre, they are commonly fortunate : for that men in thefe dayes, haue more ' refpect to the out ward fhow then the inward fubftance, where-in they imitate good Lapidaryes, who chufe the ftones that delyght the eye, meafuring the value not by the hidden vertue, but by the outwarde glifter-ing : or wife Painters, who laye their beft coulours, vpon their worft counterfeite.

And in this me thinketh Nature hath dealt indiffe-rently, that a foole whom euery one abhorreth, fhoulde haue beautie, which euery one defireth : that the excel-lencie of the one might excufe the vanitie of the other : for as we in nothing more differ from the Gods, then when we are fooles, fo in nothing doe we come neere them fo much, as when we are amiable. This caufed *Helen* to be fnatched vp for a Starre, and *Ariadne* to be placed in the Heauens, not that they were wife, but faire, fitter to adde a Maieftie to the Skie, then beare a Maieftie in Earth. *Iuno* for all hir iealoufie, be-holding *Iô*, wifhed to be no Goddeffe, fo fhe might be fo gallant. Loue commeth in at the eye, not at the eare, by feeing Natures workes, not by hearing womens words. And fuch effects [affects] and pleafure doth fight bring vnto vs, that diuers haue lyued by looking on faire and beautifull pictures, defiring no meate, nor h[e]arkning to any Mufick. What made the Gods fo often to trewant from Heauen, and mych [mich] heere on earth, but beautie ? What made men to imagine, that the Firmament was God, but the* beautie ? which is fayd to bewitch the wife, and enchaunt them that made it. *Pigmalion* for beautie, loued an Image of Iuory, *Appelles* the counterfeit of *Campafpe*, and none we haue heard off fo fenceleffe, that the name of beautie, cannot either breake or bende. It is this onely that Princes defire in their Houfes, Gardeins, Orchards, and Beddes, following *Alexander*, who more efteemed the face of *Venus*, not yet finifhed, then the Table of the nyne Mufes perf[e]éted. And I am of that minde that there can be nothing giuen vnto mortall men by the immortall Gods, eyther more noble or more neceffary then beautie. For as when the counterfeit of *Ganimedes*, was fhowen at a market, euery one would faine buye it, bicaufe *Zeuxis* had there-in fhewed his greateft cunning : fo when a beautifull woman appeareth in a multitude, euery man is drawne to fue to hir, for that the Gods (the onely Painters of beautie) haue in hir expreffed, the art of their Deitie. But I wil heere reft my felfe, knowing that if I fhould runne fo farre as Beautie would carry me, I fhoulde fooner want breath to tell hir praifes, then matter to proue them, thus I am perfwaded, yat my faire daughter fhal be wel maryed, for there is none, that will or can demaund a greater ioynter then Beautie.

My fecond childe is wittie, but yet wanton, which in my minde, rather addeth a delyght to the man, then a difgrace to the mayde, and fo lynked are thofe two qualyties together, that to be wanton without wit, is Apifhnes : and to be thought wittie without wantonnes, precifeneffe. When *Lais* being very pleafaunt, had told a merry ieft : It is pitie fayde *Ariftippus*, that *Lais* hauing fo good a wit, fhould be a wanton. Yea quoth *Lais*, but it were more pitie, that *Lais* fhoulde be a wanton and haue no good wit. *Ofyris* King of the *Aegyptians*, being much delyghted with pleafaunt conceipts, would often affirme, that he had rather haue a virgin, that could giue a quicke aunfwere that might cut him, then a milde fpeach that might claw him. When it was obieéted to a gentlewoman, yat fhe was neither faire nor fortunate, and yet quoth fhe, wife and wel fauoured, thinking it the chiefeft gift yat Nature could beftow, to haue a Nutbrowne hue, and an ex-

cellent head. It is wit yat allureth, when eueryword ſhal
haue his weight, when nothing ſhal proceed, but it ſhal
either fauour of a ſharpe conceipt, or a ſecret concluſion.
And this is the greateſt thing, to conceiue readely and
aunſwere aptly, to vnderſtand whatſoeuer is ſpoken, and
to reply as though they vnderſtoode nothing. A
Gentleman yat once loued a Lady moſt entirely,
walking with hir in a parke, with a deepe figh began ·
to ſay, O yat women could be conſtant, ſhe replyed,
O yat they could not, Pulling hir hat ouer hir head,
why quoth the gentleman doth the Sunne offend your
eyes, yea, aunſwered ſhe the ſonne of your mother,
which quicke and ready replyes, being well marked
of him, he was enforced to ſue for yat which he was
determined to ſhake off. A noble man in *Sienna*,
diſpoſed to ieſt with a gentlewoman of meane birth,
yet excellent qualities, between game and earneſt gan
thus to ſalute hir. I know not how I ſhold commend
your beautie, becauſe it is ſomwhat to[o] brown, nor
your ſtature being ſomwhat to[o] low, and of your wit I
can not iudge, no quoth ſhe, I bele[e]ue you, for none
can iudge of wit, but they that haue it, why then quoth
he, doeſt thou thinke me a foole, thought is free my
Lord quoth ſhe, I wil not take you at your word.
He perceiuing al outward faults to be recompenced
with inward fauour, choſe this virgin for his wife.
And in my ſimple opinion, he did a thing both worthy
his ſtocke and hir vertue. It is wit that flouriſheth,
when beautie fadeth : that waxeth young when age
approcheth, and reſembleth the Iuie leafe, who although it be dead, continueth greene. And bicauſe
of all creatures, the womans wit is moſt excellent,
therefore haue the Poets fained the Muſes to be
women, the Nimphes, the Goddeſſe[s] : enſamples of
whoſe rare wiſedomes, and ſharpe capacities would
nothing but make me commit Idolatry with my
daughter.

I neuer heard but of three things which argued a
fine wit, Inuention, Conceiuing, Aunſwering. Which

haue all bene found fo common in women, that
were it not I fhould flatter them, I fhould think them
fingular.

Then this fufficeth me, that my feconde daughter
fhall not lead Apes in Hell, though fhe haue not a
penny for the Prieft, bicaufe fhe is wittie, which bind-
eth weake things, and loofeth ftrong things, and
worketh all things, in thofe that haue either wit themfelues, or loue wit in others.

My youngeft though no pearle to hang at ones eare,
yet fo precious fhe is to a well difpofed minde, that
grace feemeth almoft to difdaine Nature. She is deformed in body, flowe of fpeache, crabbed in countenaunce, and almoft in all parts crooked : but in behauiour fo honeft, in prayer fo deuout, fo precife in al
hir dealings, that I neuer heard hir fpeake anye thing
that either concerned not good inftruction, or godlye
mirth.

Who neuer delyghteth in coftly apparell, but euer
defireth homely attire, accompting no brauery greater
then vertue : who beholding hir vglye fhape in a
glaffe, fmilyng fayd : This face were faire, if it were
tourned, noting that the inward motions would make
the outward fauour but counterfeit. For as ye precious
ftone *Sandaftra*, hath nothing in outward appearaunce
but that which feemeth blacke, but being broken
poureth forth beames lyke the Sunne : fo vertue
fheweth but bare to the outward eye, but being
pearced with inward defire, fhineth lyke Chriftall.
And this dare I auouch yat as the *Trogloditæ* which
digged in the filthy ground for rootes, and found the
ineftimable ftone *Topafon*, which inriched them euer
after : fo he that feeketh after my youngeft daughter,
which is deformed, fhall finde the great treafure of
pietie, to comfort him during his lyfe. Beautifull
women are but lyke the *Ermine*, whofe fkinne is
defired, whofe carcaffe is difpifed, the vertuous contrariwife, are then moft lyked, when theyr fkinne is
leafte loued.

Then ought I to take leaft care for hir, whom euerye one that is honeft will care for: fo that I will quiet my felf with this perfwafion, that euery one fhal haue a wooer fhortly. Beautie cannot liue with-out a hufband, wit will not, vertue fhall not.

NOw Gentleman, I haue propounded my reafons, for euery one I muft now afke you the queftion. If it were your chaunce to trauaile to *Sienna,* and to fee as much there as I haue tolde you here, whether would you chufe for your wife the faire foole, the witty wanton, or the crooked Saint.

When fhee had finifhed, I ftoode in a maze, feeing three hookes layed in one bayte, vncertaine to aun- fwere what myght pleafe hir, yet compelled to faye fome-what, leaft I fhould difcredit my felfe : But feeing all were whift to heare my iudgement, I replied thus.

LAdye *Iffyda,* and Gentle-woemenne all, I meane not to trauayle to *Sienna* to wooe Beautie, leaft in comming home the ayre chaunge it, and then my labour bee loft : neyther to feeke fo farre for witte, leaft fhee accompt me a foole, when I myght fpeede as well neerer hande : nor to fue to Vertue, leaft in *Italy* I be infeĉted with vice : and fo looking to gette *Iupiter* by the hand, I catch *Pluto* by the heele. But if you will imagine that great *Magnifico* to haue fent his three Daughters into England, I would thus debate with them before I would barg[a]in[e] with them. I loue Beautie wel, but I could not finde in my hart to marry a foole : for if fhe be impudent I fhal not rule hir : and if fhe be obftinate, fhe will rule me, and my felfe none of the wifeft, me thinketh it were no good match, for two fooles in one bed are too many.

Witte of all thinges fetteth my fancies on edge, but I fhould hardly chufe a wanton : for be fhe neuer fo wife, if alwayes fhe want one when fhe hath me, I had as leife [liefe] fhe fhould want me too, for of all my apparell I woulde haue my cappe fit clofe.

Vertue I cannot miflike, which hether-too I haue honoured, but fuch a crooked Apoftle I neuer brooked : for vertue may well fatte my minde, but it will neuer feede mine eie, and in mariage, as market folkes tel me, the hufband fhould haue two eies, and the wife but one : but in fuch a match it is as good to haue no eye, as no appetite.

But to aunfwere of three inconueniences, which I would chufe (although each threaten a mifchiefe) I muft needes take the wife wanton : who if by hir wantonneffe fhe will neuer want wher fhe likes, yet by her wit fhe will euer conceale whom fhe loues, and to weare a horne and not knowe it, will do me no more harme then to eate a flye, and not fee it.

Iffyda I know not whether ftong with mine anfwer, or not content with my opinion, replied in this manner.

Then *Fidus* when you match, God fend you fuch a one, as you like beft : but be fure alwaies, that your head be not higher then your hat. And thus faining an excufe departed to hir lodging, which caufed al the company to breake off their determined paftimes, leauing me perplexed with a hundred contrary imaginations.

For this *Philautus* thought I, that eyther I did not hit the queftion which fhe would, or that I hit it too full againft hir will : for to faye the trueth, wittie fhe was and fome-what merrie, but God knoweth fo farre from wantonneffe, as my felfe was from wifdome, and I as farre from thinking ill of hir, as I found hir from taking me well.

Thus all night toffing in my bedde, I determined the next daye, if anye opportunitie were offered, to offer alfo my importunate feruice. And found the time fitte, though hir minde fo froward, that to thinke of it my heart throbbeth, and to vtter it, will bleede frefhly.

The next daye I comming to the gallery where fhe was folitaryly walking, with hir frowning cloth, as fick lately of the folens [fullens], vnderftanding my father to bee gone on hunting, and al other the Gentlewomen either walked abro[a]d to take the aire, or not yet re[a]dy to come out of their chambers, I aduentured in one fhip to put all my wealth, and at this time to open my long conce[a]led loue, determining [determined] either to be a Knight as we faye, or a knitter of cappes. And in this manner I vttered my firft fpeach.

L Ady, to make a long preamble to a fhort fute, wold feeme fuperfluous, and to beginne ab-ruptly in a matter of great waight, might be thought abfurde : fo as I am brought into a doubt whether I fhould offend you with too many wordes, or hinder my felfe with too fewe. She not ftaying for a longer treatife brake me of[f] thus roundly.

Gentle-man a fhort fute is foone made, but great matters not eafily graunted, if your requeft be reafo-ble a word wil ferue, if not, a thoufand will not fuffice. Therfore if ther be any thing that I may do you pleafure in, fee it be honeft, and vfe not tedious difcourfes or colours of retorick [Rhethoricke], which though they be thought courtly, yet are they not efteemed neceffary: for the pureft Emerau[l] dfhineth bri[gh]teft when it hath no oyle, and trueth delighteth beft, when it is apparayled worft.

Then I thus replyed.

F Ayre Lady as I know you wife, fo haue I found you curteous, which two qualities meeting in one of fo rare beautie, muft forfhow fome great meruaile, and workes fuch effeＣtes [effeＣt] in thofe, that eyther haue heard of your prayfe, or feene your perfon, yat they are enforced to offer them-felues vnto your feruice, among the number of which your vaffalles, I though leaft worthy, yet moft willing, am nowe come to prof-fer both my life to do you good, and my lyuinges to be at your commaund, which franck offer proceeding

of a faythfull mynde, can neyther be refufed of you,
nor mifliked.   And bicaufe I would cut of[f] fpeaches
which might feeme to fauor either of flattery, or de-
ceipte, I conclude thus, that as you are the firft, vnto
whome I haue vowed my loue, fo you fhall be the
laft, requiring nothing but a friendly acceptaunce of
my feruice, and good-will for the rewarde of it.

Iffyda whofe right eare beganne to gloe, and both
whofe cheekes waxed read [redde], eyther with choler,
or bafhfulneffe, tooke me vp thus for ftumbling.

G Entle-man you make me blufh as much for anger
as fhame, that feeking to prayfe me, and proffer
your felfe, you both bring my good name into queftion, and your ill meaning into difdaine : fo that
thinking to prefent me with your hart, you haue
thruft into my hands the Serpent *Amphisbena*, which
hauing at e[a]ch ende a fting, hurteth both wayes. You
tearme me fayre, and ther-in you flatter, wife and
there-in you meane wittie, curteous which in other
playne words, if you durft haue vttered it, you
would haue named wanton.   Haue you ·thought me
Fidus, fo light, that none but I could fit your loofeneffe ? or am I the wittie wanton which you harped
vpon yefter-night, that would alwayes giue you the
ftynge in the head ? you are much deceyued in mee
Fidus, and I as much in you : for you fhall neuer
finde me for your appetite, and I had thought neuer
to haue tafted you fo vnplefant to mine.   If I be
amiable, I will doe thofe things that are fit for fo
good a face : if deformed, thofe things which fhall
make me faire.   And howfoeuer I lyue, I pardon
your prefumption, knowing it to be no leffe common
in Court' than foolifh, to tell a faire tale, to a foule
Lady, wherein they fharpen I confeffe their wittes,
but fhewe as I thinke fmall wifedome, and you among
the reft, bicaufe you would be accompted courtly, hau=
affayed to feele the veyne you cannot fee, wherein you
follow not the beft Phifitions, yet the moft, who feel-

ing the pulſes, doe alwayes ſay, it betokeneth an
Ague, and you ſeeing my pulſes beat, pleaſauntly
iudge me apte to fall into a fooles Feuer : which leaſte
it happen to ſhake mee heere-after, I am minded to
ſhake you off now, vſing but one requeſt, wher I ſhold
ſeeke oft to reuenge, that is, that you neuer attempt
by word or writing to ſolicite your ſute, which is no
more pleaſaunt to me, then the wringing of a ſtraight
ſhoe.

When ſhe had vttered theſe bitter words, ſhe was
going into hir chamber : but I that now had no ſtaye
of my ſelfe, began to ſtaye hir, and thus agayne to
replye.

I Perceiue *Iffida* that where the ſtreame runneth
ſmootheſt, the water is deepeſt, and where the
leaſt ſmoake is, there to be the greateſt fire : and
wher the mildeſt countenaunce is, there to be the
melancholieſt conceits. I ſweare to thee by the
Gods, and there ſhe interrupted me againe, in this
manner.

F *Idus* the more you ſweare, the leſſe I beleeue
you, for that it is a pra& tiſe in Loue, to haue as
little care of their owne oathes, as they haue of others
honors, imitating *Iupiter*, who neuer kept oath he
ſwore to *Iuno*, thinking it lawfull in loue to haue as
ſmall regard of Religion, as he had of chaſtitie. And
bicauſe I wil not feede you with delayes, nor that
you ſhould comfort your ſelfe with tryall, take this for
a flatte aunſwere, that as yet I meane not to loue any,
and if I doe, it is not you, and ſo I leaue you. But
once againe I ſtayed hir ſteppes being now throughly
heated as well with loue as with cholar, and thus I
thundered.

I F I had vſed the polycie that Hunters doe, in
catching of *Hiena*, it might be alſo, I had now
won you : but comming of the right ſide, I am en-

tangled my felfe, and had it ben on ye left fide, I fhold haue inueigled thee. Is this the guerdon for good wil, is this ye courtefie of Ladies, the lyfe of Courtiers, the foode of louers? Ah *Iffida*, little doft thou know the force of affection, and therfore thou rewardeft it lightly, neither fhewing curtefie lyke a Louer, nor giuing thankes lyke a Ladye. If I fhould compare my bloud with thy birth, I am as noble : if my wealth with thine, as rich : if confer qualities, not much inferiour : but in good wil as farre aboue thee, as thou art beyond me in pride.

Doeft thou difdaine me bicaufe thou art beautiful? why coulours fade, when courtefie flourifheth. Doeft thou reiect me for that thou art wife? why wit hauing tolde all his cardes, lacketh many an ace of wifedome, But this is incident to women to loue thofe that leaft care for them, and to hate thofe that moft defire them, making a fta[c]ke of that, which they fhould vfe for a ftomacher.

And feeing it is fo, better loft they are with a lyttle grudge, then found with much griefe, better folde for forrow, then bought for repentaunce, and better to make no accompt of loue, then an occupation : Wher all ones feruice be it neuer fo great is neuer thought inough, when were it neuer fo lyttle, it is too much. When I had thus raged, she thus re-plied.

F*Idus* you goe the wrong way to the Woode, in ma-king a gappe, when the gate is open, or in feek-ing to enter by force, when your next way lyeth by fa-uor. Where-in you follow the humour of *Aiax*, who loofing *Achilles* fhielde by reafon, thought to winne it againe by rage : but it fell out with him as it doth com-monly, with all thofe yat are cholaricke, that he hurt no man but himfelf, neither haue you moued any to offence but your felfe. And in my minde, though fimple be the comparifon, yet feemely it is, that your anger is lyke the wrangling of children, who when they cannot get what they would haue by playe, they fall to crying, and not vnlyke the vfe of foule gamefters, who hauing loft the maine by true iudgement, thinke to face it out with a falfe oath, and you miffing of my loue, which you required in fport, determine to hit [get] it by fpite. If you haue a commffion to take vp Ladyes, lette me fee it: if a priuiledge, let me know it: if a cuftome, I meane to breake it.

You talke of your birth, when I knowe there is no difference of blouds is [in] a bafen, and as lyttle doe I efteeme thofe that boaft of their aunceftours, and haue themfelues no vertue : as I doe of thofe that crake of their loue, and haue no modeftie. I knowe Nature hath prouided, and I thinke our lawes allow it, that one maye loue when they fee their time, not that they muft loue when others appoint it.

Where-as you bring in a rabble of reafons, as it were to bynde mee agaynft my will, I aunfwere that in all refpectes I thinke you fo farre to excell mee, that I cannot finde in my heart to matche with you. For one of fo great good will as you are, to encounter with one of fuch pride as I am, wer[e] neither commendable nor conuenient, no more then a patch of Fuftian in a Damafke coat.

As for my beautie and wit, I had rather make them better then they are, being now but meane by vertue, then worfe then they are, which woulde then be no[t]hing, by Loue.

Now wher-as you bring in (I know not by what proofe, for I thinke you were neuer fo much of womens counfells) that there women beft lyke, where they be leaft beloued, then ought the[y] more to pitie vs, not to oppreffe vs, feeing we haue neither free will to chufe, nor fortune to enioy. Then *Fidus* fince your eyes are fo fharpe, that you cannot onely looke through a Milftone, but cleane through the minde, and fo cunning that you can leuell at the difpofitions of women whom you neuer knew, me thinketh you fhold vfe the meane, if you defire to haue the ende, which is to hate thofe

T

whom you would faine haue to loue you, for this haue you fet for a rule (yet out of fquare) that women then loue moſt, when they be loathed moſt. And to the ende I might ſtoope to your lure, I pray [you] begin to hate me, that I may loue you.

Touching your loofing and finding, your buying and fellyng, it much ſkilleth not, for I had rather you ſhoulde loofe me fo you might neuer finde me againe, then finde me that I ſhould thinke my felfe loſt : and rather had I be folde of you for a penny, then bought for you with a pound. If you meane either to make an Art or an Occupation of Loue, I doubt not but you ſhal finde worke in the Court fufficient : but you ſhal not know the length of my foote, vntill by your cunning you get commendation. A Phrafe now there is which be-longeth to your Shoppe boorde, that is, to make loue, and when I ſhall heare of what faſhion it is made, if I like the pattorn [patterne], you ſhall cut me a partlet : fo as you cut it not with a paire of left handed ſheeres. And I doubte not though you haue marred your firſt loue in the making, yet by the time you haue made three or foure loues, you will proue an expert work-manne : for as yet you are like the Taylours boy, who thinketh to take meafure before he can handle the ſheeres.

And thus I proteſt vnto you, bicaufe you are but a younge begynner, that I will helpe you to as much cuſtome as I canne, fo as you will promyfe mee to fowe no falfe ſtitches, and when myne old loue is worne thread-bare, you ſhall take meafure of a newe.

In the meane feafon do not difcourage your felf. *Appelles* was no good Paynter the firſt day : For in euery occupation one muſt firſt endeauour to beginne. He that will fell lawne muſt learne to folde it, and he that will make loue, muſt learne firſt to courte it.

As ſhe was in this vaine very pleafaunt, fo I think ſhe would haue bene verye long, had not the Gentle-woemen called hir to walk, being fo faire a day : then taking hir leaue very curteouſly, ſhe left me alone, yet turning againe fhe faide : will you not manne vs, *Fidus*, beeing fo proper a man ? Yes quoth I, and without afking to, had you beene a proper woman.   Then fmyling fhee faide : you fhould finde me a proper woman, had you bene a proper work-man.   And fo fhe departed.

Nowe *Philautus* and *Euphues*, what a traunce was I left in, who bewailing my loue, was anfwered with hate : or if not with hate, with fuch a kind of heate, as almoft burnt the very bowels with-in me.   What gre[a]ter difcurtefie could ther poffibly reft in the minde of a Gentle-woman, then with fo many nips, fuch bitter girdes, fuch difdainfull glickes to anfwere him, that honoured hir?   What crueltie more vnfit for fo comely a Lady, then to fpurre him that galloped, or to let him bloud in the hart, whofe veine fhe fhold haue ftanched in the liuer?   But it fared with me as with the herb Bafill, the which ye more it is croufhed, the fooner it fpringeth, or the rue [Rew], which the oftner it is cutte, the better it groweth, or the poppy, which the more it is troden with the feete, the more it florifheth. For in thefe extremities, beaten as it were to the ground with difdain, my loue re[a]cheth to the top of the houfe with hope, not vnlike vnto a Tree, which though it be often felled to the hard roote, yet it buddeth againe and getteth a top.

But to make an ende both of my tale and my forrowes, I will proceede, onely crauing a little pacience, if I fall into mine old paffions : With-that *Philautus* came in with his fpoake, faying : in fayth, *Fidus*, mee thinketh I could neuer be weary in hearing this difcourfe, and I feare me the ende will be to[o] foone, although I feele in my felf the impreffion of thy fo[r]rows. Yea quoth *Euphues*, you fhall finde my friend *Philautus* fo kinde harted, that before you haue done, he will be farther in loue with hir, then you were : for as your Lady faide, *Philautus* will be bound to make loue as warden of yat occupation.   Then *Fidus*, well God graunt *Philautus* better fucceffe than I hadde, which was too badde. For my Father being returned from hunting, and the Gentle-women from walking, the table was couered, and we all fet downe to dinner, none more pleafaunt then *Iffyda*, which would not conclude hir mirth, and I not melancholie, bicaufe I would couer my fadneffe, leaft either fhe might thinke me to doat, or my Father fufpect me to defire hir. And thus we both in table talke beganne to reft. She requefting me to be hir caruer, and I not attending well to that fhe carued [craued], gaue hir falt, which when fhe receiued, fhee gan thus to reply.

I N footh Gentle-manne I feldome eate falte for feare of anger, and if you giue it mee in token that I want witte, then will you make me cholericke before I eate it : for woemen be they neuer fo foolifh, would euer be thought wife.

I ftand [ftaied] not long for mine aunfwere, but as well quickened by hir former talke, and defirous to crye quittaunce for hir prefent tongue, fayd thus.

If to eate ftore of falt caufe one to frette, and to haue no falte fignifie lacke of wit, then do you caufe me to meruaile, that eating no falte you are fo captious, and louing no falt you are fo wife, when in deede fo much wit is fufficient for a woman, as when fhe is in the raine can warne hir to come out of it.

You miftake your ayme quoth *Iffyda*, for fuch a fhowre may fall, as did once into *Danaes* lap, and then yat woman were a foole that would come out of it : but it may be your mouth is out of tafte, therfore you were beft feafon it with falt.

In deede quoth I, your aunfweres are fo frefh, that with-out falt I can hardly fwallow them. Many nips were returned that time betweene vs, and fome fo bitter, that I thought them to proceede rather of mallice, to worke difpite, then of mirth to fhewe difporte.

My Father very defirous to heare queftions afked, willed me after dinner, to vfe fome demaund, which after grace I did in this forte.

L Ady *Iffyda*, it is not vnlik[e]ly but yat you can aunfwer a queſtion as wiſely, as the laſt nyght you
aſked one wilylie, and I truſt you wil be as ready to
refolue any doubt by entreatie, as I was by commaundement.

There was a Lady in *Spaine*, who after the diſeaſe [de-
ceaſe] of hir Father hadde three futors, (and yet neuer a
good Archer) the one excelled in all giftes of the bodye,
in-fomuch that there could be nothing added to his per-
feƈtion, and fo armed in all poyntes, as his very lookes
were able to pearce the heart of any Ladie, eſpecially
of fuch a one, as ſeemed hir felfe to haue no leſſe
beautie, than ſhe had perfonage.

For that, as betweene the fimilitude of manners
there is a friendſhip in euerie refpeƈte abfolute : fo in
the compoſition of the bodye there is a certaine loue
engendred by one[s] looke, where both the bodyes refemble each other as wouen both in one lombe [loome].
The other hadde nothing to commend him but a quicke
witte, which hee hadde alwayes fo at his will, that nothing could be fpoken, but he would wreſt it to his
owne purpofe, which wrought fuch delight in this
Ladye, who was no leſſe wittie than hee, that you
woulde haue thought a mariage to be folempnized
before the match could be talked of.   For there is nothing in loue more requifite, or more deleƈtable, then
pleafaunt and wife conference, neyther canne there
aryfe any ſtorme in loue which by witte is not turned
to a calme.

The thirde was a Gentle-man of great poſſeſſions,
large reuenues, full of money, but neither the wifeſt
that euer enioyed fo much, nor ye proper[e]ſt that euer
defired fo much, he had no plea in his fute, but gyllt
which rubbed well in a hoat hand is fuch a greafe as will
fupple a very hard heart.   And who is fo ignorant that
knoweth not, gold [to] be a key for euery locke, chief-
lye with his Ladye, who hir felfe was well ſtored, and
are [as] yet infeƈted with a defyre of more, that ſhee could
not but lende him a good countenaunce in this match.

Now Lady *Iffida*, you are to determine this *Spaniſh* bargaine, or if you pleaſe, we wil make it an *Engliſh* controuerſie : ſuppoſing you to be the Lady, and three ſuch Gentlemen to come vnto you a wo[o]ing. In faith who ſhould be the ſpeeder?

Entleman (quoth *Iffida*) you may aunſwere your owne queſtion by your owne argument if you would, for if you conclude the Lady to be beautiful, wittie and wealthy, then no doubt ſhe will take ſuch a one, as ſhould haue comelyneſſe of body, ſharpeneſſe of wit, and ſtore of riches : Otherwiſe, I would condempne that wit in hir, which you ſeeme ſo much to commend, hir ſelfe excelling in three qualyties, ſhee ſhould take one, which was endued but with one : in perfeςt loue the eye muſt be pleaſed, the eare delighted, the heart comforted : beautie cauſeth the one, wit the other, wealth the third.

To loue onely for comelyneſſe, were luſt : to lyke for wit onely, madneſſe : to deſire chiefly for goods, couetouſneſſe : and yet can there be no loue with-out beautie, but we loath it : nor with-out wit, but wee ſcorne it : nor with-out riches, but we repent it. Euery floure hath his bloſſome, his fauour, his ſappe : and euery deſire ſhould haue to feede the eye, to pleaſe the wit, to maintaine the roote.

Ganimedes maye caſt an amiable countenaunce, but that feedeth not : *Vlyſſes* tell a wittie tale, but that fatteth not : *Cræſus* bring bagges of gold, and that doth both : yet with-out the ayde of beautie he cannot beſtow it, and with-out wit he knowes not how to vſe it. So that I am of this minde, there is no Lady but in hir choyce wil be ſo reſolute, that either ſhe wil lyue a virgin till ſhe haue ſuch a one, as ſhall haue all theſe three properties, or els dye for anger, if ſhe match with one that wanteth any one of them.

I perceiuing hir to ſtand ſo ſtifly, thought if I might to remoue hir footing, and replyed againe.

Ady you now thinke by pollicie to ſtart, where
you bound me to aunſwere by neceſſitie, not
ſuffering me to ioyne three flowers in one Noſegay, but
to chuſe one, or els to leaue all. The lyke muſt I
craue at your hands, that if of force you muſt conſent
to any one, whether would you haue the proper man,
the wiſe, or the rich.

She as not without an anſwere, quickly requited me.
Lthough there be no force, which may compel
me to take anye, neither a profer, where-by I
might chuſe all : Yet to aunſwere you flatly, I woulde
haue the wealthieſt, for beautie without riches, goeth
a begging, and wit with-out wealth, cheapeneth all
things in the Faire, but buyeth nothing.

Truly Lady quoth I, either you ſpeake not as you
think, or you be far ouerſhot, for me thinketh, that he
yat hath beautie, ſhal haue money of ladyes for almes,
and he that is wittie wil get it by craft : but the rich
hauing inough, and neither loued for ſhape nor ſence,
muſt either keepe his golde for thoſe he knowes not, and
[or] ſpend it on them that cares not. Well, aunſwered
Iffida, ſo many men, ſo many mindes, now you haue
my opinion, you muſt not thinke to wring me from it,
for I had rather be as all women are, obſtinate in mine
owne conccipt, then apt to be wrought to others con-
ſtructions.

My father liked hir choyce, whether it were to flatter hir, or for feare to offend hir, or that he loued money himſelfe better then either wit or beautie. And
our concluſions thus ended, ſhe accompanied with hir
gentlewomen and other hir feruaunts, went to hir Vn-
cles, hauing tar[r]ied a day longer with my father, then
ſhe appoynted, though not ſo manye with me, as ſhee
was welcome.

Ah *Philautus*, what torments diddeſt thou thinke
poore *Fidus* endured, who now felt the flame euen to
take full holde of his heart, and thinking by ſolitari-
neſſe to driue away melancholy, and by imagination to
forget loue, I laboured no otherwiſe, then he that to

haue his Horfe ftande ftill, pricketh him with the fpurre,
or he that hauing fore eyes rubbeth them with falt
water. At the laft with continual abftinence from meat,
from company, from fleepe, my body began to confume, and my head to waxe idle, infomuch that the fufte-
nance which perforce was thruft into my mouth, was
neuer difgefted, nor ye talke which came from my adle
braines liked : For euer in my flumber me thought
Iffida prefented hir felf, now with a countenance pleafaunt and merry, ftreight-waies with a colour full of
wrath and mifchiefe.

My father no leffe forrowfull for my difeafe, then
ignorant of ye caufe, fent for diuerf[e] Phifitions, among
the which thei came an *Italian,* who feeling my pulfes,
cafting my water, and marking my lookes, commaunded
the chamber to be voyded, and fhutting the doore applyed this medicine to my malady.   Gentleman, there
is none that can better heale your wound than he yat
made it, fo that you fhould haue fent for *Cupid,* not
Aefculapius, for although they be both Gods, yet will
they not meddle in each others office. *Appelles* wil
not goe about to amend *Lifippus* caruing, yet they
both wrought *Alexander* : nor *Hippocrates* bufie him-
felf with *Ouids* art, and yet they both defcribed *Venus.*
Your humour is to be purged not by the Apothecaries
confe&ions, but by the following of good counfaile.
You are in loue *Fidus*?   Which if you couer in a
clofe cheft, will burne euery place before it burft the
locke.   For as we know by Phifick that poyfon wil
difperfe it felfe into euery veyne, before it part the
hart : fo I haue heard by thofe yat in loue could fay
fomwhat, that it maimeth euerye parte, before it kill the
Lyuer.   If therefore you will make me priuie to all
your deuifes, I will procure fuch meanes, as you fhall
recouer in fhort fpace, otherwife if you feeke to conceale the partie, and encreafe your paffions, you fhall
but fhorten your lyfe, and fo loofe your Loue, for
whofe fake you lyue.

When I heard my Phyfition fo pat to hit my difeafe,

I could not diffemble with him, leaft he fhould bewray it, neither would I, in hope of remedy.

Vnto him I difcourfed the faithfull loue, which I bore to *Iffida*, and defcribed in euery perticular, as to you I haue done. Which he hearing, procured with in one daye, Lady *Iffida* to fee me, telling my Father, that my difeafe was but a confuming Feuer, which he hoped in fhort time to cure.

When my Lady came, and faw me fo altered in a moneth, wafted to the harde bones, more lyke a ghoaft then a lyuing creature, after many words of comfort (as women want none about ficke perfons) when fhe faw opportunitie, fhe afked me whether the *Italian* wer[e] my meffenger, or if he were, whether his embaf-fage were true, which queftion I thus aunfwered.

Lady to diffemble with the worlde, when I am departing from it, woulde profite me nothing with man, and hinder me much with God, to make my deathbed the place of deceipt, might haften my death, and encreafe my daunger.

I haue loued you long, and now at the length [I] muft leaue you, whofe harde heart I will not impute to difcurtefie, but deftinie, it contenteth me that I dyed in fayth, though I coulde not liue in fauour, neyther was I euer more defirous to begin my loue, then I am now to ende my life. Thinges which cannot be altered are to be borne, not blamed : follies paft are fooner remembred then redreffed, and time loft [paft] may well be repented, but neuer recalled. I. will not recount the paffions I haue fuffered, I think the effect fhow them, and now it is more behoo[ue]full for me to fall to praying for a new life, then to remember the olde : yet this I ad[de] (which though it merit no mercy to faue, it deferueth thankes of a friend) that onely I loued thee, and liued for thee, and nowe dye for thee. And fo turning on my left fide, I fetched a deepe figh.

Iffyda the water ftanding in hir eyes, clafping my

298   *Euphues and his England.*

hand in hirs, with a fadde countenaunce anfwered mee thus.

MY good *Fidus*, if the encreafing of my forrowes, might mittigate the extremitie of thy ficknes, I could be content to refolue my felfe into teares to ridde thee of trouble : but the making of a frefh wound in my body, is nothing to the healing of a feftred fore in thy bowelles: for that fuch difeafes are to be cured in the end, by the names of their originall. For as by Bafill the Scorpion is engendred, and by the meanes of the fame hearb deftroyed : fo loue which by time and fancie is bred in an idle head, is by time and fancie banifhed from the heart : or as the Salamander which being a long fpace nourifhed in the fire, at the laft quencheth it, fo affe&ion hauing taking holde of the fancie, and liuing as it were in the minde of the louer, in tra& of tyme altereth and chaungeth the heate, and turneth it to chilneffe.

It is no fmall griefe to me *Fidus*, that I fhould bee thought to be the caufe of thy languifhing, and cannot be remedy of thy difeafe. For vnto thee I will reueale more then either wifdome would allowe, or my modeftie permit.

And yet fo much, as may acquit me of vngratitude towards thee, and ridde thee of the fufpition concieued of me.

SO it is *Fidus* and my good friende, that about a two yeares paft, ther was in court a Gentleman, not vnknown vnto thee, nor I think vnbeloued of thee, whofe name I will not conceale, leaft thou fhouldeft eyther thinke me to forge, or him not worthy to be named. This Gentleman was called *Thirfus*, in all refpe&es fo well qualified as had he not beene in loue with mee, I fhould haue bene enamoured of him.

But his haftineffe preuented my heate, who began to fue for that, which I was ready to proffer [offer], whofe fweete tale although I wifhed it to be true, yet at the firft I could not beleeue it : For that men in matters of loue haue as many wayes to deceiue, as they haue wordes to vtter.

I feemed ftraight laced, as one neither accuftomed to fuch fuites, nor willing to entertaine fuch a feruant, yet fo warily, as putting him from me with my little finger, I drewe him to me with my whole hand.

For I ftoode in a great mam[m]ering, how I might behaue my felfe, leaft being too coye he might thinke mee proud, or vfing too much c[o]urtefie, he might iudge mee wanton. Thus long time I held him in a doubt, thinking there-by to haue iuft tryall of his faith, or plaine knowledge of his falfhood. In this manner I led my life almoft one yeare, vntill with often meeting and diuers conferrences, I felt my felfe fo wounded, that though I thought no heauen to my happe, yet I lyued as it were in hell till I had enioyed my hope.

For as the tree *Ebenus* though it no way be fet in a flame, yet it burneth with fweete fauors : fo my minde though it could not be fired, for that I thought my felfe wife, yet was it almoft confumed to afhes with pleafaunt delights and fweete cogitations : in-fomuch as it fared with mee, as it doth with the trees ftriken with thunder, which hauing the barkes founde, are brufed in the bodye, for finding my outwarde partes with-out blemyfhe, looking into my minde, coulde not fee it with-out blowes.

I now perceiuing it high time to vfe the Phifition, who was alwayes at hande, determined at the next meeting to conclud[e] fuch faithful and inuiolable league of loue, as neither the length of time, nor the diftance of place, nor the threatning of friendes, nor the fpight of fortune, nor the feare of death, fhould eyther alter or diminifh : Which accordingly was then finifhed, and hath hether-to bene truely fulfilled.

Thirfus, as thou knoweft hath euer fince bene beyonde the Seas, the remembraunce of whofe conſtancie is the onely comfort of my life : neyther do I reioyce in any thing more, then in the fayth of my good *Thirſus.*

Then *Fidus* I appeale in this caſe to thy honeſtie, which ſhall determine of myne honour. Wouldeſt thou haue me inconſtant to my olde friend, and ſayth-full to a newe ? Knoweſt thou not that as the Almond tree beareth moſt fruite when he is olde, ſo loue hath greateſt fayth when it groweth in age. It falleth out in loue, as it doth in Vines, for the young Vines bring the moſt wine but the olde the beſt : So tender loue maketh greateſt ſhowe of bloſſomes, but tryed loue bringeth forth ſweeteſt iuyce.

And yet I will ſay thus much, not to adde courage to thy attemptes, that I haue taken as great delight in thy company, as euer I did in anyes (my *Thirſus* onely excepted) which was the cauſe that oftentymes, I would eyther by queſtions moue thee to talke, or by quarrels incenſe thee to choller, perceiuing in thee a wit aunſwerable to my deſire, which I thought throughly to whet by ſome diſcourſe. But wert thou in comlines *Alexander,* and my *Thirſus, Therſites,* wert thou *Vlyſſes,* he *Mydas,* thou *Crœſus,* he *Codrus,* I would not forſake him to haue thee : no not if I might ther-by prolong thy life, or ſaue mine owne, ſo faſt a roote hath true loue taken in my hart, that the more it is digged at, the deeper it groweth, the oftener it is cut, the leſſe it bleedeth, and the more it is loaden, the better it beareth.

What is there in this vile earth that more com-mendeth a woman then conſtancie ? It is neyther his wit, though it be excellent that I eſteeme, neyther his byrth though it be noble, nor his bringing vppe, which hath alwayes bene courtlye, but onelye his conſtancie and my fayth, which no torments, no tyrant, not death ſhall diſſolue. For neuer ſhall it be ſaid that *Iffyda* was falſe to *Thirſus,* though *Thirſus* bee faythleſſe (which the Gods forfend) vnto *Iffyda.*

For as *Amulius* the cunning painter fo portrayed *Minerua*, that which waye fo-euer one caft his eye, fhe alwayes behelde him : fo hath *Cupid* fo exqui-fetlye drawne the Image of *Thirfus* in my heart, that what way fo-euer I glaunce, mee thinketh hee look-eth ftedfaftlye vppon me : in-fomuch that when I haue fcene any to gaze on my beautye (fimple God wotte though it bee) I haue wifhed to haue the eyes of *Auguftus Cæfar* to dymme their fightes with the fharp and fcorching beames.

Such force hath time and triall wrought, that if *Thirfus* fhoulde dye I woulde be buried with him, imitating the Eagle which *Sefta* a Virgin brought vp, who feeing the bones of the Virgin caft into the fire, threw him felfe in with them, and burnt himfelf with them. Or *Hippocrates* Twinnes, who were borne together, laughed together, wept together, and dyed together.

For as *Alexander* woulde be engrauen of no one man, in a precious ftone, but onely of *Pergotales* : fo would I haue my picture imprinted in no heart, but in his, by *Thirfus*.

Confider with thy felfe *Fidus*, that a faire woman with-out conftancie, is not vnlyke vnto a greene tree without fruit, refembling the Counterfait that *Praxitiles* made for *Flora*, before the which if one ftoode di-rectly, it feemed to weepe, if on the left fide to laugh, if on the other fide to fleepe : where-by he noted the light behauiour of hir, which could not in one conftant fhadow be fet downe.

And yet for ye great good wil thou beareft me, I can not reiect thy feruice, but I will not admit thy loue. But if either my friends, or my felfe, my goods, or my good will may ftande thee in fteede, vfe me, truft mee, commaund me, as farre foorth, as thou canft with modeftie, and I may graunt with mine honour. If to talke with me, or continually to be in thy com-pany, may in any refpect fatiffie thy defire, affure thy felfe, I wil attend on thee, as dilygently as thy Nourfe, and bee more carefull for thee, then thy Phifition. More I can not promife, without breach of my faith, more thou canſt not aſke without the fufpition of folly.

Heere *Fidus* take this Diamond, which I haue h[e]ard olde women fay, to haue bene of great force, againſt idle thoughts, vayne dreames, and phrenticke imaginations, which if it doe thee no good, aſſure thy felfe it can do thee no harme, and better I thinke it againſt fuch enchaunted fantafies, then either *Homers Moly*, or *Plinyes Centaurio*.

When my Lady had ended this ſtraunge difcourfe, I was ſtriken into fuch a maze, that for the fpace almoſt of halfe an houre, I lay as it had ben in a traunce, mine eyes almoſt ſtanding in my head without motion, my face without colour, my mouth without breath, in fo much that *Iffida* began to fcrich[e] out, and call company, which called me alfo to my felfe, and then with a faint and trembling tongue, I vttered thefe words. Lady I cannot vfe as many words as I would, bicaufe you fee I am weake, nor giue fo many thankes as I ſhould, for that you deferue infinite. If *Thirfus* haue planted the Vine, I wil[l] not gather the grapes : neither is it reafon, that he hauing fowed with payne, that I ſhould reape the ple[a]fure. This fufficeth me and delighteth me not a lit[t]le, yat you are fo faithfull, and he fo fortunate. Yet good lady, let me obtain one fmal fute, which derogating nothing from your true loue, muſt needes be lawful, that is, that I may in this my ſickneſſe enioy your company, and if I recouer, be admitted as your feruaunt : the one wil haſten my health, the other prolong my lyfe. She courteoufly graunted both, and fo carefully tended me in my ſickneſſe, that what with hir merry fporting, and good nouriſhing, I began to gather vp my crumbes, and in ſhort time to walke into a gallerie, neere adioyning vnto my chamber, wher ſhe difdained not to lead me, and fo at al times to vfe me, as though I had ben *Thirfus*. Euery euening ſhe wold put forth

either* fome pretie queſtion, or vtter fome me[r]ry conceit, to driue me from melancholy. There was no
broth that would downe, but of hir making, no meat
but of hir dreſſing, no ſleepe enter into mine eyes, but
by hir ſinging, infomuch as ſhe was both my Nurſe,
my Cooke, and my Phiſition. Being thus by hir for
the ſpace of one moneth cheriſhe[d], I waxed ſtrong
and ſo luſtie, as though I had neuer bene ſicke.

NOw *Philautus* iudge not parcially, whether was
ſhe a lady of greater conſtancie towards *Thirſus*,
or courteſie towards me?

Philautus thus aunſwered. Now furely *Fidus* in my
opinion, ſhe was no leſſe to be commended for keeping hir faith inuiolable, then to be praiſed for giuing
ſuch almes vnto thee, which good behauiour, differeth
farre from the nature of our *Italian* Dames, who if
they be conſtant they difpife al other that ſeeme to
loue them. But I long yet to heare the ende, for me
thinketh a [mat]ter begon with ſuch heate, ſhoulde not
ende with a bitter colde.

O *Philautus*, the ende is ſhort and lamentable, but
as it is haue it.

SHe after long recreating of hir ſelfe in the country,
repayred againe to the court, and ſo did I alſo,
wher I lyued, as the Elephant doth by aire, with the
fight of my Lady, who euer vſed me in all hir ſecrets
as one that ſhe moſt truſted. But my ioyes were too
great to laſt, for euen in the middle of my bliſſe, there
came tidings to *Iffida*, that *Thirſus* was ſlayn by the
Turkes, being then in paye with the King of *Spaine*,
which battaile was ſo bloody, that many gentlemen
loſt their lyues.

Iffida ſo diſtraught of hir wits, with theſe newes
fell into a phrenſie, hauing nothing in hir mouth, but
alwayes this, *Thirſus* ſlayne, *Thirſus* ſlayne, euer
d[o]ubling this ſpeach with ſuch pitiful cryes and
ſcri[t]ches, as it would haue moued the ſouldiers of *Vliſſes*

to forrow. At the laft by good keeping, and fuch mean es as by Phificke were prouided, fhe came againe to hir felfe, vnto whom I writ many letters to take patiently the death of him, whofe life could not be recalled, diuers fhe aunfwered, which I will fhewe you at my better leafure.

But this was moft ftraunge, that no fute coulde allure hir againe to loue, but euer fhee lyued all in blacke, not once comming where fhe was moft fought for. But with-in the terme of fiue yeares, fhe began a lyttle to lyften to mine old fute, of whofe faithfull meaning fhe had fuch tryall, as fhe coulde not thinke that either my loue was buylded vppon luft, or deceipt.

But deftenie cut off my loue, by the cutting off hir lyfe, for falling into a hot peftilent feuer, fhe dyed, and how I tooke it, I meane not to tell it* : but forfaking the Court prefently, I haue heere lyued euer fince, and fo meane vntill Death fhall call me.

NOw Gentlemen I haue helde you too long, I feare me, but I haue ended at the laft. You fee what Loue is, begon with griefe, continued with forrowe, ended with death. A paine full of pleafure, a ioye replenifhed with mifery, a Heauen, a Hell, a God, a Diuell, and what not, that either hath in it folace or forrowe? Where the dayes are fpent in thoughts, the nights in dreames, both in daunger, either be- guylyng us of that we had, or promifing vs that we had [haue] not. Full of iealoufie with-out caufe, and voyde of feare when there is caufe : and fo many inconue- niences hanging vpon it, as to recken them all were infinite, and to tafte but one of them, intollerable.

Yet in thefe dayes, it is thought the fignes of a good wit, and the only vertue peculyar to a courtier, For loue they fay is in young Gentlemen, in clownes it is luft, in olde men dotage, when it is in al menne, madneffe.

But you *Philautus*, whofe bloud is in his chiefeft

heate, are to take great care, leaſt being ouer-warmed with loue, it ſo inflame the liuer, as it driue you into a conſumption.

And thus the olde man brought them into dinner, wher they hauing taken their repaſt, *Philautus* aſwell in the name of *Euphues* as his own, gaue this anſwer to the old mans tale, and theſe or the like thankes for his coſt and curteſie.

Father, I thanke you, no leſſe for your talke which I found pleaſaunt, then for your counſell, which I accompt profitable, and ſo much for your great cheere and curteous entertainment as it deſerueth of thoſe that can-not deſerue any.

I perceiue in England the woemen and men are in loue conſtant, to ſtraungers curteous, and bounti-full in hoſpitalitie, the two latter we haue tryed to your coſt, the other we haue heard to your paines, and may iuſtifie them al wherſoeuer we become to your praiſes and our pleaſure. This only we craue, that neceſſitie may excuſe our boldneſſe, and for amendes we will vſe ſuch meanes, as although we can-not make you gaine much, yet you ſhall looſe little.

Then *Fidus* taking *Philautus* by the hand, ſpake thus to them both.

Entle-men and friendes, I am aſhamed to receiue ſo many thankes for ſo ſmall curteſie, and ſo farre off it is for me to looke for amends for my coſt, as I deſire nothing more then to make you ammendes for your company, and your good wills [will] in ac-compting well of ill fare : onely this I craue, that at your returne, after you ſhall be feaſted of great perſon-ages, you vo[u]chſafe to viſitte the cotage of poore *Fidus*, where you ſhall be no leſſe welcome than *Iupiter* was to *Bacchus* : Then *Euphues*.

We haue troubled you too long, and high tyme it is for poore Pilgrimes to take the daye before them, leaſt being be-nighted, they ſtraiue curteſie in an other place, and as we ſay in *Athens*, fiſhe and geſſe in three

U

dayes are ftale : Not-withftanding we will be bold to fee you, and in the meane feafon we thank you, and euer, as we ought, we will pray for you.

Thus after many farewelles, with as many welcomes of the one fide, as thankes of the other, they departed, and framed their fteppes towards London. And to driue away the time, *Euphues* began thus to inftruct *Philautus.*

THou feeft *Philautus* the curtefie of England to furpaffe, and the conftancie (if the olde Gentleman tolde the trueth) to excell, which warneth vs both to be thankfull for the benefits we receiue, and circumfpect in the behauiour we vfe, leaft being vnmindfull of good turnes, we bee accompted ingrate, and being diffolute in our liues, we be thought impudent.

When we come into London, wee fhall walke in the garden of the worlde, where amonge many flowers we fhall fee fome weedes, fweete Rofes and fharpe Nettles, pleafaunt Lillyes and pricking Thornes, high Vines and lowe Hedges. All thinges (as the fame goeth) that maye eyther pleafe the fight, or diflike the fmell, eyther feede the eye with delight, or fill the nofe with infection.

Then good *Philautus* lette the care I haue of thee be in fteede of graue counfell, and my good will towardes thee in place of wifdome.

I hadde rather thou fhouldeft walke amonge the beddes of w[h]ol[e]fome potte-hearbes, then the knottes of pleafaunt flowers, and better fhalt thou finde it to gather Garlyke for thy ftomack, then a fweete Violet for thy fences.

I feare mee *Philautus,* that feeing the amyable faces of the Englyfhe Ladyes, thou wilt caft off[f] all care both of my counfayle and thine owne credit. For wel I know that a frefh coulour doth eafily dim a quicke fight, that a fweete Rofe doth foonest pearce a fine fent, that pleafaunt firroppes doth chiefelieft infecte a delicate tafte, that beautifull woemen do firft of all allure them that haue the wantonneſt eyes and the whiteſt mouthes.

A ſtraunge tree there is, called *Alpina*, which bringeth forth the fayreſt bloſſomes of all trees, which the Bee eyther fuſpecting to be venemous, or miſliking bicauſe it is ſo glorious, neither taſteth it, nor commeth neere it.

In the like caſe *Philautus* would I haue thee to imitate the Bee, that when thou ſhalt beholde the amiable bloſſomes of the *Alpine* tree in any woemanne, thou ſhunne them, as a place infected eyther with poyſon to kill thee, or honnye to deceiue thee : For it were more conuenient thou ſhouldeſt pull out thine eyes and liue with-out loue, then to haue them cleare and be infected with luſt.

Thou muſt chuſe a woeman as the Lapidarie doth a true Saphire, who when he ſeeth it to gliſter, couereth it with oyle, and then if it ſhine, he alloweth it, if not, hee breaketh it: So if thou fall in loue with one that is beautifull, caſt ſome kynde of coulour in hir face, eyther as it were myſlyinge [miſliking] hir behauiour, or hearing of hir lightneſſe, and if then ſhee looke as fayre as before, wooe hir, win hir, and weare hir.

Then my good friende, conſider with thy ſelfe what thou art, an *Italian*, where thou art, in England, whome thou ſhalt loue if thou fall into that vaine, an Aungell : let not thy eye go beyond thy eare, nor thy tongue ſo farre as thy feete.

And thus I coniure thee, that of all thinges that thou refrayne from the hot fire of affection.

For as the precious ſtone *Autharſitis* beeing throwne into the fyre looketh blacke and halfe dead, but being caſt into the water gliſtreth like the Sunne beames : ſo the precious minde of man once put into the flame of loue, is as it were vglye, and loſeth his vertue, but ſprinckled with the water of wiſdome, and deteſtation of ſuch fond delightes, it ſhineth like the golden rayes of *Phœbus*.

And it ſhall not be amiſſe, though my Phiſicke be

fimple, to prefcribe a ftraight diot [diet] before thou fall into thine olde defeafe.

Firft let thy apparell be but meane, neyther too braue to fhew thy pride, nor too bafe to bewray thy pouertie, be as careful to keepe thy mouth from wine, as thy fingers from fyre. Wine is the glaffe of the minde, and the onely fauce that *Bacchus* gaue *Ceres* when he fell in loue : be not daintie mouthed, a fine tafte noteth the fond appetites, that *Venus* fayde hir *Adonis* to haue, who feing him to take chiefeft delight in coaftle [coftlie] cates, fmyling fayd this. I am glad that my *Adonis* hath a fweete tooth in his head, and who knoweth not what followeth. But I will not wade too farre, feeing heeretofore as wel in my cooling card, as at diuers other times, I haue giuen thee a caueat, in this vanity of loue to haue a care : and yet me thinketh the more I warne thee, the leffe I dare truft thee, for I know not how it commeth to paffe, that euery minute I am troubled in minde about thee.

When *Euphues* had ended, *Philautus* thus began.

E Vphues, I thinke thou waft borne with this word loue in thy mouth, or yat thou art bewitched with it in minde, for ther is fcarce three words vttered to me, but the third is Loue : which how often I haue aunfwered thou knoweft, and yet that I fpeake as I thinke, thou neuer beleeueft : either thinking thy felfe, a God, to know thoughts, or me worfe than a Diuell, not to acknowledge them. When I fhall giue anye oc- cafion, warne me, and that I fhould giue none, thou haft already armed me, fo that this perfwade thy felfe, I wil fticke as clofe to thee, as the foale doth to the fhoe. But truely, I muft needes commende the courtefie of *England*, and olde *Fidus* for his conftancie to his Lady *Iffida*, and hir faith to hir friende *Thirfus*, the remem- braunce of which difcourfe didde often bring into my minde the hate I bore to *Lucilla*, who loued all, and was not found faithfull to any. But I lette that paffe, leaft thou come in againe with thy fa-burthen, and hit

me in the teeth with loue, for thou haſt ſo charmed me,
that I dare not ſpeake any word that may be wreſted
to charitie, leaſt thou ſay, I meane Loue, and in truth,
I thinke there is no more difference betweene them,
then betweene a Broome, and a Beeſome.

I will follow thy dyot [diet] and thy counſayle, 1
thanke thee for thy good will, ſo that I wil now walke
vnder thy ſhadowe and be at thy commaundement: Not
ſo aunſwered *Euphues*, but if thou follow me, I dare be
thy warrant we will not offend much. Much talke
ther was in the way, which much ſhortned their way :
and at laſt they came to London, where they met
diuers ſtraungers of their friends, who in ſmall ſpace
brought them familiarly acquainted with certaine Engliſh gentlemen who much delighted in ye company of
Euphues, whom they found both ſober and wiſe, yet
ſome times mer[r]y and pleaſant. They wer brought
into al places of ye citie, and lodged at ye laſt in a
Merchaunts houſe, wher they continued till a certeine
breach. They vſed continually the Court, in ye which
Euphues tooke ſuch delyght, yat he accompted al ye
praiſes he hard of it before, rather to be enuious, then
otherwiſe, and to be parciall, in* not giuing ſo much
as it deſerued, and yet to be pardoned bicauſe they
coulde not. It happened yat theſe Engliſh gentlemen
conducted theſe two ſtraungers to a place, where diuers
gentlewomen wer : ſome courtiers, others of ye country:
Wher being welcome, they frequented almoſt euery
day for ye ſpace of one moneth, enterteining of time
in courtly paſtimes, though not in ye court, inſomuch
yat if they came not, they wer ſent for, and ſo vſed as
they had ben countrymen, not ſtraungers. *Philautus*
with this continual acceſſe and often conference with
gentlewomen, began to weane himſelfe from ye counſaile of *Euphues*, and to wed his eyes to the comelines
of Ladies, yet ſo warily as neither his friend could by
narrow watching diſcouer it, neither did he by any
wanton countenance, bewray it, but carying the Image
of Loue, engrauen in ye bottome of his hart, and the

picture of courtefie, imprinted in his face, he was
thought to *Euphues* courtly, and knowen to himfelfe
comfortleffe. Among a number of Ladyes he fixed his
eyes vpon one, whofe countenaunce feemed to promife
mercy, and threaten mifchief, intermedling a defire of
liking, with a difdain of loue : fhewing hir felfe in courtefie to be familyar with al, and with a certein com[e]ly
pride to accept none, whofe wit wold commonly taunt
without defpite, but not without difport, as one yat
feemed to abhorre loue worfe than luft, and luft worfe
then murther, of greater beautie then birth, and yet of
leffe beautie then honeftie, which gate hir more honor
by vertue then nature could by Arte, or fortune might
by promotion, fhe was redy of anfwer, yet wary : fhril
of fpeach, yet fweet : in al hir paffions fo temperate,
as in hir greateft mirth none wold think hir wanton,
neither in hir deepeft grief folum [fullen], but alwaies to
looke with fo fober cheerfulnes, as it was hardly thought
wher fhe wer more commended for hir grauitie of ye
aged, or for hir courtlines of ye youth : oftentimes delighted to heare difcourfes of loue, but euer defirous
to be inftruĉted in learning : fomwhat curious to keepe
hir beautie, which made hir com[e]ly[e], but more careful
to increafe hir credit, which made hir commendable :
not adding ye length of a haire to courtlines, yat might
detraĉt ye bredth of a haire from chaftitie : In al[l] hir
talke fo pleafant, in al hir lookes fo amiable, fo graue
modeftie ioyned with fo wittie mirth, yat they yat wer
entangled with hir beautie, wer inforced to prefer hir
wit before their wils : and they yat loued hir vertue,
wer compelled to prefer their affeĉtions before hir
wifdome : Whofe rare qualyties, caufed fo ftraunge
euents, yat the wife were allured to vanitie, and the
wantons to vertue, much lyke ye riuer in *Arabia*, which
turneth golde to droffe, and durt to filuer. In conclufion, ther wanted nothing in this Englifh Angell yat
nature might adde for perfeĉtion, or fortune could giue
for wealth, or god doth commonly beftow on mortal
creatures : And more eafie it is in ye defcription of fo

rare a perfonage, to imagine what fhe had not, then to repeat al fhe had. But fuch a one fhe was, as almoft they all are yat ferue fo noble a Prince, fuch virgins cary lights before fuch a *Vefta*, fuch Nymphes, arrowes with fuch a *Diana*. But why go I about to fet hir in black and white, whome *Philautus* is now with all colours importraying in ye Table of his hart. And furely I think by this he is half mad, whom long fince, I left in a great maze. *Philautus* viewing all thefe things, and more then I haue vttered (for yat the louers eye perceth deeper) wythdrew himfelfe fecretly into his lodging and locking his [the] dore, began to debate with himfelfe in this manner.

AH thrice vnfortunate is he that is once faithful, and better it is to be mercileffe fouldiour, then a true louer : the one liueth by an others death, ye other dyeth by his owne life. What ftraunge fits be thefe *Philautus* yat burne thee with fuch a heate, yat thou fhakeft for cold, and all thy body in a fhiuering fweat, in a flaming yce, melteth like wax and hardeneth like the Adama[n]t? Is it loue? then would it were death : for likelyer it is yat I fhould loofe my life, then win my Loue. Ah *Camilla*, but why do I name thee, when thou doft not heare me, *Camilla*, name thee I will, though thou hate me. But alas ye found of thy name doth make me found for grief. What is in me yat thou fhouldeft not difpife, and what is ther not in thee that I fhould not wonder at. Thou a woman, ye laft thing God made, and therefore ye beft. I a man yat could not liue without thee, and therefore ye worft. Al things wer made for man, as a fouereign, and man made for woman, as a flaue. O *Camilla*, woulde either thou hadft ben bred in *Italy*, or I in *England*, or wold thy vertues wer leffe then thy beautie, or my vertues greater then my affections.

I fee that *India* bringeth golde, but England breedeth goodneffe : And had not England beene thruft into a corner of the world it would haue filled ye whole world

with woe. Where fuch women are as we haue talked
of in *Italy*, heard of in *Rome*, read of in *Greece*, but
neuer found but in this Ifland: And for my part (I
fpeake foftly, bicaufe I will not heare my felfe) would
there were none fuch here, or fuch euery wher. Ah
fond *Euphues* my deere friend, but a fimple foole if
thou beleeue now thy cooling Carde, and an obftinate
foole if thou do not recant it. But it may be thou
layeft that Carde for ye eleuation of *Naples* like an
Aftronomer. If it wer fo I forgiue thee, for I muft
beleeue thee, if for the whole world. Behold *England*,
wher *Camilla* was borne, the flower of courtefie, the
picture of comelyneffe : one that fhameth *Venus*, beeing fome-what fairer, but much more vertuous, and
ftayneth *Diana* being as chaft, but much more amiable.
I but *Philautus* ye more beuti[beautie] fhe hath, ye more
pride, and ye more vertue ye more precifenes. The
Pecock is a Bird for none but *Iuno*, the Doue, for none but
Vefta : None muft wear *Venus* in a Tablet, but *Alexander*, none *Pallas* in a ring but *Vlyffes*. For as there
is but one *Phœnix* in the world, fo is there but one
tree in *Arabia*, where-in fhe buyldeth, and as there is
but one *Camilla* to be heard off, fo is ther but one
Cæfar that fhe wil[l] like off. Why then *Philautus* what
refteth for thee but to dye with patience, fe[e]ing thou
mayft not lyue with ple[a]fure. When thy difeafe is fo
daungerous yat the third letting of bloud is not able
to recouer thee, when neither *Ariadnes* thrid [threed],
nor *Sibillas* bough, nor *Medeas* feede, may remedy thy
griefe. Dye, dye, *Philautus*, rather with a fecret fcarre,
then an open fcorne. *Patroclus* can-not mafke in *Achilles*
armour without a maine [maime], nor *Philautus* in the
Englifh Court without a mocke. I but ther is no
Pearle fo hard but Viniger breaketh it, no Diamond fo
ftony, but bloud mollyfieth, no hart, fo ftif but Loue
weakeneth it. And what then? Bicaufe fhee may
loue one, is it neceffarye fhee fhould loue thee?
Bee there not infinite in *England*, who as farre exceede
thee in wealth, as fhe doth all the *Italians* in wifedome,

and are as farre aboue thee in all qualyties of the body, as ſhe is aboue them in all giftes of the mir de ? Doeſt thou not ſee euery minute the noble youth of *England* frequent the Court, with no leſſe courage than thou cowardiſe. If Courtlye brauery, may allure hir, who more gallant, then they ? If perſonage, who more valyant ? If wit [wittie,] who more ſharp, if byrth, who more noble, if vertue, who more deuoute ?

When there are all thinges in them that ſhoulde delyght a Ladye, and no one thing in thee that is in them, with what face *Philautus* canſt thou deſire, which they can-not deſerue, or with what ſeruice deſerue that, whiche ſo manye deſyre before thee ?

The more beautye *Camilla* hath, the leſſe hope ſhouldeſt thou haue : and thinke not but the bayte that caught thee, hath beguiled other Englyſhe-men or now. Infantes they canne loue, neyther ſo hard h[e]arted to deſpyſe it, nor ſo ſymple not to diſcerne it.

It is likely then *Philautus* that the Foxe will let the Grapes hange for the Gooſe, or the Engliſh-man be-queath beautie to the *Italian*? No no *Philautus* aſ-ſure thy ſelfe, there is no *Venus* but ſhe hath hir Temple, where on the one ſide *Vulcan* may knocke but *Mars* ſhall enter : no Sainte but hath hir ſhrine, and he that can-not wynne with a *Pater noſter*, muſt offer a pennye.

And.as rare it is to ſee the Sunne with-out a light, as a fayre woeman with-out a louer, and as neere is Fancie to Beautie, as the pricke to the Roſe, as the ſtalke to the rynde, as the earth to the roote.

Doeſt thou not thinke that hourely ſhee is ſerued and ſued vnto, of thy betters in byrth, thy equal[l]es in wealth, thy* inferiors in no reſpeĉt.

If then ſhe haue giuen hir fayth, dareſt thou call hir honour into ſuſpition of falſhood ?

If ſhe refuſe ſuch vaine delightes, wilt thou bring hir wiſdome into the compaſſe of folly ?

If ſhe loue ſo beautiful a peece, then wil ſhe not be vnconſtant : If ſhe vow virginitie, ſo chaſt a Lady can-not be periured : and of two thinges the one of theſe muſt be true, that eyther hir minde is alreadye ſo weaned from loue, that ſhe is not to be moued, or ſo ſettled in loue, that ſhe is not to be remoued.

I but it maye bee, that ſo younge and tender a heart hath not felte the impreſſion of Loue : I but it cannot bee, that ſo rare perfection ſhould wante that which they all wiſh, affection.

A Roſe is ſweeter in the budde, then full blowne. Young twigges are ſooner bent then olde trees. White Snowe ſooner melted then hard Yce : which proueth that the younger ſhee is, the ſooner ſhe is to bee wooed, and the fayrer ſhee is, the likelier to be wonne. Who will not run with *Atlanta*, though he be lame? Who whould not wraſtle with *Cleopatra*, though he were ſicke? Who feareth to loue *Camilla*, though he were blinde?

Ah beautie, ſuch is thy force, that *Vulcan* courteth *Venus*, ſhe for comlineſſe a Goddeſſe, he for vglineſſe a diuell, more fit to ſtrike with a hammer in his forge, then to holde a Lute in thy chamber.

Whether doſt thou wade *Philautus* in launcing the wound thou ſhouldeſt taint, and pricking the heart which aſketh a plaiſter : for in deciphering what ſhe is, thou haſt forgotten what thou thy ſelfe art, and being dafeled with hir beautie, thou ſeeſt not thine own baſeneſſe. Thou art an *Italian* poore *Philautus*, as much miſliked for the vice of thy countrey, as ſhe meruailed at for the vertue of hirs, and with no leſſe ſhame doſt thou heare, then know with griefe. How if any Engliſh-man be infected with any myſdemeanour, they ſay with one mouth, hee is Italionated : ſo odious is that nation to this, that the very man is no leſſe hated for the name, then the countrey for the manners.

O *Italy* I muſt loue thee, bicauſe I was borne in thee, but if the infection of the ayre be ſuch, as whoſoeuer breede in thee, is poyſoned by thee, then had I rather be a Baſtard to the Turke *Ottomo*, then heire to the Emperour *Nero*.

Thou which here-tofore waſt moſt famous for vic-

tories, art become moſt infamous by thy vices, as much
diſdained now for thy bea[ſt]lines in peace, as once
feared for thy battayles in warre, thy *Cæſar* being
turned to a vicar, thy Conſulles to Cardinalles, thy ſacred
Senate of three hundred graue Counſellors, to a ſhameleſſe Sinod of three thouſand greedy caterpillers. Where
there is no vice puniſhed, no vertue prayſed, where
none is long loued if he do not ill, where none ſhal be
long loued if he do well. But I leaue to name thy ſinnes,
which no Syphers[Ciphers] can number, and I would I
were as free from the infeſtion of ſome of them, as I
am far from the reckoning of all of them, or would I
were as much enuied for good, as thou art pittied
for ill.

Philautus would thou haddeſt neuer liued in *Naples*
or neuer left it. What new ſkirmiſhes doſt thou now
feele betweene reaſon and appetite, loue and wiſdome,
daunger and deſire.

Shall I go and attyre my ſelfe in coſtly apparell,
tuſhe a faire pearle in a Murrians eare cannot make
him white? Shall I ruffle in new deuices, with Chaines,
with Bracelettes, with Ringes and Robes, tuſhe the
precious Stones of *Manſolus* Sepulchre cannot make
the dead carcaſſe ſweete.

Shall I curle my hayre, coulour my face, counterfayte courtlyneſſe? tuſhe there is no paynting can make
a pyſture ſenſible. No no *Philautus*, eyther ſwallowe
the iuyce of *Mandrak[e]*, which maye caſt thee into a
dead ſleepe, or chewe the hearbe Cheruell, which may
cauſe thee to miſtake euery thing, ſo ſhalt thou either dye
in thy ſlumber, or thinke *Camilla* deformed by thy potion.

No I can-not do ſo though I would, neither* would*
I* though* I* could.* But ſuppoſe thou thinke thy ſelfe
in perſonage comely, in birth noble, in wit excellent,
in talke eloquent, of great reuenewes: yet will this only
be caſt in thy teethe as an obloquie, thou art an *Italian.*

I but all that be blacke digge not for coales, all
things that breede in the mudde, are not Euets, all that
are borne in *Italy*, be not ill. She will not think[enquire]

what most are, but enquire what I am. Euerye one
that fucketh a Wolfe is not rauening, ther is no countrey but hath fome as bad as *Italy*, many that haue
worfe, none but hath fome. And canft thou thinke
that an Englifh Gentleman wil fuffer an *Italian* to be
his Riual[l]? No, no, thou muft either put vp a quarrell
with fhame, or trye the Combat with perill. An Englifh man hath three qualyties, he can fuffer no partner
in his loue, no ftraunger to be his equal, nor to be
dared by any. Then *Philautus* be as wary of thy life,
as careful for thy loue : thou muft at *Rome*, reuerence
Romulus, in *Boetia Hercules*, in *Englande* thofe that
dwell there, els fhalt thou not lyue there.

Ah Loue what wrong doeft thou me, which once
beguildeft me with yat I had, and now beheaddeft me
for that I haue not. The loue I bore to *Lucilla* was cold
water, the loue I owe *Camilla* hoate fire, the firfte was
ended with defame, the laft muft beginne with death.

I fee now that as the refiluation of an Ague is defperate, and the fecond opening of a veyne deadly, fo
the renuing of loue is, I know not what to terme it,
worfe then death, and as bad, as what is worft. I perceiue at the laft the punifhment of loue is to liue.
Thou art heere a ftraunger without acquaintance, no
friend to fpeake for thee, no one to care for thee, *Euphues* will laugh at thee if he know it, and thou wilt
weepe if he know it not. O infortunate *Philautus*, born
in the wane of the Moone, and as lykely [like] to
obtain thy wifh, as the Wolfe is to catch [eate] the
Moone. But why goe I about to quench fire with a
fword, or with affection to mortifie my loue ?

O my *Euphues*, would I had thy wit, or thou my
wil. Shall I vtter this to thee, but thou art more likely
to correct my follyes with counfaile, then to comfort
me with any pretie conceit. Thou wilt fay that fhe is
a Lady of great credit, and I heere of no countenaunce.
I but *Euphues*, low trees haue their tops, fmal fparkes
their heat, the Flye his fplene, ye Ant hir gall, *Philautus* his affection, which is neither ruled by reafon, nor

led by appointment. Thou broughteſt me into *Eng-lande Euphues* to fee and am blynde, to feeke aduen-tures, and I haue loſt myſelf, to remedy loue, and I am now paſt cure, much like *Seriphuis* ye [that] ole drudge in *Naples*, who coueting to heale his bleard eye, put it out. My thoughts are high, my fortune low, and I refemble that foolifh Pilot, who hoyfeth vp all his fayles, and hath no winde, and launc[h]eth out his fhip, and hath no water. Ah Loue thou takeſt away my taſt, and prouokeſt mine appetite, yet if *Euphues* would be as willing to further me now, as he was once wily to hin-der me, I fhold think my felf fortunate and all yat are not amorous to be fooles. There is a ſtone in the floud of *Thracia*, yat whofoeuer findeth it, is neuer after grieued, I would I had yat ſtone in my mouth, or that my body were in yat Riuer, yat either I might be with-out griefe, or without lyfe. And with thefe wordes, *Euphues* knocked at the dore, which *Philautus* opened pretending droufineſſe, and excuſing his abfence by Idleneſſe, vnto whom *Euphues* fayd.

What *Philautus* doeſt thou fhunne the Courte, to ſleepe in a corner, as one either cloyed with delight, or hauing furfetted with defire, beleeue me *Philautus* if the winde be in that doore, or thou fo deuout to fall from beautie to thy beads, and to forfake ye court to lyue in a Cloiſter, I cannot tel whether I fhould more wonder at thy fortune, or prayfe thy wifedome, but I feare me, if I liue to fee thee fo holy, I fhall be an old man before I dye, or if thou dye not before thou be fo pure, thou fhalt be more meruayled at for thy years, then eſteemed for thy vertues. In footh my good friende, if I fhould tarry a yeare in *England*, I could not abide an houre in my chamber, for I know not how it com-meth to paſſe, yat in earth I thinke no other Paradife, fuch varietie of delights to allure a courtly eye, fuch rare puritie to draw a well difpofed minde, yat I know not whether they be in *Englande* more amorous or ver-tuous, whether I fhoulde thinke my time beſt beſtowed, in viewing goodly Ladies, or hearing godly leſſons. I

had thought no woman to excel *Liuia* in ye world, but
now I fee yat in *England* they be al as good, none
worfe, many better, infomuch yat I am enforced to
thinke, yat it is as rare to fee a beautifull woman in
England without vertue, as to fee a faire woman in
Italy without pride. Curteous they are without coynes, but not without a care, amiable without pride, but
not without courtlines : mer[r]y without curiofitie, but
not without meafure, fo yat conferring ye Ladies of
Greece, with ye ladies of *Italy*, I finde the beft but indifferent, and comparing both countries with ye Ladies
of *England*, I accompt them al ftark naught. And
truly *Philautus* thou fhalt not fhriue me like a ghoftly
father, for to thee I will confeffe in two things my ex-
treme folly, ye one in louing *Lucilla*, who in comparifon of thefe had no fpark of beautie, ye other for
making a cooling card againft women, when I fee thefe
to haue fo much vertue, for yat in the firft I muft
acknowledge my iudgement raw, to difcerne fhadowes,
and rafh in the latter to giue fo peremtory fentence, in
both I thinke my felfe, to haue erred fo much, that I
recant both, beeing ready to take any penaunce thou
fhalt enioyne me, whether it be a faggot for Herefie,
or a fine for Hipocrifie. An Hereticke I was by mine
inuectiue againft women, and no leffe then an Hipo-
crite for diffembling with thee, for nowe *Philautus* I
am of that minde that women, but *Philautus* taking
holde of this difcourfe, interrupted him with a fodaine
reply, as followeth.

STaye *Euphues*, I can leuell at the thoughtes of thy
heart by the words of thy mouth, for that commonly the tongue vttereth the minde, and the out-ward
fpeach bewrayeth ye inward fpirit. For as a good roote
is knowen by a faire bloffome, fo is the fubftaunce of
the heart noted by ye fhew of the countenaunce. I can
fee day at a little hole, thou muft halt cunningly if thou
beguile a Cripple, but I cannot chufe but laugh to fee
thee play with the bayt, that I feare thou haft fwallowed,

thinking with a Myſt, to make my ſight blynde, bicauſe
I ſhold not perceiue thy eyes bleared, but in faithe
Eupheus, I am nowe as well acquainted with thy conditions as with thy perſon, and vſe hath made me ſo
expert in thy dealyngs, that well thou mayeſt iuggle
with the world, but thou ſhalt neuer deceiue me.

A burnt childe dreadeth the fire, he that ſtumbleth
twice at one ſtone is worthy to breake his ſhins, thou
mayſt happely forſweare thy ſelfe, but thou ſhalt neuer
delude me. I know thee now as readely by thy viſard
as thy viſage: It is a blynde Gooſe that knoweth
not a Foxe from a Fearne-buſh, and a fooliſh fellow
that cannot diſcerne craft from conſcience, being once
couſened. But why ſhould I lament thy follyes with
griefe, when thou ſeemeſt to colour them with deceite.
Ah *Euphues* I loue thee well, but thou hateſt thy ſelfe,
and ſeekeſt to heape more harms on thy head by a
little wit, then thou ſhalt euer claw of by thy great wiſdom, al fire is not quenched by water, thou haſt not loue
in a ſtring, affeċtion is not thy ſlaue, you [thou] canſt not
leaue when thou liſteſt. With what face *Euphues* canſt
thou returne to thy vomit, ſeeming with the greedy
hounde to lap vp that which thou diddeſt caſt vp. I
am aſhamed to rehearſe the tearmes that once thou
diddeſt vtter of malice againſt women, and art thou not
aſhamed now again to recant them? they muſt needs
think thee either enuious vpon ſmal occaſion, or amou-
rous vpon a light cauſe, and then will they all be as
ready to hate thee for thy ſpight, as to laugh at thee
for thy looſeneſſe. No *Euphues* ſo deepe a wound cannot be healed with ſo light a playſter, thou maiſt by
arte recouer the ſkin, but thou canſt neuer couer the
ſkarre, thou maiſt flatter with fooles bicauſe thou art
wiſe, but the wiſe will euer marke thee for a foole.
Then ſure I cannot ſee what thou gaineſt if the ſimple
condemne thee of flatterie, and the graue of folly. Is
thy cooling Carde of this propertie, to quench fyre in
others, and to kindle flames in thee? or is it a whetſtone to make thee ſharpe and vs blunt, or a ſword to

cut wounds in me and cure them in *Euphues*? Why
didſt thou write that agaynſt them thou neuer thoughteſt,
or if thou diddeſt it, why doeſt thou not follow it?
But it is lawfull for the Phiſition to ſurfet, for the
ſhepheard to wander, for *Euphues* to preſcribe what he
will, and do what he lyſt.

The ſick patient muſt keepe a ſtraight diot [dyet], the
ſilly ſheepe a narrow folde, poore *Philautus* muſt beleeue
Euphues and all louers (he onelye excepted) are cooled
with a carde of teene [tenne], or rather fooled with a
vaine toy.   Is this thy profeſſed puritie to crye *peccaui*?
thinking it as great ſinne to be honeſt, as ſhame not to
be amorous, thou that diddeſt blaſpheme the noble ſex
of women with-out cauſe, doſt thou now commit Idolatrie with them with-out care? obſeruing as little
grauitie then in thine vnbrideled furie, as you [thou] doſt
now reaſon by thy diſordinate fancie.   I ſee now that
there is nothing more ſmooth then glaſſe, yet nothing
more brittle, nothing more faire then ſnow, yet nothing
les firm, nothing more fine then witte, yet nothing
more fickle.   For as *Polypus* vpon what rock ſoeuer
he liketh, turneth himſelfe into the ſame likeneſſe, or
as the bird *Piralis* ſitting vpon white cloth is white,
vpon greene, greene, and changeth hir coulour with
euery cloth, or as our changeable ſilk, turned to ye
Sunne hath many coulours, and turned backe the contrary, ſo wit ſhippeth it ſelf to euery conceit being
conſtant in nothing but inconſtancie.   Wher is now
thy conference with *Atheos*, thy deuotion, thy Diuinitie? Thou ſayeſt that I am fallen from beautie to my
beades, and I ſee thou art come from thy booke to
beaſtlines, from coting of ye ſcriptures, to courting with
Ladies, from *Paule* to *Ouid*, from the Prophets to
Poets, reſembling ye wanton *Diophantus*, who refuſed
his mothers bleſſing, to heare a ſong, and thou for-
ſakeſt Gods bleſſing to ſit in a warme Sunne.   But
thou *Euphues* thinkeſt to haue thy prerogatiue (which
others will not graunt thee for a priuiledge) that vnder the
couler [colour] of wit, thou maiſt be accounted wiſe and,

being obſtinate, thou art to be thought ſinguler. There is no coyne good ſiluer, but thy half-penny, if thy glaſſe gliſter it muſt needs be gold, if you [thou] ſpeak a ſentence it muſt be a law, if giue a cenſer an oracle, if dreame a Prophecie, if conieçture a truth : inſo-much, yat I am brought into a doubt, whether I ſhould more lament in thee, thy want of gouer[ne]ment, or laugh at thy fained grauity : But as that rude Poette *Cherilus* hadde nothing to be noted in his verſes, but onely the name of *Alexander*, nor that rurall Poet *Daretus* any thing to couer his deformed ape, but a white curtain, ſo *Euphues* hath no one thing to ſha-dow his ſhameleſſe wickednes, but onely a ſhew of wit. I ſpeake al this *Euphues*, not that I enuie thy eſtate, but that I pitty it, and in this I haue diſcharged the duetye of a friend, in that I haue not wincked at thy folly. Thou art in loue *Euphues*, contrarie to thine o[a]th, thine honor, thine honeſtie, neither would any profeſſing that thou doeſt, liue as thou doeſt, which is no leſſe grief to me than ſhame to thee : excuſe thou maiſt make to me, bicauſe I am credulous, but amends to the world thou canſt not frame, bicauſe thou art come out of *Greece*, to blaſe thy vice in *England*, a place too honeſt for thee, and thou too diſhoneſt for any place. And this my flat and friendly de[a]ling if thou wilt not take as I meane, take as thou wilt : I feare not thy force, I force not thy friendſhip : And ſo I ende.

Euphues not a little amaſed with the diſcurteous ſpeach of *Philautus*, whome he ſawe in ſuch a burn-ing feuer, did not applye warme clothes to continue his ſweate, but gaue him colde drink to make him ſhake, eyther thinking ſo ſtraunge a maladie was to be cured with a deſperate medicine, or determining to vſe as little arte in Phiſicke, as the other did honeſtie in friendſhippe, and therfore in ſteede of a pyll to purge his hotte bloud, he gaue him a choake-peare to ſtoppe his breath, replying as followeth.

I had thought *Philautus*, that a wounde healing ſo

x

faire could neuer haue bred to a Fiſtula, or a bodye
kept ſo well from drinke, to a dropſie, but I well perceiue that thy fleſhe is as ranke as the wolues, who as
ſoone as he is ſtricken recouereth a ſkinne, but rank-
leth inwardly vntill it come to the lyuer, and thy
ſtomacke as queſie as olde *Neſtors,* vnto whome pappe
was no better then poyſon, and thy body no leſſe diſ-
tempered then *Hermogineus,* whom abſtinence from
wine, made oftentimes dronken. I ſee thy humor is
loue, thy quarrell ie[a]louſie, the one I gather by thine
addle head, thy other by they ſuſpicious nature : but
I leaue them both to thy will and thee to thine owne
wickedneſſe.   Pretily to cloake thine own folly, thou
calleſt me theeſe firſt, not vnlike vnto a curſt wife,
who deſeruing a check, beginneth firſt to ſcolde.

There is nothing that can cure the kings Euill, but
a Prince, nothing eaſe a pluriſie but letting bloud,
nothing purge thy humour, but that which I cannot
giue thee, nor thou gette of any other, libertie.

Thou ſeemeſt to coulour craft by a friendly kindnes,
taking great care for my bondage, that I might not
diſtruſt thy follies, which is, as though the Thruſh in
the cage ſhould be ſory for the Nightingale which
ſingeth on the tree, or the Bear at the ſtake lament
the miſhap of the Lion in the foreſt.

But in trueth *Philautus* though thy ſkin ſhewe thee
a fox, thy little ſkil tryeth thee a ſheep.   It is not the
coulour that commendeth a good painter, but the good
countenance, nor the cutting that valueth the Diamond, but the vertue, nor the gloſe of the tongue that
tryeth a friend, but ye faith.   For as al coynes are not
good yat haue the Image of *Cæſar,* nor al golde that
are coyned with the kinges ſtampe, ſo all is not trueth
that beareth the ſhew of godlines, nor all friends that
beare a faire face, if thou pretende ſuch loue to *Euphues,* carrye thy heart on the backe of thy hand, and
thy tongue in the plame, that I may ſee what is in
thy minde, and thou with thy fingers claſpe thy mouth.
Of a ſtraunger I canne beare much, bicauſe I know

not his manners, of an enimy more, for that al proceedeth of malice, all things of a friend, if it be to
trye me, nothing if it be to betray me : I am of *Sci-
pios* minde, who had rather that *Hannibal* fhould eate
his hart with falt, then *Lælius* grieue it with vnkinde-
neffe : and of the lyke with *Lælius*, who chofe rather to
bee flayne with the *Spaniards*, then fufpecfted of *Scipio*.

I can better take a blifter of a Nettle, then a prick
of a Rofe : more willing that a Rauen fhould pecke
out mine eyes, then a Turtle pecke at them. To dye
of the meate one lyketh not, is better then to furfet
of that he loueth : and I had rather an enemy fhoulde
bury me quicke, then a friende belye me when I am
dead.

But thy friendfhip *Philautus* is lyke a new fafhion,
which being vfed in the morning, is accompted olde
before noone, which varietie of chaunging, being oftentimes noted of a graue Gentleman in [of] *Naples*, who
hauing bought a Hat of the neweft fafhion, and beft
block in all *Italy*, and wearing but one daye, it was
tolde him yat it was ftale, he hung it vp in his ftudie,
and viewing al forts, al fhapes, perceiued at ye laft,
his olde Hat againe to come into the new fafhion,
where-with fmiling to himfelfe he fayde, I haue now
lyued compaffe, for *Adams* olde Apron, muft make
Eue a new Kirtle : noting this, that when no new
thing could be deuifed, nothing could be more new
then ye olde. I fpeake this to this ende *Philautus*,
yat I fee thee as often chaunge thy head as other[s] do
their Hats, now beeing friend to *Aiax*, bicaufe he
fhoulde couer thee with his buckler, now to *Vlyffes*,
that he may pleade for thee with his eloquence, now
to one, and nowe to an other, and thou dealeft with
thy friendes, as that Gentleman did with his felt, for
feeing not my vaine, aunfwerable to thy vanities, thou
goeft about (but yet the neereft way) to hang me vp
for holydayes, as one neither fitting thy head nor
pleafing thy humor, but when *Philautus* thou fhalt fee
that chaunge of friendfhips fhal make thee a fat Calfe,

and a leane Cofer, that there is no more hold in a new
friend then a new fafhion, yat Hats alter as faft as the
Turner can turne his block, and harts as foone as one
can turne his back, when feeing euery one return to
his olde wearing, and finde it ye beft, then compelled
rather for want of others, then good wil of me, thou
wilt retire to *Euphues*, whom thou laydft by ye wals,
and feeke him againe as a new friend, faying to thy
felf, I haue lyued compaffe, *Euphues* olde faith muft
make *Philautus* a new friend. Wherein thou refem-
bleft thofe yat are the firft comming of new Wine, leaue
ye olde, yet finding that grape more pleafaunt then
wholefome, they begin to fay as *Cal[l]if[h]ines* did to
Alexander, yat he had rather carous olde grains with *Diogenes* in his difh, then new grapes with *Alexander* in his
ftanding Cup, for of al Gods fayd he, I loue not
Aefculapius. But thou art willing to chaunge, els
wouldeft thou be vnwilling to quarrel, thou keepeft
only company out of my fight, with *Reynaldo* thy
country-man, which I fufpecting, concealed, and now
prouing it do not care, if he haue better deferued ye
name of a friend then I, god knoweth, but as *Achilles*
fhield being loft on ye feas by *Vliffes*, was toft by ye
fea to ye Tombe of *Aiax*, as a manifeft token of his
right : fo thou being forfaken of *Reynaldo*, wilt bee
found in *Athens* by *Euphues* dore, as ye true owner.
Which I fpeak not as one loth to loofe thee, but careful thou loofe not thy felfe. Thou thinkeft an Apple
maye pleafe a childe, and euery odde aunfwere appeafe
a friend. No *Philautus*, a plaifter is a fmall amends
for a broken head, and a bad excufe, will not purge
an ill accufer. A friend is long a getting, and foone
loft, like a Merchants riches, who by tempeft loofeth
as much in two houres, as he hath gathered together
in twentie yeares. Nothing fo faft knit as glaffe, yet
once broken, it can neuer be ioyned, nothing fuller of
mettal then fteele, yet ouer heated it wil neuer be
hardned, friendfhip is ye beft pearle, but by difdain
thrown into vineger, it burfteth rather in peeces, then

it wil bow to any foftnes. It is a falt fifh yat water
cannot make frefh, fweet honny yat is not made bitter
with gall, harde golde yat is not to bee mollified with
fire, and a miraculous friend yat is not made an enimy
with contempt. But giue me leaue to examine ye
caufe of thy difcourfe to ye quick, and omitting ye
circumftance, I wil[l] to ye fubftance. The onely thing
thou layeft to my charge is loue, and that is a good
ornament, ye reafons to proue it, is my praifing of
women, but yat is no good argument. Am I in loue
Philautus? with whom it fhold be thou canft not coniecture, and that it fhold not be with thee, thou giueft
occafion. *Priamus* began to be iealous of *Hecuba,*
when he knew none did loue hir, but when he loued
many, and thou of me, when thou art affured I loue
none, but thou thy felf euery one. But whether I
loue or no, I cannot liue in quiet, vnleffe I be fit for
thy diet, wherin thou doft imitate *Scyron* and *Pro-
cuftes,* who framing a bed of braffe to their own bignes, caufed it to be placed as a lodging for all paffen-
gers, infomuch yat none could trauel yat way, but he
was enforced to take meafure of their fheets : if he
wer to[o] long for ye bed, they cut off his legs for catching cold, it was no place for a longis [lungis], if to fhort
they racked him at length, it was no pallet for a dwarfe :
and certes *Philautus,* they are no leffe to be difcom-
mended for their crueltie, then thou for thy folly. For
in like manner haft thou built a bed in thine owne
brains, wherin euery one muft be of thy length, if he loue
you [thou] cutteft him fhorter, either with fome od[de]
deuife, or graue counfel, fwearing (rather then thou
wo[u]ldft not be beleued) yat *Protagenes* portrai[e]d
Venus with a fponge fprinkled with fweete water, but if
once fhe wrong it, it would drop bloud : that hir Iuorie
Combe would at the firft tickle the haires, but at the
laft turne all the haires into Adders : fo that nothing
is more hatefull than Loue. If he loue not, then*
[thou] ftretcheft out lyke a Wyre-drawer, making a wire
as long as thy finger, longer then thine arme, pullyng

on with the pincers with the fhoemaker a lyttle fhoe
on a great foote, till thou crack thy credite, as he doth
his ftitches, alleadging that Loue followeth a good
wit, as the fhadowe doth the body, and as requifite
for a Gentleman, as fteele in a weapon.

A wit fayeft thou with-out loue, is lyke an Egge
with-out falte, and a Courtier voyde of affection, like
falt without fauour.   Then as one pleafing thy felfe in
thine owne humour, or playing with others for thine
owne pleafure, thou rolleft all thy wits to fifte Loue
from Luft, as the Baker doth the branne from his
flower, bringing in *Venus* with a Torteyfe vnder hir
foote, as flowe to harmes : hir Chariot drawen with
white Swannes, as the cognifance of *Vefta*, hir birds
to be Pigeons, noting pietie : with as many inuentions
to make *Venus* currant, as the Ladies vfe flights in
Italy to make themfelues counterfaite.   Thus with the
Aegyptian thou playeft faft or loofe, fo that there is
nothing more certeine, then that thou wilt loue, and
nothing more vncerteine then when, tourning at one
time thy tayle to the winde, with the Hedge-hogge,
and thy nofe in the winde, with the Weather-cocke,
in one gale both hoyfing fayle and weighing Anker,
with one breath, making an Alarme and a Parly, dif-
charging in the fame inftaunt, both a Bullet and a falfe
fire.   Thou haft rackte me, and curtalde me, fometimes I was too long, fometimes to[o] fhorte, now to[o]
bigge, then too lyttle, fo that I muft needes thinke
thy bed monftrous, or my body, eyther thy brains out
of temper, or my wits out of tune : infomuch as I can
lyken thy head to *Mercuri[e]s* pipe, who with one ftop
caufed *Argus* to ftare and winke.   If this fault bee in
thy nature, counfel canne do little good, if in thy difeafe, phificke can do leffe : for nature will haue hir courfe,
fo that perfwafions are needeleffe, and fuch a mallady in
the Marrowe, will neuer out of the bones, fo that
medicines are booteleffe.

Thou fayeft that all this is for loue, and that I beeing thy friend, thou art loth to wink at my folly : truly

I fay with *Tully*, with faire wordes thou fhalt yet perfwade me: for experience teacheth me, that ftraight trees
haue crooked rootes, fmooth baites fharpe hookes, that
the fayrer the ftone is in the Toades head, the more peftilent the [her] poyfon is in hir bowelles, that talk the more
it is feafoned with fine phrafes, the leffe it fauoreth of
true meaning. It is a mad Hare yat wil be caught with
a Taber, and a foolifh bird that ftaieth the laying falt
on hir taile, and a blinde Goofe that commeth to the
Foxes fermon, *Euphues* is not entangled with *Philautus*
charmes. If all were in ieft, it was to broad weighing
the place, if in earneft to bad, confidering the perfon,
if to try thy wit, it was folly to bee fo hot, if thy friendfhip, mallice to be fo haftie: Haft thou not read fince
thy comming into *England* a pretie difcourfe of one
Phialo, concerning the rebuking of a friende? Whofe
reafons although they wer but few, yet were they fufficient, and if thou defire more, I could rehearfe infinite. But thou art like the *Epicure*, whofe bellye is
fooner filled then his eye: For he coueteth to haue
twentie difhes at his table, when hee can-not difgeft one
in his ftomacke, and thou defireft manye reafons to
bee brought, when one might ferue thy turne, thinking
it no Rayne-bowe that hath al coulours, nor auncient
armoury, that are not quartered with fundry co[a]tes, nor
perfect rules yat haue not [a] thoufand reafons, and of al
the reafons would thou wouldeft follow but one, not to
checke thy friende in a brauerie, knowing that rebuckes
ought not to weigh a graine more of falt then fuger:
but to be fo tempered, as like pepper they might be
hoat in the mouth, but like treacle wholfom[e] at the
heart: fo fhal they at ye firft make one blufhe if he
were pale, and well confidered better, if he were not
paft grace.

If a friende offend he is to be whipped with a good
Nurfes rodde, who when hir childe will not be ftill,
giueth it together both the twigge and the teate, and
bringeth it a fleepe when it is waywarde, afwell with
rocking it as rating it.

The admonition of a true friend fhould be like the practife of a wife Phifition, who wrappeth his fharpe pils in fine fugar, or the cunning Chirurgian, who launcing ye wound with an yron, immediatly applyeth to it foft lint, or as mothers deale with their children for worms, who put their bitter feedes into fweete reafons, if this order had beene obferued in thy difcourfe, that enterlaching [interlafing] fowre tauntes with fugred counfell, bearing afwell a gentle raine, as vfing a hard fnaffle, thou mighteft haue done more with the whifke of a wand, then now thou canft with the prick of the fpur, and auoyded that which now thou maift not, extream[e] vnkindneffe. But thou art like that kinde Iudge, which *Propertius* noteth, who condempning his friend, caufed him for the more eafe to be hanged with a filken twift. And thou like a friend cutteft my throat with a Rafor, not with a hatchet for my more honor. But why fhould I fet downe the office of a friend, when thou like our *Athenians*, knoweft what thou fhouldeft doe, but like them, neuer doft it. Thou faieft I eat mine own words in prayfing women, no *Philautus* I was neuer eyther fo wicked, or fo witleffe, to recant truethes, or miftake coulours. But this I fay, that the Ladyes in *England* as farre excell all other countryes in vertue, as *Venus* doth all other woemen in beautie. I flatter not thofe of whome I hope to reape benefit, neyther yet fo prayfe them, but that I think them women : ther is no fword made of fteele but hath yron, no fire made of wood but hath fmoake, no wine made of grapes but hath leefe, no woeman created of flefh but hath faultes : And if I loue them *Philautus*, they deferue it.

But it grieueth not thee *Philautus* that they be fayre, but that they are chafte, neyther doft thou like mee the worfe for commending theyr beautie, but thinkeft they will not loue thee well, bicaufe fo vertuous, where-in thou followeft thofe, who better efteeme the fight of the Rofe, then the fauour, preferring fayre weedes before good hearbes, chufing rather to weare a painted flower in their bofomes, then to haue a wholfome roote in their broathes, which refembleth the fafhion of your Maydens in *Italy*, who buy that for the beft cloth yat wil weare whiteft, not that wil laft longeft. There is no more praife to be giuen to a faire face then to a falfe glaffe, for as the one flattereth vs with a vaine fhaddow to make vs proud in our own conceits, fo ye other feedeth vs with an idle hope to make vs peeuifh in our owne contemplations.

Chirurgians affyrme, that a white vaine beeing ftriken, if at the fyrft there fpringe out bloud, it argu- eth a good conftitution of bodye : and I think if a fayre woeman hauing heard the fuite of a Louer, if fhe hlufh at ye firft brunt, and fhew hir bloud in hir face, fhew- eth a well dyfpofed minde : fo as vertuous woemenne I confeffe are for to bee chofen by the face, not when they blufhe for the fhame of fome finne committed. but for feare fhe fhould comitte any, al women fhal be as *Cæfar* would haue his wife, not onelye free from finne, but from fufpition : If fuch be in the Englyfh courte, if I fhould not prayfe them, thou wouldeft [then wouldeft thou] faye I care not for their vertue, and now I giue them their commendation, thou fweareft I loue them for their beautie : So that it is no leffe labour to pleafe thy mind, then a fick mans mouth, who can realifh no- thing by the tafte, not that the fault is in the meat, but in his malady, nor thou like of any thing in thy he[a]d, not that ther is any diforder in my fayings, but in thy fences. Thou doft laft of all obiecte yat which filence might well refolue, that I am fallen from Prophets to Poets, and returned againe with the dog to my vomit, which GOD knoweth is as farre from trueth as I knowe thou art from wifdome.

What haue I done *Philautus*, fince my going from *Naples* to *Athens*, fpeake no more then the trueth, vtter no leffe, flatter me not to make me better then I am, be-lye me not to make me worfe, forge nothing of malice, conceale nothing for loue : did I euer vfe any vnfeemelye talke to corrupt youth ? tell me where : did I euer deceiue thofe that put me in truft ? tell mee whome : haue I committed any fact worthy eyther of death or defame? thou canst not recken what. Haue I abufed my felfe towardes my fuperiors, equalles, or inferiors? I thinke thou canst not deuife when : But as there is no wool fo white but the Diar can make blacke, no Apple fo fweete but a cunning grafter can chaunge into a Crabbe : fo is there no man fo voyde of cryme that a fpightful[fpitefull]tongue cannot make him to be thought a caitife, yet commonly it falleth out fo well that the cloth weareth the better being dyed, and the Apple eateth pleafaunter beeing grafted, and the innocente is more efteemed, and thriueth fooner being enuied for vertue, and belyed for malice. For as he that ftroke *Iafon* on the ftomacke, thinking to kill him, brake his impoftume with ye blow, wherby he cured him : fo oftentimes it fareth with thofe that deale malitioufly, who in fteed of a fword apply a falue, and thinking to be ones Prieft, they become his Phifition. But as the Traytour that clyppeth the coyne of his Prince, maketh it lyghter to be wayed, not worfe to be touched : fo he that by finifter reports, feemeth to pare the credite of his friend, may make him lighter among the common fort, who by weight often-times are deceiued with counterfaites, but nothing empayreth his good name with the wife, who trye all gold by the touch·ftone.

A Straunger comming into the *Capitol* of *Rome* feeing all the Gods to be engrauen, fome in one ftone, fome in an other, at the laft he perceiued *Vulcan*, to bee wrought in Iuory, *Venus* to be carued in Ieate, which long time beholding with great delyght, at the laft he burft out in thefe words, neither can this white Iuory *Vulcan*, make thee a white Smith, neither this faire woman Ieat, make thee a faire ftone. Where-by he noted that no cunning could alter the nature of the one, nor no Nature tranfforme the colour of the other. In lyke manner fay I *Philautus*, although thou haue fhadowed my guiltleffe life, with a defamed counterfait, yet fhall not thy black *Vulcan* make either thy accu-

fations of force, or my innocencie faultie, neither fhal
the white *Venus* which thou haft portrayed vpon the
blacke Ieat of thy malyce, make thy conditions amiable, for *Vulcan* cannot make Iuory blacke, nor *Venus*
chaunge the coulour of Ieat, the one hauing receiued
fuch courfe by Nature, the other fuch force by Vertue.

What caufe haue I giuen thee to fufpect me, and
what occafion haft thou not offered me to deteft thee?
I was neuer wife inough to giue thee counfaile, yet euer
willing to wifh thee well, my wealth fmall to do thee
good, yet ready to doe my beft: Infomuch as thou
couldeft neuer accufe me of any difc[o]urtefie, vnleffe it
were in being more carefull of thee, then of my felfe.
But as all floures [flowers] that are in one Nofegay, are
not of one nature, nor all Rings that are worne vppon
one hande, are not of one fafhion : fo all friendes that af-
fociate at bedde and at boord, are not one of difpofition. *Scipio* muft haue a noble minde, *Lælius* an humble fpirite : *Titus* muft luft after *Sempronia*, *Gifippus*
muft leaue hir : *Damon* muft goe take order for his
lands, *Pithias* muft tarry behinde, as a Pledge for his
life : *Philautus* muft doe what he will, *Euphues* not
what he fhould. But it may be that as the fight of
diuers colours, make diuers beafts madde : fo my prefence doth driue thee into this melancholy. And feeing
it is fo, I will abfent my felfe, hier [hire] an other lodging in *London*, and for a time giue my felfe to my booke,
for I haue learned this by experience, though I be young,
that Bauins be knowen by their bands, Lyons by their
clawes, Cockes by their combes, enuious mindes by
their manners. Hate thee I will not, and truft thee I
may not : Thou knoweft what a friende fhoulde be, but
thou wilt neuer liue to trye what a friend is. Farewell *Philautus*, I wil not ftay to heare thee replye, but
leaue thee to thy lyft, [luft] *Euphues* carieth this Pofie
written in his hande, and engrauen in his heart. *A
faithfull friend, is a wilfull foole.* And fo I taking leaue,
till I heare thee better minded, *England* fhall be my abode
for a feafon, depart when thou wilt, and againe fare-well.

Euphues in a great rage departed, not fuffering *Philautus* to aunfwere one word, who ftood in a maze, after the fpeache of *Euphues*, but taking courage by loue, went immediatelye to the place where *Camilla* was dauncing, and ther wil I leaue him, in a thoufand thoughts, hammering in his head, and *Euphues* feeking a new chamber, which by good friends he quickly got, and there fell to his *Pater nofter*, wher a while I will not trouble him in his prayers.

NOw you fhall vnderftand that *Philautus* fur-thered as well by the opportunitie of the time, as the requefts of certeine Gentlemen his friends, was entreated to make one in a Mafque, which *Philautus* perceiuing to be at the Gentlemans houfe where *Camilla* laye, affented as willyngly to goe, as he defired to fpeede, and all things beeing in a readineffe, they went with fpeede : where beeing welcommed, they daunced, *Philautus* taking *Camilla* by the hande, and as time ferued, began to boord hir in this manner.

IT hath ben a cuftome faire Lady, how commend-able I wil not difpute, how common you know, that Mafquers do therfore couer their faces that they may open their affections, and vnder ye colour of a daunce, difcouer their whole defires : the benefit of which priueledge, I wil not vfe except you graunt it, neither can you refufe, except you break it.   I meane only with queftions to trye your wit, which fhall neither touch your honour to aunfwere, nor my honeftie to afke.

Camilla tooke him vp fhort, as one not to feeke how to reply, in this manner.

GEntleman, if you be leffe, you are too bolde, if fo, too broade, in clayming a cuftome, where there is no prefcription.   I knowe not your name, bicaufe you feare to vtter it, neither doe I defire it, and you feeme to be afhamed of your face, els would you not hide it, neither doe I long to fee it : but as for any cuf-

tome, I was neuer fo fuperftitious, that either I thought
it treafon to breake them, or reafon to keepe them.

As for the prouing of my witte, I had rather you
fhould accompt me a foole by filence, then wife by
aunfwering? For fuch queftions in thefe affemblyes,
moue fufpition where there is no caufe, and therefore
are not to be refolued leaft there be caufe.

Philautus, who euer as yet but played with the bait,
was now ftroke with the hooke, and no leffe delyghted
to heare hir fpeake, then defirous to obtaine his fuite,
trayned hir by the bloud in this fort.

I F the patience of men were not greater then the
peruerfeneffe of women, I fhould then fall from a
queftion to a quarrell, for that I perceiue you draw
the counterfaite of that I would fay, by the conceit of
that you thinke others haue fayd : but whatfoeuer the
colour be, the picture is as it pleafeth the Paynter :
and whatfoeuer were pretended, the minde is as the
h[e]art doth intend. A cunning Archer is not knowen
by his arrow but by his ayme : neither a friendly affection by the tongue, but by the faith. Which if it be
fo, me thinketh common courtefie fhould allow that,
which you feeke to cut off by courtly coyneffe, as one
either too young to vnderftand, or obftinate to ouer-
thwart, your yeares fhall excufe the one, and my humour pardon[the] other.

And yet Lady I am not of that faint minde, that
though I winke with a flafh of lyghtening, I dare not
open mine eyes againe, or hauing once fuffered a repulfe, I fhould not dare to make frefh affault, he that
ftriketh fayle in a ftorme, hoyfeth them higher in a
calm, which maketh me the bolder to vtter that, which
you difdaine to heare, but as the Doue feemeth angry,
as though fhe had a gall, yet yeeldeth at the laft to delight : fo Ladyes pretende a great fkyrmifhe at the firft,
yet are boorded willinglye at the laft.

I meane therefore to tell you this, which is all, that
I loue you : And fo wringing hir by the hand, he
ended : fhe beginning as followeth

Entleman (I follow my firſt tearme) which ſheweth rather my modeſtie then your defart, feeing you refemble thofe which hauing once wet their feete, care not how deepe they wade, or thofe that breaking the yce, weigh not how farre they ſlippe, thinking it law-full, if one ſuffer you to treade awry, no ſhame to goe ſlipſhad [ſlippeſhood]: if I ſhould ſay nothing then would you vaunt that I am wonne : for that they that are ſilent feeme to confent, if any thing, then would you boaft that I would be woed, for that [ye] caſtles that come to parlue [*parle*], and woemen that delight in courting, are willing to yeelde : So that I muſt eyther heare thofe thinges which I would not, and feeme to be taught by none, or to holde you talke, which I ſhould not, and runne into the ſufpition of others. But certainlye if you knewe how much your talke difpleaſeth me, and how lit[t]le it ſhould profit you, you would think the time as vainely loſt in beginning your talke, as I accompt ouer long, vntill you ende it.

If you build vpon cuſtome that Maſkers haue liber-tie to fpeake what they ſhould not, you ſhall know that woemen haue reafon to make them heare what they would not, and though you can vtter by your vifarde what-foeuer it be with-out bluſhing, yet cannot I heare it with-out ſhame. But I neuer looked for a better tale of fo ill a face, you ſay a bad coulour maye make a good countenaunce, but he that conferreth your difordered difcourfe, not your deformed attyre, may rightly faye, that he neuer fawe fo crabbed a vifage, nor hearde fo crooked a vaine. An archer faye you is to be knowne by his ayme, not by his arrowe : but your ayme is fo ill, that if you knewe how farre wide from the white your ſhaft ſticketh, you would here-after rather break your bow, then bend it. If I be too young to vnderſtand your deſtinies, it is a figne I can-not like [looke], if too obſtinate, it is a token I will not : therefore for you to bee difpleafed, it eyther needeth not, or booteth not. Yet goe you farther, thinking to make a great vertue of your little valure, feeing that lightning may caufe you wincke, but it fhall not ftricke you blinde, that a ftorme may make you ftrycke fayle, but neuer cut the maft, thaf a hotte fkyrmifhe may caufe you to retyre, but neuer to runne away : what your cunning is, I knowe not, and likely it is your courage is great, yet haue I heard, that he that hath efcaped burning with lightning, hath beene fpoyled with thunder, and one that often hath wifhed drowning, hath beene hanged once for al, and he that fhrinketh from a bullette in the maine bat-taile, hath beene ftriken with a bil in the rerewarde. You fall from one thing to an other, vfing no decorum, except this, that you ftudy to haue your difcourfe as farre voyde of fence, as your face is of fauor, to the ende, that your diffigured countenaunce might fup-plye the diforder of your ill couched fentences, amonge the which you bring in a Doue with-out a gall, as farre from the matter you fpeake off, as you are from the maft[e]rye you would haue, who although fhe can-not be angry with you in that fhe hath no gall, yet can fhe laugh at you for that fhe hath a fpleene.

I will ende where you beganne, hoping you will beginne where I end, you let fall your queftion which I looked for, and pickt a quarrell which I thought not of, and that is loue : but let hir that is difpofed to aunfwere your quarrell, be curious to demaund your queftion.

And this[thus] Gentle-manne I defire you, all queftions and other quarrelles fet aparte, you thinke me as a friende, fo farre forth as I can graunt with modeftie, or you require with good manners, and as a friende I wifhe you, that you blowe no more this fire of loue, which will wafte you before it warme mee, and make a colde [coale] in you, before it can kindle in me : If you think otherwife I may [canne] afwell vfe a fhift to driue you off, as you did a fhewe to drawe me on. I haue aunfwered your cuftome, leaft you fhould argue me of coynes, no otherwife then I might mine honour faued, and your name vnknowen.

By this time entered an other Mafque, but almoft

after the fame manner, and onely for *Camillas* loue,
which *Philautus* quickly efpyed, and feeing his *Camilla*
to be courted with fo gallant a youth, departed : yet
with-in a corner, to the ende he might decipher the
Gentle-man whom he found to be one of the braueft
youthes in all *England*, called *Surius*, then wounded
with griefe, hee founded with weakneffe, and going to
his chamber beganne a frefhe to recount his miferies
in this forte.

Ah myferable and accurfed *Philautus*, the verye
monfter of Nature and fpeĉtacle of fhame, if thou liue
thou fhalt be defpyfed, if thou dye not myffed, if wo[o]e
poynted at, if win lo[a]thed, if loofe laughed at, bred
either to liue in loue and be forfaken, or die with loue
and be forgotten.

Ah *Camilla* would eyther I had bene born without
eyes not to fee thy beautie, or with-out eares not to
heare thy wit, the one hath enflamed me with the
defire of *Venus*, the other with the giftes of *Pallas*,
both with the fire of loue : Loue, yea loue *Philautus*,
then the which nothing canne happen vnto man more
miferable.

I perceiue now that the Chariotte of the Sunne
is for *Phœbus*, not for *Phaeton*, that *Bucephalus* will
ftoupe to none but *Alexander*, that none can founde
Mercurius pipe but *Orpheus*, that none fhall win *Camillas*
liking but *Surius* a Gentlemanne. I confeffe of greater
byrth then I, and yet I dare fay not of better [greater]
faith. It is he *Philautus* that will fleete all the fat from
thy [the] beard, in-fomuch as fhe will difdaine to looke
vpon thee, if fhe but once thinke vppon him. It is he
Philautus that hath wit to trye hir, wealth to allure hir,
perfonage to entice hir, and all thinges that eyther
Nature or Fortune can giue to winne hir.

For as the *Phrigian* Harmonie being moued to the
Calenes maketh a great noyfe, but being moued to
Apollo it is ftill and quiet : fo the loue of *Camilla*
defired of mee, mooueth I knowe not how manye dif-
cordes, but proued of *Surius*, it is calme, and confenteth.

It is not the fweete flower that Ladyes defyre, but the fayre, whiche maketh them weare that in theyr heades, wrought forth with the needle, not brought forth by Nature : And in the lyke manner they accompte of that loue, whiche arte canne coulour, not that the heart dooth confeffe, where-in they imitate the Maydens as (*Euphues* often hath told mee) of *Athens*, who tooke more delight to fee a frefhe and fine coulour, then to taft a fweete and wholfome firrop.

I but howe knoweft thou that *Surius* fayth is not as great as thine, when thou art affured thy vertue is no leffe then his ? He is wife, and that thou feeft : valyaunt, and that thou feareft : rich, and that thou lackeft : fit to pleafe hir, and difplace thee : and without fpite be it fayd, worthye to doe the one, and willing to attempt the other.

Ah *Camilla, Camilla,* I know not whether I fhould more commend thy beautie or thy wit, neither can I tell whether thy lookes haue wounded me more or thy words, for they both haue wrought fuch an alteration in my fpirites, that feeing thee filent, thy comelyneffe maketh me in a maze, and hearing thee fpeaking, thy wifedome maketh me ftarke madde.

I but things aboue thy height, are to be looked at, not reached at. I but if now I fhould ende, I had ben better neuer to haue begon [begun]. I but time muft weare away loue, I but time may winne it. Hard ftones are pearced with foft droppes, great Oakes hewen downe with many blowes, the ftonieft heart mollyfied by continuall perfwafions, or true perfeueraunce.

If deferts can nothing preuaile, I will practife deceipts, and what faith cannot doe, coniuring fhall. What faift thou *Philautus,* canft thou imagine fo great mifchiefe againft hir thou loueft ? Knoweft thou not, that Fifh caught with medicines, and women gotten with witchcraft are neuer wholefom[e]? No, no, the Foxes wiles fhal neuer enter into ye Lyons head, nor *Medeas* charmes into *Philautus* heart. I, but I haue h[e]ard that extremities are to be vfed, where the meane will not

Y

ferue, and that as in loue ther is no meafure of griefe,
fo there fhould be no ende of guile, of two mifchiefes
the leaft is to be chofen, and therefore I thinke it
better to poyfon hir with the fweet bait of loue, then
to fpoile my felfe with the bitter fting of death.

If fhe be obftinate, why fhould not I be defperate? if
fhe be voyd of pitie, why fhoulde I not be voyde of pietie?
In the ruling of Empires there is required as great policie as prowes [proweffe], in gouerning an Eftate, clofe
crueltie doth more good then open clemencie, for ye
obteining of a kingdome, af well mifchiefe as mercy,
is to be practife[d]. And then in the winning of my Loue,
the very Image of beautie, courtefie and wit, fhall I
leaue any thing vnfought, vnattempted, vndone? He
that defireth riches, muft ftretche the ftring that will
not reach, and practife all kindes of getting. He that
coueteth honour, and can-not clymbe by the ladder,
muft vfe al colours of luftineffe: He that thirfteth for
Wine, muft not care how he get it, but wher he maye
get it, nor he that is in loue, be curious, what meanes
he ought to vfe but re[a]dy to attempt any: For flender
affection do I think that, which either the feare of Law,
or care of Religion may diminifh.

Fye *Philautus*, thine owne wordes condempne thee
of wickedneffe: tufh the paffions I fuftaine, are neither to be quieted with counfaile, nor eafed by reafon:
therefore I am fully refolued, either by Arte to winne
hir loue, or by defpayre to loofe mine owne lyfe.

I haue hearde heere in *London* of an *Italian*, cun-
ning in Mathematicke named *Pfellus*, of whome in
Italy I haue hearde in fuche cafes canne doe much by
Magicke, and will doe all thinges for money, him will
I affaye, as well with golde as other good tournes, and
I thinke there is nothing that can be wrought, but fhal
be wrought for gylt, or good wil, or both.

And in this rage, as one forgetting where hee
was, and whome hee loued, hee went immediately to
feeke Phificke for that, which onely was to bee found
by Fortune.

Ere Gentlemen you maye fee, into what open finnes the heate of Loue driueth man, efpecially where one louing is in difpayre, either of his owne imperfection or his Ladyes vertues, to bee beloued againe, which caufeth man to attempt thofe thinges, that are contrarie to his owne minde, to Religion, to honeftie.

What greater villany can there be deuifed, then to enquire of Sorcerers, South-fayers, Coniurers, or learned Clearkes for the enioying of loue? But I will not refell that heere, which fhall bee confuted heere-after.

Philautus hath foone founde this Gentleman, who conducting him into his ftudie, and demaunding of him the caufe of his comming, *Philautus* beginneth in this manner, as one paft fhame to vnfold his fute.

After *Pfellus* (and Countrey-man,) I neyther doubt of your cunning to fatiffie my requeft, nor of your wifedome to conceale it, for were either of them wanting in you, it might tourne mee to trouble, and your felfe to fhame.

I haue hearde of your learning to be great in Magicke, and fomewhat in Phificke, your experience in both to be exquifit, which caufed me to feeke to you for a remedie of a certeine griefe, which by your meanes maye be eafed, or els no wayes cured.

And to the ende fuch cures may be wrought, God hath ftirred vp in all times Clearkes of greate vertue, and in thefe our dayes men of no fmall credite, among the which, I haue hearde no one, more commended then you, which althoughe happelye your modeftye will denye, (for that the greateft Clearkes doe commonlye diffemble their knowledge) or your precifeneffe not graunt it, for that cunning men are often [more] daungerous, yet the worlde doth well know it, diuers haue tryed it, and I muft needes beleeue it.

Pfellus not fuffering him to raunge, yet defirous to know his arrant, aunfwered him thus.

GEntleman and countryman as you fay, and I be leeue, but of that heereafter : if you haue fo great confidence in my cunning as you protefl, it may bee your ftrong imagination fhall worke yat in you, which my Art cannot, for it is a principle among vs, yat a vehement thought is more auayleable, then ye vertue of our figures, formes, or charecters. As for keeping your counfayle, in things honeft, it is no matter, and in caufes vnlawful, I will not meddle. And yet if it threaten no man harme, and maye doe you good, you fhall finde my fecrecie to be great, though my fcience be fmal, and therefore fay on.

THere is not farre hence a Gentlewoman whom I haue long time loued, of honeft parents, great vertue, and fingular beautie, fuch a one, as neither by Art I can defcribe, nor by feruice deferue : And yet bicaufe I haue heard many fay, that wher cunning muft worke, the whole body muft be coloured, this is hir fhape.

She is a Virgin of the age of eighteene yeares, of ftature neither too high nor too low, and fuch was *Iuno* : hir haire blacke, yet comely, and fuch had *Læda* : hir eyes hafill, yet bright, and fuch were the lyghtes of *Venus*.

And although my fkill in Phifognomie be fmall, yet in my iudgement fhe was borne vnder *Venus*, hir forhead, nofe, lyppes, and chinne, fore-fhewing (as by fuch rules we geffe) both a defire to lyue, and a good fucceffe in loue. In compleftion of pure fanguine, in condition a right Sainte, feldome giuen to play, often to prayer, the firft letter of whofe name (for that alfo is neceffary) is *Camilla.*

THis Lady I haue ferued long, and often fued vnto, in-fomuch that I haue melted like wax againft the fire, and yet liued in the flame with the flye *Pyraufta.* O *Pfellus* the tormentes fuftained by hir prefence, the griefes endured by hir abfence, the pyning thoughtes in the daye, the pinching dreames in the night, the dying life, the liuing death, the ie[a]loufie at all times, and the difpaire at this inftant, can neyther be vttered of me with-out fl[o]udes of teares, nor heard of thee with-out griefe.

No *Pfellus* not the tortures of hell are eyther to be compared, or fpoken of in the refpeɛt of my tor-mentes : for what they all had feuerally, all that and more do I feele ioyntly : In-fomuch that with *Syfiphus* I rolle the ftone euen to the toppe of the Hill, when it tumbleth both it felfe and me into the bottome of hell : yet neuer ceafing I attemp[t]e to renewe my labour, which was begunne in death, and can-not ende in life.

What dryer thirft could *Tantalus* endure then I, who haue almoft euerye houre the drinke I dare not tafte, and the meate I can-not ? In-fomuch that I am torne vpon the wheele with *Ixion*, my lyuer gnawne of the Vultures and Harpies : yea my foule troubled euen with the vnfpeakeable paines of *Megæra, Tifi-phone, Aleɛto* : whiche fecrete forrowes although it were more meete to enclofe them in a Laborinth, then to fette them on a Hill : Yet where the minde is paft hope, the face is paft fhame. It fareth with me *Pfellus* as with the *Auftrich* [Oftridge], who pricketh none but hir felfe, which caufeth hir to runne when fhe would reft : or as it doth with the *Pelicane*, who ftricketh bloud out of hir owne bodye to do others good : or with the Wood Culuer, who plucketh of hir [his] fe[a]thers in winter to keepe others from colde : or as with the Storke, who when fhe is leaft able, carrieth the greateft burthen. So I praɛtife all thinges that may hurt mee to do hir good that neuer regardeth my paynes, fo farre is fhee from rewarding them.

For as it is impoffible for the beft *Adamant* to drawe yron vnto it if the *Diamond* be neere it, fo is it not to bee looked for, that I with all my feruice, fuite, defartes, and what els fo-euer that may draw a woe-manne, fhould winne *Camilla*, as longe as *Surius*, a precious ftone in hir eyes, and an eye fore in mine, bee prefent, who loueth hir I knowe too wel, and fhee

him I feare me, better, which loue wil breed betweene
vs fuch a deadly hatred, that beeing dead, our bloud
cannot bee mingled together like *Florus* and *Aegithus*,
and beeing burnt, the flames fhall parte like *Polinices*
and *Eteocles*, fuch a mortall enmitie is kindled, that
nothing can quench it but death: and yet death fhall
not ende [it].

What counfell canne you giue me in this cafe? what
comfort? what hope?

When *Acontius* coulde not perfwade *Cydippe* to loue,
he practifed fraude. When *Tarquinius* coulde not
winne *Lucretia* by prayer, hee vfed force.

When the Gods coulde not obtaine their defires by
fuite, they turned them-felues into newe fhapes, leauing
nothing vndonne, for feare, they fhould bee vndonne.

The defeafe of loue *Pfellus*, is impatient, the defire
extreame, whofe affaultes neyther the wife can refifl
by pollicie, nor the valiaunt by ftrength.

Iulius Cæfar a noble Conquerour in warre, a graue
Counfaylour in peace, after he had fubdued *Fraunce*,
Germanie, Britaine, Spaine, Italy, Thefalay [*Theffalia*],
Aegipt, yea entered with no leffe puiffaunce then good
fortune into *Armenia*, into *Pontus*, into *Africa*, yeelded
in his chiefeft victories to loue, *Pfellus*, as a thing
fit for *Cæfar*, who conquered all thinges fauing him-felfe,
and a deeper wound did the fmall Arrowe of *Cupid*
make, then all the fpeares of his enimies.

Hannibal lot leffe valiaunt in armes, nor more fortunate in loue, hauing fpoyled *Ticinum, Trebia, Traf-
mena* and *Cannas*, fubmitted him-felfe in *Apulia* to ye
loue of a woman, whofe hate was a terrour to all men,
and became fo bewitched, that neyther the feare of
death, nor the defire of glorye coulde remoue him
from the lappe of his louer.

I omitte *Hercules*, who was conftrained to vfe a diflaffe
for the defire of his loue. *Leander*, who ventured to croffe
the Seaes for *Hero*. *Hyphus* [*Iphis*] that hanged himfelfe, *Pyramus* that killed him-felfe and infinite more,
which coulde not refifl the hot fkyrmifhes of affection.

And fo farre hath this humour crept into the minde, that *Biblis* loued hir Brother, *Myrr[h]a* hir Father, *Canace* hir nephew : In-fomuch as ther is no reafon to be giuen for fo ftraung[e] a griefe, nor no remedie fo vnlawefull, but is to bee fought for fo monftrous a defeafe. My defeafe is ftraung[e], I my felfe a ftraunger, and my fuite no leffe ftraunge then my name, yet leaft I be tedious in a thing that requireth hafte, giue eare to my tale.

I Haue hearde often-tymes that in Loue there are three thinges for to bee vfed, if time ferue, violence, if wealth be great, golde, if neceffitie compel, forcerie.

But of thefe three but one can ftand me in fteede, the laft, but not the leaft, whiche is able to worke the mindes of all woemen like wax, when the others can fcarfe wind them like[a] with. Medicines there are that can bring it to paffe, and men ther are that haue, fome by potions, fome by verfes, fome by dreames, all by deceite, the enfamples were tedious to recite, and you knowe them, the meanes I come to learne, and you can giue them, which is the onely caufe of my comming, and may be the occafion of my pleafure, and certainlye the waye both for your prayfe and profit.

Whether it be an enchaunted leafe, a verfe of *Pythia*, a figure of *Amphion*, a Charecter of *Ofchanes*, an Image of *Venus*, or a braunch of *Sybilla*, it fkilleth not.

Let it be eyther the feedes of *Medea*, or the bloud of *Phillis*, let it come by Oracle of *Apollo*, or by Prophecie of *Tyrefias*, eyther by the intrayles of a Goat, or what els foeuer I care not, or by all thefe in one, to make fure incantation and fpare not.

If I winne my loue, you fhall not loofe your labour, and whether it redound or no to my greater perill, I will not yet forget your paines.

Let this potion be of fuch force, that fhe may doat in hir defire, and I delight in hir diftreffe.

And if in this cafe you eyther reueale my fuite or denye it, you fhall foone perceyue that *Philautus* will

dye as defperatelye in one minute, as he hath liued
this three monethes carefully, and this your ftudie
fhall be my graue, if by your ftudye you eafe not my
griefe.

When he had thus ended he looked fo fternly vpon
Pfellus, that he wifhed him farther off, yet taking him
by the hande, and walking into his chamber, this good
man began thus to aunfwere him.

Entleman, if the inward fpirite be aunfwerable to
the outward fpeach, or the thoughtes of your
heart agreeable to the words of your mouth, you fhal
breede to your felfe great difcredite, and to me no
fmall difquyet. Doe you thinke Gentleman that the
minde being created of God, can be ruled by man, or
that anye one can moue the heart, but he that made
the heart? But fuch hath bene the fuperftition of olde
women, and fuch the folly of young men, yat there
could be nothing fo vayne but the one woulde inuent,
nor anye thing fo fenceleffe but the other would beleeue : which then brought youth into a fooles Paradife, and hath now caft age into an open mockage.

What the force of loue is, I haue knowen, what the
effects haue bene I haue heard, yet could I neuer
learne that euer loue could be wonne, by the vertues
of hearbes, ftones or words. And though many there
haue bene fo wicked to feeke fuch meanes, yet was
there neuer any fo vnhappy to finde them.

Parrhafius painting *Hopplitides*, could neither make
him that ranne to fweate, nor the other that put off his
armour to breathe, adding this as it were for a note,
No further then colours : meaning that to giue lyfe
was not in his Pencil, but in the Gods.

And the like may be faid of vs that giue our mindes
to know the courfe of the Starres, the Plannets, the
whole Globe of heauen, the Simples, the Compounds,
the bowels of the Earth, that fomething we may geffe
by the out-ward fhape, fome-thing by the natiuitie :
but to wreft the will of man, or to wreath his heart to

oui humours, it is not in the compaffe of Arte, but in the power of the moft higheft.

But for bicaufe there haue bene manye with-out doubt, that haue giuen credit to the vayne illufions of Witches, or the fonde inuentions of idle perfons, I will fet downe fuch reafons as I haue heard, and you wil laugh at, fo I hope, I fhal both fatiffie your minde and make you a lyttle merry, for me thinketh there is nothing that can more delyght, then to heare the things which haue no weight, to be thought to haue wrought wonders.

If you take Pepper, the feede of a Nettle, and a pretie quantitie of *Pyretum*, beaten or pounded alto-gether, and put into Wine of two yeares olde, when-foeuer you drinke to *Camilla*, if fhe loue you not, you loofe your labour. The coft is fmall, but if your be-liefe be conftant you winne the goale, for this Receipt ftandeth in a ftrong conceipt.

Egges and Honnye, blended with the Nuts of a Pine tree, and laid to your left fide, is of as great force when you looke vppon *Camilla* to bewitch the minde, as the *Quinteffence* of Stocke-fifh, is to nourifh the body.

An hearbe there is, called *Anacamforitis*, a ftrange name and doubtleffe of a ftraunge nature, for whofo-euer toucheth it, falleth in loue, with the perfon fhee next feeth. It groweth not in *England*, but heere you fhal haue that which is not halfe fo good, that will do as much good, and yet truly no more.

The Hearbe *Carifium*, moyftened with the bloude of a Lyfarde, and hanged about your necke, will caufe *Camilla* (for hir you loue beft) to dreame of your fer-uices, fuites, defires, defertes, and whatfoeuer you would wifh hir to thinke of you, but beeing wakened fhe fhall not remember what fhee dreamed off. And this Hearbe is to be founde in a Lake neere *Boetia*, of which water who fo drinketh, fhall bee caught in Loue, but neuer finde the Hearbe : And if hee drincke not, the Hearbe is of no force.

There is in the Frogges fide, a bone called *Apocy-con*, and in the heade of a young Colte, a bounch

named *Hippomanes*, both fo effectuall, for the obteining of loue, that who fo getteth either of them,
fhall winne any that are willyng, but fo iniurioullye
both crafte and Nature dealt with young Gentlemen
that feeke to gaine good will by thefe meanes, that the
one is lycked off before it can be gotten, the other
breaketh as foone as it is touched. And yet vnleffe
Hippomanes be lycked, it can-not worke, and except
Apocycon be found it is nothing worth.

I omit the Thiftle *Eryngium*, the Hearbes *Cata-
nenci* and *Pyteuma*, *Iuba* his *Charito blæpheron*, and
Orpheus Staphilinus, all of fuch vertue in cafes of loue,
that if *Camilla* fhoulde but taft any one of them in hir
mouthe, fhe woulde neuer lette it goe downe hir
throate, leafte fhee fhoulde bee poyfoned, for well you
knowe Gentleman, that Loue is a Poyfon, and therefore by Poyfon it muft be mayntayned.

But I will not forgette as it were the Methridate of
the Magitians, the Beaft *Hiena*, of whom there is no
parte fo fmall, or fo vyle, but it ferueth for their purpofe : Infomuch that they accompt *Hyena* their God
that can doe al, and their Diuel that will doe all.

If you take feauen hayres of *Hyenas* lyppes, and carrye them fixe dayes in your teeth, or a peece of hir
fkinne nexte your bare hearte, or hir bellye girded to
hir [your] left fide, if *Camilla* fuffer you not to obtaine
your purpofe, certeinely fhe can-not chufe, but thanke
you for your paines.

And if you want medicines to winne women, I haue
yet more, the lungs of a *Vultur*, the afhes of *Stellio*,
the left ftone of a Cocke, the tongue of a Goofe, the
brayne of a Cat, the laft haire of a Wolues taile.
Thinges eafie to be hadde, and commonly practifed,
fo that I would not haue thee ftande in doubte of thy
loue, when either a young Swallow famifhed, or the
fhrowding fheete of a deere friend, or a waxen Taper
that burnt at his feete, or the enchaunted Needle that
Medea hid in *Iafons* fleeue, are able not onely to make
them defire loue, but alfo dye for loue.

How doe you now feele your felfe *Philautus* ? If the leaft of thefe charmes be not fufficient for thee, all exorcifmes and coniurations in the world will not ferue thee.

You fee Gentleman, into what blynde and grofe errours in olde time we were ledde, thinking euery olde wiues tale to be a truth, and euery merry word, a very witchcraft. When the *Aegyptians* fell from their God to their Priefts of *Memphis*, and the *Grecians*, from their Morall queftions, to their difputations of *Pirrhus*, and the *Romaines* from religion, to polycie : then began all fuperftition to breede, and all impietie to blo[o]me, and to be fo great, they haue both growen, that the one being then an Infant, is nowe an Elephant, and the one beeing then a Twigge, is now a Tree.

They inuented as many Enchauntments for loue, as they did for the Tooth-ach, but he that hath tryed both will fay, that the beft charme for a Toothe, is to pull it out, and the beft remedie for Loue, to weare it out. If incantations, or potions, or amorous fayings could haue preuailed, *Circes* would neuer haue loft *Vlyffes*, nor *Phædra Hippolitus*, nor *Phillis Demophoon*.

If Coniurations, Characters, Circles, Figures, F[i]endes, or Furies might haue wrought anye thing in loue, *Medea* woulde not haue fuffered *Iafon* to alter his minde.

If the firropes of *Micaonias*, or the Verfes of *Aen[ea]s* or the *Satyren* of *Dipfas* were of force to moue the minde, they all three would not haue bene martired with the torments of loue.

No no *Philautus* thou maift well poyfon *Camilla* with fuch drugges, but neuer perfwade hir : For I confeffe that fuch hearbes may alter the bodye from ftrength to weakeneffe, but to thinke that they can moue the minde from vertue to vice, from chaftitie to luft, I am not fo fimple to beleeue, neither would I haue thee fo finful as to doubt [doe] it.

L *Vcilla* miniftring an amorous potion vnto hir hufband *Lucretius*, procured his death, whofe life fhe onely defired.

Ariſtotle noteth one that beeing inflamed with the beautie of a faire Ladye, thought by medicine to procure his bliſſe, and wrought in the ende hir bane : So was *Caligula* ſlaine of *Cæſonia*, and *Lucius Lucullus* of *Caliſtine* .

Perſwade thy ſelfe *Philautus* that to vſe hearbes to winne loue will weaken the body, and to think that hearbes can further, doth hurt the ſoule : for as great force haue they in ſuch caſes, as noble men thought them to haue in the olde time. *Achimeni[u]s* the hearbe was of ſuch force, that it was thought if it wer thrown into the battaile, it would make all the ſoldiers tremble: but where was it when the *Humbri* and *Tentoni* were exiled by warre, wher grewe *Achiminis*[*Achimenius*], one of whoſe leaues would haue ſaued a thouſand liues?

The Kinges of *Perſia* gaue their ſouldiers the plant *Latace*, which who ſo hadde, ſhoulde haue plentye of meate and money, and men and al things : but why did the ſoldiers of *Cæſar* endure ſuch famine · in *Pharſalia*, if one hearbe might haue eaſed ſo many heartes.

Where is *Balis* that *Iuba* ſo commendeth, the which coulde call the dead to lyfe, and yet hee himſelfe dyed ?

Democritus made a confe(tion, that who-ſoeuer dranke it ſhould haue a faire, a fortunate, and a good childe. Why did not the *Perſian* Kinges ſwill this Nectar, hauing ſuch deformed and vnhappy iſſue ?

Cato was of that minde, that three enchaunted wordes coulde heale the eye-ſight : and *Varro*, that a verſe of *Sybilla* could eaſe the goute, yet the one was fayne to vſe running water, which was but a colde medicine, the other patience, which was but a drye playſter.

I would not haue thee thinke *Philautus* that loue is to bee obteined by ſuch meanes, but onely by fayth, vertue, and conſtancie.

Philip King of *Macedon* caſting his eye vppon a fayre Virgin became enamoured, which *Olympias* his wife perceiuing, thought him to bee enchaunted, and caufed one of the feruauntes to bring the Mayden vnto hir, whome fhee thought to thruft both to exile and fhame : but vieweing hir fayre face with-out blemyfhe, hir chafte eyes with-out glauncinge, hir modeft countenaunce, hir fober and woemanlye behauiour, finding alfo hir vertues to be no leffe then hir beautie, fhee fayde, in my felfe there are charmes, meaning that there was no greater enchauntment in loue, then temperaunce, wifdome, beautie and chaftitie. Fond therefore is the opinion of thofe that thinke the minde to be tyed to Magick, and the practife of thofe filthy, that feeke thofe meanes.

Loue dwelleth in the minde, in the will, and in the hearts, which neyther Coniurer canne alter nor Phificke. For as credible it is, that *Cupid* fhooteth his A[r]rowe and hytteth the heart, as that hearbes haue the force to bewitch the heart, onelye this difference there is, that the one was a fiction of poetrie, the other of fuperftition. The will is placed in the foule, and who canne enter there, but hee that created the foule ?

No no Gentle-man what-foeuer you haue heard touching this, beleeue nothing : for they in myne opinion which imagine that the mynde is eyther by incantation or excantation to bee ruled, are as far from trueth, as the Eaft from the Weft, and as neere impietie againft God, as they are to fhame among men, and fo contrary is it to the profeffion of a Chriftian, as *Paganifme.*

Suffer not your felfe to bee lead with that vile conceypte, practife in your loue all kinde of loyaltie. Be not mute, nor full of bab[b]le, bee fober, but auoyde follenneffe, vfe no kinde of ryotte eyther in banqueting, which procureth furfeites, nor in attyre, which hafteth beggerye.

If you thinke well of your witte, be alwayes pleafaunt, if yll bee often filent : in the one thy talke fhal proue thee fharpe, in the other thy modeftie, wife.

All fyfhe are not caught with Flyes, all woemenne are not allured with perfonage. Frame letters, ditties, Muficke, and all meanes that honeftie may allowe : For he wooeth well, that meaneth no yll, and hee fpeedeth fooner that fpeaketh what he fhould, then he that vttereth what he will. Beleeue me *Philautus* I am nowe olde, yet haue I in my head a loue tooth, and in my minde there is nothing that more pearceth the heart of a beautifull Ladye, then writinge, where thou mayft fo fette downe thy paffions and hir perfection, as fhee fhall haue caufe to thinke well of thee, and better of hir felfe : but yet fo warilye, as neyther thou feeme to prayfe hir too much, or debafe thy felfe too lowelye : for if thou flatter them with-out meane they loath it*, and if thou make of thy felfe aboue reafon they laugh at it, temper thy wordes fo well, and place euerye fentence fo wifelye, as it maye bee harde for hir to iudge, whether thy loue be more faythfull, or hir beautie amiable.

Lions fawne when they are clawed, Tygers ftoupe when they are tickled, *Bucephalus* lyeth downe when he is curryed, woemen yeelde when they are courted.

This is the poyfon *Philautus*, the enchauntment, the potions that creepeth by fleight into the minde of a woeman, and catcheth hir by affuraunce, better then the fonde deuices of olde dreames, as an *Apple* with an *Aue Marie*, or a hafill wand of a yeare olde croffed with fix Charactors, or the picture of *Venus* in Virgin Wax, or the Image of *Camilla* vppon a Moulwarpes fkinne.

It is not once mencioned in the Englifhe Courte, nor fo much as thought of in any ones confcience, that Loue canne bee procured by fuch meanes, or that anye canne imagine fuche myfchiefe, and yet I feare mee it is too common in our Countrey, where-by they incurre hate of euerye one, and loue of none.

Touching my cunning in any vile deuices of Magick it was neuer my ftudie, onely fome delyght, I tooke in the Mathematicks which made me knowen of more then I would, and of more then thinke well of me, although I neuer did hurt any, nor hindred.

But be thou quiet *Philautus*, and vfe thofe meanes that may winne thy loue, not thofe that may fhorten hir lyfe, and if I can any wayes ftande thee in fteade, vfe me as thy poore friend and countrey-man, harme I will doe thee none, good I cannot. My acquaintance in Court is fmall, and therefore my dealyngs about the Courte fhall be fewe, for I loue to ftande aloofe from *loue* and lyghtning. Fire giueth lyght to things farre off, and burneth that which is next to it. The Court fhineth to me that come not there, but fingeth thofe that dwell there. Onely my counfayle vfe, that is in writing, and me thou fhalt finde fecret, wifhing thee alwayes fortunate, and if thou make me pertaker of thy fucceffe, it fhall not tourne to thy griefe, but as much as in mee lyeth, I will further thee.

When he had finifhed his difcourfe, *Philautus* liked very well of it, and thus replyed.

WEll *Pfeilus*, thou haft wrought that in me, which thou wifheft, for if the baites that are layde for beautie be fo ridiculous, I thinke it of as great effeĉt in loue, to vfe a Plaifter as a Potion.

I now vt[t]erly diffent from thofe that imagine Magicke to be the meanes, and confent with thee, that thinkeft letters to be, which I will vfe, and howe I fpeede I will tell thee, in the meane feafon pardon me, if I vfe no longer aunfwere, for well you know, that he that hath the fit of an Ague vpon him, hath no luft to talke but to tumble, and Loue pinching me I haue more defire to chew vpon melancholy, then to difpute vpon Magicke, but heereafter I will make repaire vnto you, and what I now giue you in thankes, I will then requite with amends.

Thus thefe two country-men parted with certeine *Italian* embracings and termes of courtefie, more then common. *Philautus* we fhal finde in his lodging,

Pfellus we will leaue in his ftudi_e, the one mufing of
his loue, the other of his learning.

Ere Gentlewomen you may fee, how iuftly men
feeke to entrap you, when fcornefully you goe
about to reiect them, thinking it not vnlawfull to vfe
Arte, when they perc[e]iue you obftinate, their dealings I wil[l] not allow, neither can I excufe yours,
and yet what fhould be the caufe of both, I can
geffe.

When *Phydias* firft paynted, they vfed no colours,
but blacke, white, redde, and yeolow : *Zeuxis* added
greene, and euery one inuented a new fhadowing. At
the laft it came to this paffe, that he in painting deferued moft prayfe, that could fette downe moft con-
lours : wherby ther was more contention kindeled
about the colour, then the counterfaite, and greater
emulation for varietie in fhew, then workmanfhip in
fubftaunce.

In the lyke manner hath it fallen out in Loue, when
Adam wo[o]ed there was no pollycie, but playne dealyng,
no colours but blacke and white.   Affection was meafured by faith, not by fancie : he was not curious, nor
Eue cruell : he was not enamoured of hir beautie, nor
fhe allured with his perfonage : and yet then was fhe
the faireft woman in the worlde, and he the propereft
man.   Since that time euery louer hath put too a
lynke, and made of a Ring, a Chaine, and an odde
Corner, and framed of a playne Alley, a crooked knot,
and *Venus* Temple, *Dedalus* Laborinth.   One curleth
his hayre, thinking loue to be moued with faire lockes,
an other layeth all his lyuing vppon his backe, iudging that women are wedded to brauerie, fome vfe difcourfes of Loue, to kindle affection, fome ditties to
allure the minde, fome letters to ftirre the appetite,
diuers fighting to proue their manhoode, fundry figh-
ing to fhew their maladyes, many attempt with fhowes
to pleafe their Ladyes eyes, not few with Muficke to
entice the eare : Infomuch that there is more ftrife

now, who fhal be the fineft Louer, then who is the faithfulleft.

This caufeth you Gentlewomen, to picke out thofe that can court you, not thofe that loue you, and hee is accompted the beft in your conceipts, that vfeth moft colours, not that fheweth greateft courtefie.

A playne tale of faith you laugh at, a picked difcourfe of fancie, you meruayle at, condempning the fimpli-citie of truth, and preferring the fingularitie of deceipt, where-in you refemble thofe fifhes that rather fwallow a faire baite with a fharpe hooke, then a foule worme breeding in the mudde.

Heere-off it commeth that true louers receiuing a floute for their fayth, and a mocke for their good mean-ing, are enforced to feeke fuch meanes as might com-pell you, which you knowing impoffible, maketh you the more difdainefull, and them the more defperate. This then is my counfaile, that, you vfe your louers lyke friends, and chufe them by their faith, not by the fhew, but by the found, neither by the waight, but by the touch, as you do golde: fo fhall you be prayfed, as much for vertue as beautie. But retourne we againe to *Philautus* who thus beganne to debate with himfelfe.

WHat haft thou done *Philautus*, in feeking to wounde hir that thou defireft to winne? With what face canft thou looke on hir, whome thou foughteft to loofe? Fye, fye *Philautus*, thou bringeft thy good name into queftion, and hir lyfe into hazard, hauing neither care of thine owne credite, nor hir honour. Is this the loue thou pretendeft which is worfe then hate? Diddeft not thou feeke to poyfon hir, that neuer pinched thee?

But why doe I recount thofe thinges which are paft, and I repent, I am now to confider what I muft doe, not what I would haue done? Follyes paft, fhall be worne out with faith to come, and my death fhal fhew my defire. Write *Philautus*, what fayeft thou? write, no, no thy rude ftile wil bewray thy meane eftate, and

z

thy rafh attempt, will purchafe thine ouerthrow.   *Venus* delyghteth to heare none but *Mercury*, *Pallas* wil
be ftolne of none but *Vlyffes*, it muft bee a fmoothe
tongue, and a fweete tale that can enchaunt *Vefta.*

Befides that I dare not truft a meffenger to carye it,
nor hir to reade it, leaft in fhewing my letter fhee difclofe my loue, and then fhall I be pointed at of thofe
that hate me, and pitied of thofe that lyke me, of hir
fcorned, of all talked off.   No *Philautus*, be not thou
the bye word of the common people, rather fuffer death
by filence, then derifion by writing.

I, but it is better to reueale thy loue, then conceale it, thou knoweft not what bitter poyfon lyeth in
fweet words, remember *Pfellus*, who by experience hath
tryed, that in loue one letter is of more force, then a
thoufand lookes.   If they lyke writings they read them
often, if diflyke them runne them ouer once, and this
is certeine that fhe that readeth fuche toyes, will alfo
aunfwere them.   Onely this be fecret in conueyaunce,
which is the thing they chieflyeft defire.   Then write
Philautus write, he that feareth euery bufh, muft neuer
goe a birding, he that cafteth all doubts, fhal neuer be
refolued in any thing.   And this affure thy felfe that
be thy letter neuer fo rude and barbarous, fhee will
reade it, and be it neuer fo louing fhe will not fhewe
it, which weare a thing contrary to hir honor, and the
next way to call hir honeftie into queftion.   For thou
haft heard, yea and thy felfe knoweft, that Ladyes that
vaunt of their Louers, or fhewe their letters, are accompted in *Italy* counterfait, and in *England* they are
not thought currant.

Thus *Philautus* determined, hab, nab, to fende his
letters, flattering him-felfe with the fucceffe which he
to him-felfe faigned : and after long mufing, he thus
beganne to frame the minifter of his loue.

¶ *To the fayreft, Camilla.*

H Ard is the choyce fayre Ladye, when one is compelled eyther by filence to dye with griefe, or

... ok .?okgo

by writing to liue with fhame: But fo fweete is the defire of lyfe, and fo fharpe are the paffions of loue, that I am enforced to preferre an vnfeemely fuite, before an vntimely death. Loth I haue bin to fpeake, and in difpayre to fpeede, the one proceeding of mine own cowardife, the other of thy crueltie. If thou enquire my name, I am the fame *Philautus*, which for thy fake of late came difguifed in a Mafke, pleading cuftome for a priuiledge, and curtefie for a pardon. The fame *Philautus* which then in fecret tearmes coloured his loue, and now with bitter teares bewrayes it. If thou nothing efteeme the brynifh water that falleth from mine eyes, I would thou couldeft fee the warme bloud that droppeth from my heart. Oftentimes I haue beene in thy company, where eafily thou mighteft haue perceiued my wanne cheekes, my hol[l]ow eies, my fcalding fighes, my trembling tongue, to forfhew yat then, which I confeffe now. Then confider with thy felf *Camilla*, the plight I am in by defire, and the perill I am like to fall into by deniall.

To recount the forrowes I fuftaine, or the feruice I haue vowed, would rather breede in thee an admiration, then a belief: only this I adde for the time, which the ende fhall trye for a trueth, that if thy aunfwer be fharpe, my life wil be fhort, fo farre loue hath wrought in my pyning and almoft confumed bodye, that thou onely mayft breath into me a new life, or bereaue mee of the olde.

Thou art to weigh, not how long I haue loued thee, but how faythfully, neyther to examine the worthyneffe of my perfon, but the extremitie[s] of my paffions: fo preferring my defarts before the length of time, and my defeafe, before the greatnes of my byrth, thou wilt eyther yeelde with equitie, or deny with reafon, of both the which, although the greateft be on my fide, yet the leaft fhall not diflike me: for yat I haue alwayes found in thee a minde neyther repugnaunt to right, nor void of re[a]fon. If thou wouldft but permit me to talke with thee, or by writing fuffer me at large to difcourfe with thee, I doubt not but yat, both the caufe of my loue wo[u]ld be beleeued, andt he extremitie rewarded, both proceeding of thy beautie and vertue, the one able to allure, the other ready to pittie. Thou muft thinke that God hath not beftowed thofe rare giftes vpon thee to kyll thofe that are caught, but to cure them. Thofe that are ftunge with the Scorpion, are healed with the Scorpion, the fire that burneth, taketh away the heate of the burn, the Spider *Phalangium* that poyfoneth, doth with hir fkinne make a playfter for poyfon, and fhall thy beautie which is of force to winne all with loue, be of the crueltie to wound any with death? No *Camilla*, I take no leffe delight in thy fayre face, then pleafure in thy good conditions, affuring my felfe that for affection with-out luft, thou wilt not render malyce with-out caufe.

I commit [omit] my care to thy confideration, expecting thy Letter eyther as a Cullife to preferue, or as a fworde to deftroy, eyther as *Antidotum*, or as *Auconitum* : If thou delude mee, thou fhalt not long triumphe ouer mee lyuing, and fmall will thy glory be when I am dead. And I ende. *Thine euer, though he be neuer thine. Philautus.*

THis Letter beeing coyned, hee ftudyed how hee myght conueie it, knowing it to be no leffe perrilous to truft thofe hee knewe not in fo weightye a cafe, then dyffycult for him-felfe to haue opportunitie to delyuer it in fo fufpitious a company : At the laft taking out of his clofette a fayre Pomegranet, and pullyng all the kernelles out of it, hee wrapped his Letter in it, clofing the toppe of it finely, that it could not be perceyued, whether nature agayne hadde knitte it of purpofe to further him, or his arte had ouercome natures cunning. This Pomegranet hee tooke, beeing him-felfe both meffenger of his Letter, and the mayfter, and infinuating him-felfe into the companie of the Gentlewoemen, amonge whom was alfo *Camilla*, hee

was welcommed as well for that he had beene long tyme
abfent, as for that hee was at all tymes pleafaunt,
much good communication there was touching manye
matters, which heere to infert were neyther conuenient,
feeing it doth not concern the Hyftorie, nor expedient,
feeing it is nothing to the delyuerie of *Philautus* Letter.
But this it fell out in the ende, *Camilla* whether longing for fo faire a Pomegranet, or willed to afke it, yet
loth to require it, fhe fodeinlye complayned of an old
defeafe, wherwith fhee manye times felt hirfelfe grieued,
which was an extreame heate in ye ftomack, which
aduantage *Philautus* marking, would not let flip, when
it was purpofely fpoken, that he fhould not giue them
the flippe : and therefore as one gladde to haue fo conuenient a time to offer both his duetie and his deuotion,
he beganne thus.

I Haue heard *Camilla*, of Phifitions, that there is
nothing eyther more comfortable, or more profitable for the ftomack or enflamed liuer, then a Pomgranet, which if it be true, I am glad that I came in
fo good tyme with a medicine, feeing you were in fo ill
a time fupprifed with your maladie : and verily this will
I faye, that there is not one Kernell but is able both to
eafe your paine, and to double your pleafure, and with
that he gaue it hir, defiring that as fhe felte the working
of the potion, fo fhee would confider of the Phifition.

Camilla with a fmyling countenaunce, neyther fufpecting the craft, nor the conueyer, anfwered him with
thefe thankes.

I thank you Gentleman as much for your counfell
as your curtefie, and if your cunning be anfwerable to
eyther of them, I will make you amendes for all of
them : yet I wil not open fo faire a fruite as this is,
vntill I feele the payne that I fo much feare.   As you
pleafe quoth *Philautus*, yet if euery morning you take
one kernell, it is the way to preuent your difeafe, and
me thinketh that you fhould be as carefull to worke
meanes before it come, that you haue it not, as to vfe
meanes to expell it when you haue it.

I am content, aunfwered *Camilla*, to trye your phifick, which as I know it can do me no great harme, fo it may doe me much good.

In truth fayd one of the Gentlewomen then prefent, I perceiue this Gentleman is not onely cunning in Phificke, but alfo very carefull for his Patient.

It beho[o]ueth, quoth *Philautus*, that he that miniftreth to a Lady, be as defirous of hir health, as his owne credite, for that there redoundeth more prayfe to the Phifition that hath a care to his charge, then to him that hath only a fhow of his Art.   And I truft *Camilla* will better accept of the good will I haue to ridde hir of hir difeafe, then the gift, which muft worke the effect.

Otherwife quoth *Camilla*, I were verye much to blame, knowing that in manye the behauiour of the man, hath wrought more then the force of the medicine.   For I would alwayes haue my Phifition, of a cheerefull countenaunce, pleafauntlye conceipted, and well proportioned, that he might haue his fharpe Potions mixed with fweete counfayle, and his fower drugs mitigated with merry difcourfes.   And this is the caufe, that in olde time, they paynted the God of Phificke, not lyke *Saturne* but *Aefculapius* : of a good complection, fine witte, and excellent conftitution.

For this I know by experience, though I be but young to learne, and haue not often bene ficke, that the fight of a pleafant and quicke witted Phifitian, hath remoued that from my heart with talke, that he could not with all his Triacle.

That might well be, aunfwered *Philautus*, for the man that wrought the cure, did perchaunce caufe the difeafe, and fo fecret might the griefe be, that none could heale you, but he that hurte you, neither was your heart to be eafed by any in-ward potion, but by fome outward perfwafion : and then it is no meruaile if the miniftring of a few wordes, were more auayleable then Methridate.

Wel Gentleman faid *Camilla*, I wil[l] neither difpute in Phifick, wherin I haue no fkill, neither aunfwere you, to your laft furmife[s], which you feeme to leuell at, but thanking you once againe both for your gift and good will, we wil vfe other communication, not forgetting to afke for your friend *Euphues*, who hath not long time be[e]ne, where he might haue bene welcommed at all times, and that he came not with you at this time, we both meruayle, and would faine know.

This queftion fo earneftlye afked of *Camilla*, and fo hardlye to be aunfwered of *Philautus*, nipped him in the head, notwithftanding leaft he fhold feeme by long filence to incurre fome fufpition, he thought a bad ex- cufe better then none at all, faying that *Euphues* now a dayes became fo ftudious (or as he tearmed it, fuper- fticious) that he could not himfelfe fo much, as haue his company.

Belike quoth *Camilla*, he hath either efpyed fome new faults in the women of *England*, where-by he feeketh to abfent himfelfe, or fome olde haunt that will caufe him to fpoyle himfelfe.

Not fo fayd *Philautus*, and yet that it was fayd fo I will tell him.

Thus after much conference, many queftions, and long time fpent, *Philautus* tooke his leaue, and beeing in his chamber, we will ther[e] leaue him with fuch cogi- tations, as they commonly haue, that either attende the fentence of lyfe or death at the barre, or the aun- fwere of hope or difpaire of their loues, which none can fet downe but he that hath them, for that they are not to be vttered by the coniecture of one that would imagine what they fhould be, but by him that knoweth what they are.

Camilla the next morning opened the Pomegranet, and faw the letter, which reading, pondering and peruf- ing, fhe fell into a thoufande contrarieties, whether it were beft to aunfwere it or not, at the laft, inflamed with a kinde of cholar, for that fhe knew not what be- longed to the perplexities of a louer, fhe requited his frawd and loue, with anger and hate, in thefe termes,

To Philautus.

I Did long time debate with my felfe *Philautus,*
whether it might ftand with mine honour to fend
thee an aunfwere, for comparing my place with thy
perfon, me thought thy boldnes more, then either good
manners in thee wo[u]ld permit, or I with modeftie could
fuffer. Yet at ye laft, cafting with my felfe, yat the heat
of thy loue might clean be razed with ye coldnes of
my letter, I thought it good to commit an inconuenie-
ence, yat I might preuent a mifchiefe, chufing rather
to cut thee off fhort by rigour, then to giue thee any
iot of hope by filence. Greene fores are to be dreffed
roughly, leaft they fefter, tet[t]ars to be drawn in the beginning leaft they fpread, ring wormes to be anoynted
when they firft appeare, leaft they compaffe ye whole
body, and the affa[u]lts of loue to be beaten back at ye
firft fiege, leaft they vndermine at ye fecond. Fire is
to be quenched in ye fpark, weedes are to be rooted
in ye bud, follyes in ye bloffome. Thinking this
morning to trye thy Phifick, I perceiued thy frawd,
infomuch as the kernel yat fhoulde haue cooled my
ftomack with moiftnes, hath kindled it with cholar,
making a flaming fire, wher it found but hot imbers,
conuerting like the Spider a fweet floure [flower], into a
bitter poyfon. I am *Philautus* no *Italian* Lady, who
commonly are woed with leafings, and won with luft,
entangled with deceipt, and enioyed with delight, caught
with finne, and caft off with fhame.

For mine owne part, I am too young to knowe the
paffions of a louer, and too wife to beleeue them, and
fo farre from trufting any, that I fufpect all : not that
ther is in euery one, a practife to deceiue, but that ther
wanteth in me a capacitie to conceiue.

Seeke not then *Philautus* to make the tender twig
crooked by Arte, which might haue growen ftreight by
Nature. Corne is not to be gathered in the budde, but
in the eare, nor fruite to be pulled from the tree when
it is greene, but when it is mellow, nor Grapes to bee

cut for the preſſe, when they firſt riſe, but when they are full ripe: nor young Ladies to be ſued vnto, that are fitter for a rodde then a huſbande, and meeter to beare blowes then children.

You muſt not think of vs as of thoſe in your own countrey, that no ſooner are out of the cradell, but they are ſent to the court, and wo[o]ed ſome-times before they are weaned, which bringeth both the Nation and their names, not in queſtion onely of diſhoneſtie, but into obliquie.

This I would haue thee to take for a flat aunſwere, that I neither meane to loue thee, nor heereafter if thou follow thy ſute to heare thee. Thy firſt practiſe in the Maſque I did not allow, the ſeconde by thy writing I miſlyke, if thou attempt the third meanes, thou wilt enforce me to vtter that, which modeſtie now maketh me to conceale.

If thy good will be ſo great as·thou telleſt, ſeeke to mitigate it by reaſon or time, I thanke thee for it, but I can-not requit it, vnleſſe either thou wert not *Philautus*, or I not *Camilla.* Thus pardoning thy boldnes vppon con-dition, and reſting thy friend if thou reſt thy ſute, I ende.

Neither thine, nor hir owne,
Camilla.

His letter *Camilla* ſtitched into an *Italian* petrack [Petracke] which ſhe had, determining at the next comming of *Philautus*, to deliuer it, vnder the pretence of aſking ſome queſtion, or the vnderſtanding of ſome worde. *Philautus* attending hourelye ye ſucceſſe of his loue, made his repaire according to his accuſtomable vſe, and finding the Gentlewomen ſitting in an herbor, ſaluted them curteouſly, not forgetting to be inquiſitiue how *Camilla* was eaſed by his Pomgranet, which often-times aſking of hir, ſhe aunſwered him thus.

In faith *Philautus*, it had a faire coat, but a rotten kernell, which ſo much offended my weake ſtomacke, that the very ſight cauſed me to lo[a]th it, and the ſent to throw it into the fire.

I am fory quoth *Philautus* (who fpake no leffe then trueth) that the medicine could not worke that, which my mind wifhed, and with that ftoode as one in a traunce, which *Camilla* perceiuing, thought beft to rub no more on that gall, leaft the ftanders by fhould efpy where *Philautus* fhooe wronge him.

Well faid *Camilla* let it goe, I muft impute it to my ill fortune, that where I looked for a reftoritie, I found a confumption : and with that fhe drew out hir petrarke [Petracke], requefting him to confter hir a leffon, hoping his learning would be better for a fcholemaifter, then his lucke was for a Phifition. Thus walking in the all[e]y, fhe liftned to his conftruction, who turning the booke, found where the letter was enclofed, and diffembling that he fufpected, he faide he would keepe hir petracke vntill the morning, do you quoth *Camilla*. With yat the Gentlewomen cluftred about them both, eyther to hear how cunningly *Philautus* could confter, or how readily *Camilla* could conceiue. It fell out that they turned to fuch a place, as turned them all to a blanke, where it was reafoned, whether loue came at the fodeine viewe of beautie, or by long experience of vertue, a long difputation was like to enfue, had not *Camilla* cut it off before they could ioyne iffue, as one not willing in ye company of *Philautus* eyther to talke of loue, or thinke of loue, leaft eyther hee fhould fufpect fhe had beene wooed, or might be won, which was not done fo clofelye, but it was perceiued of *Philautus*, though diffembled. Thus after many words, they went to their dinner, where I omit their table talke, leaft I loofe mine.

After their repaft, *Surius* came in with a great train, which lightened *Camillas* hart, and was a dagger to *Philautus* breaft, who taried no longer then he had leyfure[leafure] to take his leaue, eyther defirous to read his Ladyes aunfwer, or not willing to enioy *Surius* his companie, whome alfo I will now forfake, and followe *Philautus*, to heare how his minde is quieted with *Camillas* curtefie.

Philautus no fooner ent[e]red his chamber, but he read hir letter, w[h]ich wrought fuch fkirmifhes in his minde, that he had almoft forgot reafon, falling into the old evaine of his rage, in this manner.

Ah cruell *Camilla* and accurfed *Philautus,* I fee now that it fareth with thee, as it doth with the Hare Sea, which hauing made one aftonied with hir fayre fight, turneth him into a ftone with hir venemous fauo[u]r, and with me as it doth with thofe that view the *Bafilike,* whofe eyes procure delight to the looker at the firft glymfe, and death at the fecond glaunce.

Is this the curtefie of *England* towardes ftraungers, to entreat them fo difpightfullye ? Is my good will not onely reiefted with-out caufe, but alfo difdained with-out coulour ? I but *Philautus* prayfe at the [thy] parting, if fhe had not liked thee, fhe would neuer haue aun-fwered thee. Knoweft thou not that wher they loue much, they diffemble moft, that as fayre weather com-meth after a foule ftorme, fo fweete tearmes fucceede fowre [fower] taunts ?

Affaye once againe *Philautus* by Letters to winne hir loue, and followe not the vnkinde hounde, who leau-eth the fent bycaufe hee is rated, or the baftarde Span-yell, which beeing once rebuked, neuer retriueth his game. Let *Atlanta* runne neuer fo fwiftelye, fhee will looke backe vpon *Hyppomanes,* let *Medea* bee as cruell as a f[i]ende to all Gentle-men, fhee will at the laft re-peft *Iafon.* A denyall at the firft is accompted a graunt, a gentle aunfwere a mockerie. Ladyes vfe their Louers as the Storke doth hir young ones, who pecketh [picketh] them till they bleed with hir bill, and then healeth them with hir tongue. *Cupid* him-felf muft fpend one arrowe, and thinkeft thou to fpeede with one Letter ? No no *Philautus,* he that looketh to haue cleere water muft digge deepe, he that longeth for fweete Muficke, muft fet his ftringes at the hygheft, hee that feeketh to win his loue muft ftretch his labo[u]r, and hafard his lyfe. *Venus* blifleth [blefleth] Lions in the fold, and Lambes in the chamber, Eagles at the affaulte,

and Foxes in counfayle, fo that thou muft be hardy
in the purfuit, and meeke in victory, venterous in
obtaining, and wife in concealing, fo that thou win that
with prayfe, which otherwife thou wilt loofe with peeuifh-
neffe. Faint hart *Philautus* neither winneth Caftell
nor Lady: ther[e]fore endure all thinges that fhall happen with patience, and purfue with diligence, thy
fortune is to be tryed, not by the accedents [accidents]
but by the end.

Thus Gentlewoemen, *Philautus*, refembleth the
Viper, who beeing ftricken with a reede lyeth as he
were dead, but ftricken the fecond tyme, recouereth his
ftrength: hauing his anfwer at the firft in ye [a] mafque,
he was almoft amafed, and nowe againe denied, he is
animated, prefuming thus much vpon ye good difpofition and kindneffe of woemen, that the higher they fit,
the lower they looke, and the more they feeme at the
firft to lo[a]th, the more they loue at the laft. Whofe
iudgement as I am not altogether to allow, fo can I
not in fome refpect miflike. For in this they refemble
the Crocodile, who when one approcheth neere vnto
him, gathereth vp him-felf into the roundneffe of a
ball, but running from him, ftretcheth him-felf into the
length of a tree. The willing refiftance of women was
ye caufe yat made *Arelius* (whofe arte was only to draw
women) to paynt *Venus Cnydia* catching at the ball
with hir hand, which fhe feemed to fpurn at with hir
foote. And in this poynt they are not vnlike vnto the
Mirt [Mirre] Tree, which being hewed [hewen], gathereth in his fappe, but not moued, poureth it out like
firrop. Woemen are neuer more coye then when they
are beloued, yet in their mindes neuer leffe conftant,
feeming to tye themfelu[e]s to the maft of the fhippe with
Vlyffes, when they are wooed, with a ftrong Cable:
which being well difcerned is a twine threed: throwing
a ftone at the head of him, vnto whome they immediately caft out an ap[p]le, of which their gentle nature
Philautus being perfwaded, followed his fuit againe
in this manner.

Philautus to the faire, Camilla.

I Cannot tell (*Camilla*) whether thy ingratitude be greater, or my miffortune, for perufing the few lynes thou gaueft me, I found as fmall hope of my loue as of thy courtefie. But fo extreame are the paf- fions of loue, that the more thou feekeft to quench them by difdayne, the greater flame thou encreafeft by defire. Not vnlyke vnto *Iupiters* Well, which extin- guifheth a firie [fire] brande, and kindleth a wet ficke. And no leffe force, hath thy beautie ouer me, then the fire hath ouer *Naplytia*, which leapeth into it, wherfo- euer it feeth it.

I am not he *Camilla* that will leaue the Rofe, bicaufe I [it] pricked my finger, or forfake the golde that lyeth in the hot fire, for that I burnt my hande, or refufe the fweete Chefnut, for that it is couered with fharpe hufkes. The minde of a faithfull louer, is neither to be daunted with defpite, nor afrighted with daunger. For as the Load-ftone, what winde foeuer blowe, tourneth alwayes to the North, or as *Ariftotles Quad- ratus*, which way foeuer you tourne it, is alwayes conftant : fo the faith of *Philautus*, is euermore ap- plyed to the loue of *Camilla*, neither to be remoued with any winde, or rolled with any force. But to thy letter.

Thou faift greene wounds are to be dreffed roughly leaft they fefter : certeinly thou fpeakeft lyke a good Chyrurgian, but dealeft lyke one vnfkilfull, for making a great wound, thou putteft in a fmall tent, cutting the flefh that is found, before thou cure the place that is fore : ftriking the veyne with a knife, which thou fhouldeft ftop with lynt. And fo haft thou drawn my tettar [tetter], (I vfe thine owne terme) that in feek- ing to fpoyle it in my chinne, thou haft fpreade it ouer my body.

Thou addeft thou art no *Italyan* Lady, I anfwer, would thou wert, not that I would haue thee wooed, as thou fayft they are, but that I might win thee as thou now art : and yet this I dare fay, though not to excufe al, or to difgrace thee, yat fome there are in *Italy* too wife to be caught with leafings, and too honeft to be entangled with luft, and as wary to efchue finne, as they are willing to fuftaine fhame, fo that what-foeuer the moft be, I would not haue thee thinke ill of the beft.

Thou alleadgeft thy youth and alloweft thy wife-dome, the one not apt to know ye impreffions of loue, the other fufpitious not to beleeue them. Truely *Camilla* I haue heard, that young is the Goofe yat wil eate no Oates, and a very ill Cocke that will not crow before he be olde, and no right Lyon, that will not feede on hard meat, before he taft fweet milke, and a tender Virgin God knowes it muft be, that meafureth hir affections by hir age, when as naturally they are enclyned (which thou perticularly putteft to our countrey) to play the brides, before they be able to dreffe their heades.

Many fimilytudes thou bringeft in to excufe youth, thy twig, thy corne, thy fruit, thy grape, and I know not what, which are as eafelye to be refelled, as they are to be repeated.

But my good *Camilla*, I am as vnwillyng to confute any thing thou fpeakeft, as I am thou fhouldft vtter it : infomuch as I would fweare the Crow were white, if thou fhouldeft but fay it.

My good will is greater than I can expreffe, and thy courtefie leffe then I deferue : thy counfayle to expell it with time and reafon, of fo lyttle force, that I haue neither the will to vfe the meane, nor the wit to con-ceiue it. But this I fay, that nothing can break off my loue but death, nor any thing haften my death, but thy difcourtefie. And fo I attend thy finall fentence, and my fatall deftenie. *Thine euer, though he*
be neuer thine,
Philautus.

THis letter he thought by no meanes better to be conueyed, then in the fame booke he receiued hirs, fo omitting no time, leaft the yron fhould coole before he could ftrike, he prefently went to *Camilla*, whome he founde in gathering of flowers, with diuers other Ladyes and Gentlewomen, which came afwell to recreate themfelues for pleafure, as to vifite *Camilla*, whom they all loued. *Philautus* fomewhat boldened by acquaintaunce, courteous by nature, and courtly by countenance, faluted them al with fuch te[a]rmes, as he thought meete for fuch perfonages, not forgetting to call *Camilla* his fchollar, when fhe had fchooled him being hir mafter.

One of the Ladies who delighted much in mirth, feing *Philautus* behold *Camilla* fo ftedfaftly, faide vnto him.

Entleman, what floure [flower] like you beft in all this border, heere be faire Rofes, fweete Violets, fragrant primrofes, heere wil be Iilly-floures, Carnations, fops in wine, fweet Iohns, and what may either pleafe you for fight, or delight you with fauour : loth we are you fhould haue a Pofie of all, yet willing to giue you one, not yat which fhal[l] looke beft, but fuch a one as you fhal[l] lyke beft. *Philautus* omitting no opportuni[t]ie, yat might either manifeft his affeftion or commend his wit, aunfwered hir thus.

Lady, of fo many fweet floures [flowers] to chufe the beft, it is harde, feeing they be all fo good, if I fhoulde preferre the faireft before the fweeteft you would happely imagine that either I were ftopped in the nofe, or wanton in the eyes, if the fweetneffe before the beautie, then would you geffe me either to lyue with fauours, or to haue no iudgement in colours, but to tell my minde (vpon correftion be it fpoken) of all flowers, I loue a faire woman.

In deede quoth *Flauia* (for fo was fhe named) faire women are fet thicke, but they come vp thinne, and when they begin to budde, they are gathered as though they wer blowne, of fuch men as you are Gentleman, who thinke greene graffe will neuer be drye Hay, but when ye flower of their youth (being flipped too young) fhall fade before they be olde, then I dare faye, you would chaunge your faire flower for a weede, and the woman you loued then, for the worft violet you refufe now.

Lady aunfwered *Philautus*, it is a figne that beautie was no niggard of hir flippes in this gardein, and very enuious to other grounds, feing heere are fo many in one Plot, as I fhall neuer finde more in all *Italy*, whether the reafon be the heate which killeth them, or the country that cannot beare them. As for plucking them vp foone, in yat we fhew the defire we haue to them, not the malyce. Where you coniecture, that men haue no refpect to things when they be olde, I cannot confent to your faying for well doe they know that it fareth with women as it doth with the Mulbery tree, which the elder it is, the younger it feemeth, and therfore hath it growen to a Prouerb in *Italy*, when on[e] feeth a woman ftriken in age to looke amiable, he faith fhe hath eaten a Snake : fo that I muft of force follow mine olde opinion, that I loue frefh flowers well, but faire women better.

Flauia would not fo leaue him, but thus replyed to him.

YOu are very amorous Gentleman, otherwife you wold not take the defence of that thing which moft men contemne, and women will not confeffe. For where-as you goe about to currey fauour, you make a fault, either in prayfing vs too much, which we accompt in *Englande* flatterye, or pleafing your felfe in your owne minde, which wife men efteeme as folly. For when you endeauour to proue that woemen the older the[y] are, the fayrer they looke, you thinke them eyther very credulous to beleeue, or your talke verye effectuall to perfwade. But as cunning as you are in your *Pater nofter*, I will add one Article more to your *Crede*, that is, you may fpeak in matters of loue what you will, but women will beleeue but what they lyft, and in extolling their beauties, they giue more credit to their owne glaffes, then mens glofes.

But you haue not yet aunfwered my requeſt touch-ing what flower you moſt defire : for woemen doe not refemble flowers, neyther in ſhew nor fauour.

Philautus not ſhrinking for an Aprill ſhowre, followed the chace in this manner.

Lady, I neither flatter you nor pleafe my felfe (al-though it pleafeth you fo to coniecture) for I haue al-wayes obſerued this, that to ſtand too much in mine owne conceite would gaine me little, and to claw thofe of whome I fought for no benefite, woulde profit me leffe : yet was I neuer fo ill brought vp, but that I could when time and place ſhould ſerue, giue eeury one I* lyked* their iuſt commendation, vnleffe it were among thofe that were with-out comparifon : offending in nothing but in this, that beeing too curious in praif-ing my Lady, I was like to the Painter *Protogenes*, who could neuer leaue when his worke was well, which faulte is to be excufed in him, bicaufe hee would make it better, and may be borne with in mee, for that I wiſh it excellent. Touching your firſt demaund which you feeme againe to vrge in your laſt difcourfe, I fay of al[l] flowers I loue the Rofe beſt, yet with this con-dition, bicaufe I wil not eate my word, I like a faire Lady well. Then quoth *Flauia* fince you wil[l] needes ioyne the flower with the woman, amonge all vs (and fpeake not partially) call hir your Rofe yat you moſt regarde, and if ſhe deny that name, we will enioyne hir a penance for hir pride, and rewarde you with a violet for your paynes.

Philautus being driuen to this ſhift wiſhed him felfe in his chamber, for this he thought that if he ſhoulde choofe *Camilla* ſhe woulde not accept it, if an other, ſhe might iuſtly reiect him. If he ſhoulde difcouer his loue, then woulde *Camilla* thinke him not to be fe-create, if conce[a]le it, not to be feruent : befides all, the Ladyes woulde efpie his loue and preuent it, or *Ca-milla* defpife his offer, and not regarde it. While he was thus in a deepe meditation, *Flauia* wakened him faying, why Gentleman are you in a dreame, or is there

A A

none heere worthy to make choyce of, or are wee all
so indifferent, that there is neuer a good.

Philautus seeing this Lady so curteous, and louing
Camilla so earneſtly, coulde not yet reſolue with him-
selfe what to doe, but at the laſt, loue whiche neither
regardeth what it ſpeaketh, nor where, he replied thus
at all aduentures.

L Adyes and Gentlewomen, I woulde I were so fortunate that I might chooſe euery one of you for
a flower, and then would I boldely affirme that I coulde
ſhewe the faireſt poeſie in the worlde, but follye it is
for me to wiſh that being a ſlaue, which none can hope
for, that is an Emperour.   If I make my choyfe I ſhall
ſpeede so well as he that enioyeth all *Europe.* And with
that gathering a rose he gaue it to *Camilla,* whose
coulour so encreaſed as one would haue iudged al hir
face to haue been a Rose, had it not beene ſtayned with a
naturall whit[e]neſſe, which made hir to excell the Rose.

Camilla with a ſmiling countenance as though nothing greeued, yet vexed inwardly to the heart, refuſed
the gifte flatly, pretending a re[a]dy excuſe, which was,
that *Philautus* was either very much ouer seene to take
hir before the Ladie *Flauia,* or els diſpoſed to giue hir
a mocke aboue the reſt in the companie.

Well quoth *Flauia* to *Philautus,* (who nowe ſtoode
like one that had beene beſmered) there is no harme
done, for I perceiue *Camilla* is otherwiſe ſpedde, and
if I be not much deceiued, ſhe is a flower for *Surius*
wearing, the penance ſhee ſhall haue is to make you
a Noſegay which ſhee ſhall not denye thee, vnleſſe ſhee
defie vs, and the rewarde thou ſhalt haue, is this, while
you tarrie in Englande my neece ſhal be your Violet.

This Ladyes couſin was named *Frauncis,* a fayre
Gentlewoman and a wife, young and of very good conditions, not much inferiour to *Camilla,* ſequall [equall]
ſhee could not be.

Camilla who was lo[a]th to be accompted in any company coye, endeuoured in the preſence of the Ladie

Flauia to be very curteous, and gathered for *Philautus* a pofie of all the fineſt flowers in the Garden, ſaying thus vnto him, I hope you will not be offended *Philautus* in that I coulde not be your Roſe, but imputing the faulte rather to deſtinie then diſcurtefie.

Philautus plucking vp his ſpirits, gaue hir thanks for hir paynes, and immediately gathered a violet, which he gaue miſtres *Frauncis*, which fhe c[o]urteoufly receiued, thus all partes were pleafed for that time.

Philautus was inuited to dinner, ſo that he could no longer ſtay, but pulling out the booke wherein his letter was encloſed, he deliuered it to *Camilla*, taking his humble leaue of the Lady *Flauia* and the reſt of the Gentlewomen.

When he was gone there fell much talke of him be‑ tween the Gentlewomen, one commending his wit, an other his perfonage, ſome his fauour, all his good con‑ ditions infomuch that the Lady *Flauia* bound it with an othe, that fhe thought him both wife and honeſt.

When the company was diſſolued, *Camilla* not think‑ ing to receiue an aunfwere, but a lecture, went to hir Italian booke where fhee founde the letter of *Philau‑ tus*, who without any further aduife, as one very much offended, or in a great heate, ſent him this bone to gnawe vppon.

To Philautus.

Ufficed it not thee *Philautus* to bewraie thy follies and moue my pacience, but thou muſt alfo pro‑ cure in me a minde to reuenge, and to thy felfe the meanes of a farther perill? Where diddeſt thou learne that being forbidden to be bold, thou fhouldeſt growe impudent? or being ſuffered to be familiar thou fhouldeſt waxe haile fellowe? But to ſo malepert boldnes is the demeanor of young Gentlemen come, that where they haue bene once welcome for curtefie, they thinke themfelues worthie to court any Lady by cuſtomes: wherin they imagine they vfe finguler au‑ dacitie which we can no otherwife terme then fauci‑

neffe, thinking women are to be drawen by their coyned and counterfait conceipts, as the ftraw is by the *Aumber*, or the yron by ye Loadftone, or the gold by the minerall *Chryfocolla*.

But as there is no ferpent that can breede in the Box tree for the hardneffe, nor wil build in the Cypres tree for the bitterneffe, fo is there no fond or poyfoned louer that fhall enter into my heart which is hardned like the Adamant, nor take delight in my words, which fhalbe more bitter then Gall.

It fareth with thee *Philautus* as with the droone[Drone], who hauing loft hir [his] owne wings, feekes to fpoile the Bees of theirs, and thou being clipped of thy libertie, goeft about to bereaue me of mine, not farre differing from the natures of Dragons, who fucking bloud out of the Elephant, kill him, and with the fame poyfon themfelues : and it may be that by the fame meanes that thou takeft in hande to inueigle my minde, thou entrap thine owne : a iuft reward, for fo vniuft dealing, and a fit reuenge for fo vnkinde a regard. But I truft thy purpofe fhall take no place, and that thy mallice fhall want might, wherein thou fhalt refemble the ferpent *Porphirius*, who is full of poyfon, but being toothleffe he hurteth none but himfelfe, and I doubt not but thy minde is as ful of deceipt, as thy words are of flatterie, but hauing no toothe [teeth] to bite, I haue no caufe to feare.

I had not thought to haue vfed fo fower words, but where a wande cannot rule the horfe, a fpurre muft. When gentle medicines, haue no force to purge, wee muft vfe bitter potions : and where the fore is neither to be diffolued by plaifter, nor to be broken, it is requifite, it fhould be launced.

Hearbes that are the worfe for watering, are to be rooted out, trees that are leffe fertile for the lopping, are to hewen downe. Hawkes that waxe haggard by manning, are to be caft off, and fonde louers, that encreafe in their follyes when they be reiected, are to bee difpifed.

But as to be without haire, amongſt ye *Mycanions*, is accompted no ſhame, bicauſe they be al borne balde, ſo in *Italy* to lyue in loue, is thought no fault, for that there they are all giuen to luſt, which maketh thee to con- iecture, that we in *England* recken loue as ye [to be the] chiefeſt vertue, which we abhorre as ye greateſt vice, · which groweth lyke the Iuie about the trees, and killeth them by cullyng them. Thou arte alwayes talking of Loue, and applying both thy witte and thy wealth in that idle trade : only for that thou thinkeſt thy ſelfe amiable, not vnlyke vnto the Hedgehogge, who euer- more lodgeth in the thornes, bicauſe he himſelfe is full of prickells.

But take this both for a warning and an aunſwer, that if thou proſecute thy ſuite, thou ſhalt but vndoe thy ſelfe, for I am neither to be wo[o]ed with thy paſſions, whileſt thou liueſt, nor to repent me of my rigor when thou art dead, which I wold not haue thee think to proceede of anye hate I beare thee, for I malyce none, but for loue to mine honour, which neither *Italian* ſhal violate, nor Engliſh man diminiſh. For as the precious ſtone *Chalazias*, being throwen into the fire keepeth ſtil his coldneſſe, not to be warmed with any heate, ſo my heart although dented at with ye arrowes of thy burning affections, and as it were enuironed with the fire of thy loue, ſhall alwayes keepe his hard- neſſe, and be ſo farre from being mollyfied, that thou ſhalt not perceiue it moued.

The Violet Ladie *Flauia* beſtowed on thee, I wiſhe thee, and if thou lyke it, I will further thee, otherwiſe if thou perſiſt in thine olde follyes, wherby to encreaſe my new griefes, I will neither [neuer] come where thou art, nor ſhalt thou haue acceſſe to the place where I am. For as little agreement ſhal there be betweene vs, as is betwixt the Vine and the Cabiſh, the Oke and the Olyue tree, the Serpent and the Aſh tree, the yron and *Theamedes*.

And if euer thou diddeſt loue me, manifeſt it in this, that heereafter thou neuer write to mee, ſo ſhall I both be perfwaded of thy faith, and eafed of mine owne feare. But if thou attempt againe to wring water out of the Pommice, thou fhalt but bewraye thy falfhoode, and augment thy fhame, and my feueritie.

For this I fweare, by hir whofe lyghts can neuer dye, *Vefta*, and by hir whofe heafts are not to be broken, *Diana*, that I will neuer confent to loue him, whofe fight (if I may fo fay with modeftie) is more bitter vnto me then death.

If this aunfwere will not content thee, I wil fhew thy letters, difclofe thy loue, and make thee afhamed to vndertake that, which thou canneft neuer bring to paffe. And fo I ende, thine, if thou leaue to be mine.
Camilla.

C*Amilla* difpatched this letter with fpeede, and fent it to *Philautus* by hir man, which *Philautus* hauing read, I commit the plyght he was in, to the confideration of you Gentlemen that haue ben in the like : he tare his haire, rent his clothes, and fell from the paffions of a Louer to the pang[u]es of phrenfie, but at the laft callying his wittes to him, forgetting both the charge *Camilla* gaue him, and the contents of hir Letter, he greeted hir immedia[t]lye agayne, with an aunfwere by hir owne Meffenger in this manner.

To the cruell Camilla,
greeting.

I F I were as farre in thy bookes to be beleeued, as thou art in mine to be beloued, thou fhouldeft either foone be made a wife, or euer remaine a Virgin, the one would ridde me of hope, the other acquit mee of feare.

But feeing there wanteth witte in mee to perfwade, and will in thee to confent : I meane to manifeft the beginning of my Loue, by the ende of my lyfe, the affects of the one fhal appeare by the effects of the other.

When as neither folempne oath nor found perfwafion, nor any reafon can worke in thee a remorfe, I meane by death to fhew my defire, the which the fooner it commeth, the fweeter it fhalbe, and the fhortnes of the force, fhal abate the fharpnes of the forrow. I cannot tel whether thou laugh at my folly, or lament my phrenfie, but this I fay, and with falt teares. trickling down my cheekes, I fwe[a]re, yat thou neuer foundft more ple[a]fure in reiecting my loue, then thou fhalt feele paine in remembring my loffe, and as bitter fhal lyfe be to thee, as death to me, and as forrowfull fhal my friends be to fee thee profper, as thine glad to fee me perifh.

Thou thinkeft all I write, of courfe, and makeft all I fpeake, of fmall accompt : but God who reuengeth the periuries of the diffembler, is witneffe of my truth, of whom I defire no longer to lyue, then I meane fimply to loue.

I will not vfe many wordes, for if thou be wife, few are fufficient, if froward, fuperfluous : one lyne is inough, if thou be courteous, one word too much, if thou be cruell. Yet this I adde and that in bitternes of foule, that neither my hande dareth write that, which my heart intendeth, nor my tongue vtter that, which my hande fhall execute. And fo fare-well, vnto whom onely I wifh well.

> *Thine euer, though*
> *fhortly neuer.*
> *Philautus.*

THis Letter beeing written in the extremitie of his rage, he fent by him that brought hirs. *Camilla* perceiuing a frefh reply, was not a little melancholy, but digefting it with company, and burning the letter, fhe determined neuer to write to him, nor after yat to fee him, fo refolute was fhe in hir opinion, I dare not fay obftinate leaft you gentlewomen fhoulde take pepper in the nofe, when I put but falt to your mouthes. But this I dare boldly affirme, that Ladies are to be woed with *Appelles* pencill, *Orpheus* Harpe, *Mercuries* tongue, *Adonis* beautie, *Crœfus* we[a]lth, or els neuer tc bewon[n]e, for their bewties [beauties] being blafed, their eares tickled, their mindes moued, their eyes pleafed, there appitite fatiffied, their coffers filled, when they haue al thinges they fhoulde haue and would haue, then men neede not to ftande in doubt of their comming, but of their conftancie.

But let me followe *Philautus*, who nowe both loathing his life and curfing his lucke, called to remembrance his old friend *Euphues*, whom he was wont to haue alwayes in mirth a pleafant companion, in griefe a comforter, in al his life the only ftay of his lybertie, the difcurtefie which hee offered him fo encreafed his greefe, that he fell into thefe termes of rage, as one either in an extafcie, or in a lunacie.

Nowe *Philautus* difpute no more with thy felfe of thy loue, but be defparate to ende thy life, thou haft caft off thy friende, and thy Lady hath forfaken thee, thou deftitute of both, canft neither haue comfort of *Camilla*, whom thou feeft obftinate, nor counfaile of *Euphues*, whom thou haft made enuious.

Ah my good friende *Euphues*, I fee nowe at length, though too late, yat a true friend is of more price then a kingdome, and that the faith of thee is to be preferred, before the beautie of *Camilla*.

For as falfe [fafe] being is it in the company of a truftie mate, as fleeping in the graffe Trifole, where there is no ferpent fo venemous that dare venture.

Thou waft euer carefull of my eftate, and I careleffe for thine, thou diddeft alwayes feare in me the fire of loue, I euer flattered my felfe with the bridle of wifedome, when thou waft earneft to giue me counfaile, I waxed angrie to heare it, if thou diddeft fufpect me vpon iuft caufe, I fel[l] out with thee for euery light occafion, nowe now *Euphues* I fee what it is to want a friend, and what it is to loofe one, thy wordes are come to paffe which once I thought thou fpakeft in fport, but nowe I finde them as a prophecie, that I fhould be conftrayned to ftande at *Euphues* dore as the true owner.

What ſhal I do in this extremitie? which way ſhal I turne me? of whom ſhal I ſeeke remedie? *Euphues* wil reiect me, and why ſhoulde he not? *Camilla* hath reiected me, and why ſhould ſhe? the one I haue offended with too much griefe, the other I haue ſerued with too great good will, the one is loſt with loue, the other with hate, he for that I cared not for him, ſhe becauſe I cared for hir. I but though *Camilla* be not to be moued, *Euphues* may be mollified. Trie him *Philautus*, ſue to him, make friends, write to him, leaue nothing vndone that may either ſhew in thee a ſorrowful heart, or moue in him a minde that is pitifull. Thou knoweſt he is of nature curteous, one that hateth none, that loueth thee, that is tractable in al things, Lions ſpare thoſe yat couch to them, the Tygreſſe biteth not when ſhee is clawed, *Cerberus* barketh not if *Orpheus* pipe ſweetly, aſſure thy ſelf that if thou be penitent, he will bee pleaſed: and the old friendſhip wilbe better then the newe.

Thus *Philautus* ioying nowe in nothing but onely in the hope he had to recouer the friendſhip with repentance, which he had broken off by raſhneſſe, determined to greet his friend *Euphues*, who al this while loſt no time at his booke in London, but howe he imployed it, he ſhall himſelfe vtter, for that I am neither of his counſaile nor court, but what he hath done he will not conceale, for rather he wiſheth to bewray his ignorance, then his ydlenes, and willinger you ſhall find him to make excuſe of rudeneſſe then laſineſſe. But thus *Philautus* ſaluted him.

Philautus to Euphues.

THe ſharpe Northeaſt winde (my good *Euphues*) doth neuer laſt three dayes, tempeſtes haue but a ſhort time, and the more violent the thunder is, the leſſe permanent it is. In the like man[n]er it falleth out with ye iarres and croſſings of friends which begun in a minuit [minute], are ended in a moment.

Neceffary it is that among fri[e]nds there fhould bec
fome ouerthwarting, but to continue in anger not conuenient, the Camill firft troubleth the water before he
drinke, the Frankenfence is burned before it fmell,
friendes are tryed before they are* to* be trufted, leaft
fhining like the Carbuncle as though they had fire, they
be found being touched, to be without fire.

Friendfhippe fhould be like the wine which *Homer*
much commending, calleth *Maroneum*, whereof one
pient[pinte] being mingled with fiue quartes of water,
yet it keepeth his old ftrength and vertue, not to be
qualified by any difcurtefie. Where falt doth grow
nothing els can breede, where friendfhip is built, no
offence can harbour.

Then good *Euphues* let the falling out of fri[e]nd[e]s
be a renewing of affection, that in this we may refemble
the bones of the Lyon, which lying ftil and not moued
begin to rot, but being ftriken one againft another
break out like fire, and wax greene.

The anger of friends is not vnlike vnto the phifitions *Cucurbitæ* which drawing al ye infection in ye body
into one place, doth purge al difeafes, and the rages
[iarres] of friendes, reaping vp al the hidden malices, or
fufpicions, or follyes that lay lurking in the minde
maketh the knot more durable : For as the bodie being
purged of melancholy waxeth light and apt to all labour, fo the minde as it were fcoured of miftruft,
becommeth fit euer after for beleefe.

But why doe I not confeffe that which I haue committed, or knowing my felfe guilty, why vfe I to glofe,
I haue vniuftly my good *Euphues*, picked a quarrel
againft thee, forgetting the counfell thou gaueft [giueft]
me, and defpifing that which I nowe defire. Which as
often as I call to my minde, I cannot but blufh to my
felfe for fhame, and fall out with my felfe for anger.
For in falling out with thee, I haue done no otherwife
then he that defiring [defireth] to faile falfely [fafely]
killeth him at the helme, refembling him that hauing
neede to alight fpurreth his horfe to make him ftande

ftill, or him that fwimming vpon anothers backe, feeketh to ftoppe his breath.

It was in thee *Euphues* that I put all my truft, and yet vppon thee that I powred out all my mallice, more cruel then the Crocadile, who fuffereth the birde to breede in hir mouth, yat fcoureth hir teeth, and nothing fo gentle as the princely Lyon, who faued his life, that helped his foot. But if either thy good nature can for-get, that which my ill tongue doth repent, or thy ac-cuftomable kindneffe forgiue, that my vnbridled furie did commit, I will hereafter be as willing to be thy feruant, as I am now defirous to be thy friend, and as re[a]die to take an iniurie, as I was to giue an offence.

What I haue done in thine abfence I will certifie at thy comming, and yet I doubt not but thou canneft geffe by my condition, yet this I add, that I am as ready to die as to liue, and were I not animated with the hope of thy good counfell, I would rather haue fuffered the death I wifh for, then fuftain the fhame I fought for. But nowe in thefe extremities repofing both my life in thy hands, and my feruice at thy commaundement, I attend thine aunfwere, and reft thine to vfe more then his owne.

Philautus.

THis letter he difpatched by his boye, which *Eu-phues* reading, could not tell whether he fhoulde more reioyce at his friends fubmiffion, or miftruft his fubtiltie, therefore as one not refoluing himfelfe to de-termine any thing, as yet, aunfwered him thus imme-diately by his owne meffenger.

Euphues to him, that was his Philautus.

I Haue receiued thy letter, and know the man : I read it and perceiued the matter, which I am as farre from knowing how to aunfwere, as I was from looking for fuch an errand.

Thou beginneſt to inferre a neceſſitie that friends
ſhould fall out, when as I can-not allowe a[n] [in]conue-
nience. For if it be among ſuch as are faithfull, there
ſhould be no cauſe of breach : if betweene diſſemblers,
no care of reconciliation.

The Camel ſaiſt thou, loueth water, when it is troubled, and I ſay, the Hart thirſteth for the cleare ſtreame :
and fitly diddeſt thou bring it in againſt thy ſelfe (though
applyed it, I know not how aptlye for thy ſelfe) for
ſuch friendſhip doeſt thou lyke, where braules maye
be ſtirred, not quietneſſe ſought.

The wine *Maroneum* which thou commendeſt, and
the ſalt ground which thou inferreſt, ye one is neither
fit for thy drinking, nor the other for thy taſt, for ſuch
ſtrong Wines will ouercome ſuch lyght wits, and ſo good
ſalt cannot relyſh in ſo vnſauory a mouth, neither as
thou deſireſt to applye them, can they ſtande thee in
ſteede. For often-times haue I found much water in
thy deedes, but not one drop of ſuch wine, and the
ground where ſalte ſhould grow, but neuer one corne
that had ſauour.

After many reaſons to conclude, that iarres were
requiſit[e], thou falleſt to a kinde of ſubmiſſion, which I
meruayle at : For if I gaue no cauſe, why diddeſt thou
picke a quarrell : if any, why ſhouldeſt thou craue a
pardon ? If thou canſt defie thy beſt friend, what wilt
thou doe to thine enemie ? Certeinly this muſt needes
enſue, that if thou canſt not be conſtant to thy friend,
when he doth thee good, thou wilt neuer beare with him,
when hee ſhall do thee harme : thou that ſeekeſt to ſpil
the bloud of the innocent, canſt ſhew ſmall mercye
to an offender : thou that treadeſt a Worme on ye
taile, wilt cruſh a Waſpe on the head : thou that art
angry for no cauſe, wilt I thinke runne madde for a
light occaſion.

Truly *Philautus*, that once I loued thee, I can-not
deny, that now I ſhould againe doe ſo, I refuſe : For
ſmal confidence ſhal I repoſe in thee, when I am
guiltie, that can finde no refuge in innocencie.

The malyce of a friend, is like the fling of an Afpe,
which nothing can remedie, for being pearced in the
hande it muft be cut off, and a friend thruft to the
heart it muft be pulled out.

I had as liefe *Philautus* haue a wound that inwardly
might lyghtly grieue me, then a fcar that outwardly.
fhould greatly fhame me.

In that thou feemeft fo earneft to craue attonement
thou caufeft me ye more to fufpeĉt thy truth : for either
thou art compelled by neceffitie, and then it is not
worth thankes, or els difpofed againe to abufe me, and
then it deferueth reuenge. Eeles cannot be helde in a
wet hande, yet are they ftayed with a bitter Figge leafe,
the Lamprey is not to be killed with a cudgel, yet is
fhe fpoiled with a cane, fo friends that are fo flipperie,
and wauering in all their dealyngs are not to be kept
with fayre and fmooth talke, but with rough and fharp
taunts : and contrariwife, thofe which with blowes,
are not to be reformed, are oftentimes wonne with
light perfwafions.

Which way I fhould vfe thee I know not, for now a
fharpe word moued thee, when otherwhiles a fword wil
not, then a friendly checke killeth thee, when a rafor
cannot rafe thee.

But to conclude *Philautus*, it fareth with me now,
as with thofe, that haue bene once bitten with ye
Scorpion, who neuer after feele[th] anye fting, either of
the Wafpe, or the Hornet, or the Bee, for I hauing
bene pricked with thy falfehoode fhall neuer I hope
againe be touched with any other diffembler, flatterer,
or fickle friend.

Touching thy lyfe in my abfence, I feare me it hath
bene too loofe, but feeing my counfell is no more welcome vnto thee then water into a fhip, I wil[l] not waft
winde to inftruĉt him, that wafteth himfelfe to deftroy
others.

Yet if I were as fully perfwaded of thy conuerfion,
as thou wouldeft haue mee of thy confeffion, I might
happely doe that, which now I will not.

And so fare-well *Philautus,* and though thou lyttle
esteeme my counsayle, yet haue respect to thine owne
credite : So in working thine owne good, thou shalt
keepe me from harme.

Thine once,
Euphues.

This letter pinched *Philautus* at the first, yet trust-
ing much to ye good disposition of *Euphues,* he determined to perseuer both in his sute and amend[e]ment,
and ther[e]fore as one beating his yron that he might
frame it while it were hoat, aunswered him in this
manner.

To mine onely friend,
Euphues.

THere is no bone so hard but being laid in vineger,
it might [may] be wrought, nor Iuory so tough,
but seasoned with *Zutho* it may be engrauen, nor Box so
knottie, that dipped in oyle can-not be carued, and can
ther[e] be a heart in *Euphues,* which neither will yeelde
to softnesse with gentle perswasions, nor true perse-
ueraunce ? What canst thou require at my hande, that
I will deny thee ? haue I broken the league of friend-
ship ? I confesse it, haue I misufed thee in termes, I
will not deny it.   But being sorrowfull for either, why
shouldest not thou forgiue both.
   Water is prayfed for that it sauoureth of nothing,
Fire, for that it yeeldeth to nothing : and such should
the nature of a true friend be, that it should not sauour
of any rigour, and such the effect, that it may not be
conquered with any offence : Otherwise, faith put into
the breast that beareth grudges, or contracted with him
that can remember griefes, is not vnlyke vnto Wine
poured into Firre vessels, which is present death to the
drinker.
   Friends must be vfed, as the Musitians tune their

ſtrings, who finding them in a diſcorde, doe not breake them, but either by intention or remiſſion, frame them to a pleaſant conſent : or as Riders handle their young Coltes, who finding them wilde and vntraᶜtable, bring them to a good pace, with a gentle rayne, not with a ſharp ſpurre, or as the *Scithians* ruled their ſlaues not with cruell weapons, but with the ſhewe of ſmall whippes. Then *Euphues* conſider with thy ſelfe what I may be, not what I haue beene, and forſake me not for that I deceiued thee, if thou doe, thy diſcurteſie wil breede my deſtruᶜtion.

For as there is no beaſt that toucheth the hearbe whereon the Beare hath bre[a]thed, ſo there is no man that will come neere him, vpon whom the ſuſpicion of deceipt is faſtened.

Concerning my life paſſed, I conceale it, though to thee I meane hereafter to confeſſe it : yet hath it not beene ſo wicked yat thou ſhouldeſt be aſhamed, though ſo infortunate, that I am greeued. Conſider we are in England, where our demeanour will be narrowly marked if we treade a wrie, and our follyes mocked if [we] vſe wrangling, I thinke thou art willing that no ſuch thing ſhoulde happen, and I knowe thou art wiſe to preuent it.

I was of late in the company of diuers gentlewomen, among whom *Camilla* was preſent, who meruailed not a little, that thou ſoughteſt either to abſent thy ſelfe of ſome conceiued iniurie, where there was none giuen, or of ſet purpoſe, bicauſe thou wouldeſt giue one.

I thinke it requiſite as well to auoyd the ſuſpicion of malice, as to ſhunne ye note of ingratitude, that thou repayre thither, both to purge thy ſelfe of the opinion, may be conceiued, and to giue thankes for the benefits receiued.

Thus aſſuring my ſelfe thou wilt aunſwere my expeᶜtation, and renue our olde amitie, I ende, thine aſſured to commaunde.

Philautus.

PHilautus did not fleepe about his bufines, but prefently fent this letter, thinking that if once he
could faften friendfhippe againe vppon *Euphues*, that
by his meanes he fhould compaffe his loue with *Camilla*, and yet this I durft affirme, that *Philautus* was
both willing to haue *Euphues*, and forrowfull that he
loft him by his owne lauifhnes.

Euphues perufed this letter oftentimes being in a
mammering what to aunfwere, at the laft he determined once againe to lie a loofe, thinking that if *Philautus* meant faithfully, he woulde not defift from his
fuite, and therefore he returned falutations in this
manner.

Euphues to Philautus.

THere is an hearbe in India *Philautus* of plefaunt
fmell, but who fo commeth to it feeleth prefent
fmart, for that there breede in it a number of fmall
ferpents. And it may be that though thy letter be full
of fweete words, there breed in thy heart many bitter
thoughts, fo that in giuing credite to thy letters, I may
be deceiued with thy leafings.

The Box tree is alwayes greene, but the feede is
poyfon : *Tilia* hath a fweete rinde and a ple[a]fa[u]nt
leafe, but ye fruit fo bitter that no beaft wil bite it, a diffembler hath euer-more Honnye in his mouth, and
Gall in his minde, whiche maketh me to fufpecte their
wiles, though I cannot euer preuent them.

Thou fetteft downe the office of a friend, which if
thou couldft as well performe as thou canft defcribe,
I woulde be as willing to confirme our olde league,
as I am to beleeue thy newe lawes. Water that
fauoureth nothing (as thou fayeft) may be heated and
fcald thee, and fire whiche yealdeth to nothing may be
quenched, when thou wouldeft warme thee.

So the friende in whome there was no intent to
offende, may thorowe the finifter dealings of his fellowe

bee turned to heate, beeing before colde, and the faith which wrought like a flame in him, be quenched and haue no fparke.

The powring of Wine into Firre veffels ferueth thee to no purpofe, for if it be good Wine, there is no man fo foolifh to put into Firre, if bad, who woulde power [poure it] into better then Firre.

MuftieCafkes are fitte for rotten Grapes,a barrel[l] of poyfoned Iuie is good ynough for a tunne of ftinking Oyle, and crueltie too milde a medicine for crafte.

Howe Mufitions tune their inftruments I knowe, but how a man fhould temper his friend I cannot tel, yet oftentimes the ftring breaketh that the Mufition feeketh to tune, and the friend cracketh which good counfell fhoulde tame, fuch coltes are to be ridden with a fharpe fnafle, not with a pleafant bitte, and little will the Sithian whippe be regarded, where the fharpnes of the fword is derided.

If thy lucke haue beene infortunate, it is a figne thy liuing hath not beene Godly, for commonly there commeth an yll ende where there was a naughtie beginning.

But learne *Philautus* to liue hereafter as though thou fhouldeft not liue at all, be conftant to them that truft thee, and truft them that thou haft tried, dif-femble not with thy friend, either for feare to difpleafe him, or for malice to deceiue him, know this yat the beft fimples are very fimple, if the phifition could not applie them, that precious ftones were no better then Pebble[s], if Lapidaries did not knowe them, that the beft friende is worfe then a foe, if a man doe not vfe him.

Methridate muft be taken inwardly, not fpread on plaifters, purgations muft be vfed like drink, not like bathes, the counfaile of a friend muft be faftened to the minde, not the eare, followed, not prayfed, em-ployed in good liuing, not talked off in good meaning.

I know *Philautus* we are in England, but I would we wer[e] not, not yat the place is too bafe, but that we are too bad, and God graunt thou haue done nothing which may turne thee to difcredite, or me to difpleafure. Thou fayeft thou werte of late with *Camilla*, I feare me too late, and yet perhaps too foone, I haue alwayes tolde thee, that fhe was too high for thee to clymb, and too faire for others to catch, and too vertuous for any to inueigle.

But wilde horfes breake high hedges, though they cannot leap ouer them, eager Wolues bark at ye Moone though they cannot reach it, and *Mercurie* whifteleth for *Vefta*, though he cannot winne hir.

For abfenting my felfe, I hope they can take no caufe of offence, neither that I knowe haue I giuen any. I loue not to be bold, yet would I be welcome, but geftes [guefts] and fifh fay we in *Athens* are euer ftale within three dayes, fhortly I will vifite them, and excufe my felfe, in the meane feafon I thinke fo well of them, as it is poffible for a man to thinke of women, and how well that is, I appeale to thee who alwayes madeft them no worfe then faints in heauen, and fhrines in no worfe place then thy heart.

For aunfwering thy fuite I am not yet fo haftie, for accepting thy feruice I am not fo imperious, for in friendefhip there muft be an equalitie of eftates, and be* that may bee in vs, alfo a fimilitude of [diuers] manners, and that* cannot, vnleffe thou learne a newe leffon, and leaue the olde, vntill which time I leaue thee, wifhing thee well as to my felfe.

Euphues.

THis Letter was written in haft, fent with fpeed, and aunfwered againe in poft. For *Philautus* feeing fo good counfaile could not proceede of any ill conceipt, thought once againe to follicite his friend, and that in fuch tearmes as he might be moft agreeable to *Euphues* tune. In this manner.

To Euphues health in body, and quietneſſe in minde.

IN Muſicke there are many diſcords, before there can be framed a *Diapaſon*, and in contracting of good will, many iarres before there be eſtabliſhed a friendſhip, but by theſe meanes, the Muſicke is more ſweet, and the amitie more ſound. I haue receiued thy letter, where-in there is as much good counſaile conteined as either I would wiſh, or thou thy ſelfe couldeſt giue : but euer thou harpeſt on that ſtring, which long ſince was out of tune, but now is broken, my inconſtancie.

Certes my good *Euphues*, as I can-not but commend thy wiſedome in making a ſtaye of reconciliation, (for that thou findeſt ſo lyttle ſtay in me) ſo can I not but meruayle at thy incredulytie in not beleeuing me, ſince that thou ſeeſt a reformation in me.

But it maye be thou dealeſt with me, as the Philoſopher did with his knife, who being many yeares in making of it, alwayes dealyng by the obſeruation of the ſtarres, cauſed it at the laſt to cut the hard whetſtone, ſaying that it ſkilled not how long things were a doing, but how well .they were done. And thou holdeſt me off with many delayes, vſing I knowe not what obſeruations, thinking thereby to make me a friend at the laſt, that ſhall laſte : I prayſe thy good meaning, but I miſlyke thy rigour.

Me, thou ſhalt vſe in what thou wilt, and doe that with a ſlender twiſt, that none can doe with a tough wyth. As for my being with *Camilla*, good *Euphues*, rubbe there no more, leaſt I winch, for deny I wil not that I am wroung on the withers.

This one thing touching my ſelfe I ſaye, and before him that ſeeth all things I ſweare, that heereafter I wil neither diſſemble to delude thee, nor pick quarrells to fall out with thee, thou ſhalt finde me conſtant to one, faithleſſe to none, in prayer deuout, in manners reformed, in lyfe chaſt, in words modeſt: not framing my fancie to the humour of loue, but my deedes to the ruleof zeale: And fuch a man as heere-tofore mer[r]ilye thou faideft I was, but now truly thou fhalt fee I am, and as I know thou art.

Then *Euphues* appoint the place where we maye meete, and reconcile the mindes, which I confeffe by mine owne follies were feuered. And if euer after this, I fhall feeme iealous ouer thee, or blynded towards my felfe, vfe me as I deferue, fhamefully.

Thus attending thy fpeedy aunfwere, for that de-layes are perilous, efpecially as my cafe now ftandeth. I ende thine euer to vfe as thine [his] owne.

Philautus.

E*Vphues* feeing fuch fpeedye retourne of an other aunfwere, thought *Philautus* to be very fharp fet, for to recouer him, and weighing with himfelfe, that often in mar[r]iages, ther[e] haue fallen out braules, wher the chiefeft loue fhould be, and yet againe reconcilia-tions, that none ought at any time fo to loue, that he fhould finde in his heart, at any time to hate: Fur-thermore, cafting in his minde the good he might doe to *Philautus* by his friendfhip, and the mifchiefe that might enfue by his fellowes follye, aunfwered him thus agayne fpeedely, afwell to preuent the courfe hee might otherwife take, as alfo to prefcribe what way he fhould take.

Euphues to his friend,
Philautus.

NEttells *Philautus* haue no prickells [prickles], yet they fting, and wordes haue no points, yet they pearce : though out-wardlye thou proteft great amende-ment, yet often-times the foftneffe of Wooll, which the *Seres* fende, fticketh fo faft to the fkinne, that when one looketh it fho[u]ld keepe him warme, it fetcheth bloud, and thy fmooth talke, thy fweete promifes, may when I fhal thinke to haue them perfourmed to delight me, be a corrofiue to deftroy me.

But I w[i]ll not caſt beyonde the Moone, for that in all things I know there muſt be a meane.

Thou ſweareſt nowe that thy lyfe ſhall be leade by my lyne, that thou wilt giue no cauſe of offence, by thy diſorders, nor take anye by my good meaning, which if it bee ſo, I am as willyng to bee thy friend, as I am to be mine owne.

But this take for a warning, if euer thou iarre, when thou ſhouldeſt ieſt, or follow thine owne will, when thou art to heare my counſayle, then will I depart from thee, and ſo diſplay thee, as none that is wiſe ſhall truſt thee, nor any that is honeſt ſhall lyue with thee.

I now am reſolued by thy letter, of that which I was almoſt perſwaded off, by mine owne conieĉture, touching *Camilla.*

Why *Philautus* art thou ſo mad without acquaintaunce of thy part, or familiaritie of hirs, to attempt a thing which will not onely be a diſgrace to thee, but alſo a diſcredite to hir ? Thinkeſt thou thy ſelfe either worthy to wooe hir, or ſhe willyng to wedde thee ? either thou able to frame thy tale to hir content, or ſhee ready to giue ears to thy concluſions ?

No, no *Philautus,* thou art to[o] young to wooe in *England,* though olde inough to winne in *Italy,* for heere they meaſure more the man by the qualyties of his [the] minde, then the proportion of his body. They are too experte in loue, hauing learned in this time of their long peace, euery wrinckle that is to* be* ſeene or imagined.

It is neither an ill tale wel tolde, nor a good hiſtory made better, neither inuention of new fables, nor the reciting of olde, that can eyther allure in them an appetite to loue, or almoſt an attention [intention] to heare.

It fareth not with them as it doth with thoſe in *Italy,* who preferre a ſharpe wit, before found wiſdome, or a proper man before a perfeĉt minde : they lyue not by ſhaddowes, nor feede of the ayre, nor luſte after winde. Their loue is not tyed to Art but reaſon, not to the precepts of *Ouid*, but to the perfwafions of honeftie.

But I cannot but meruayle at thy audacitie, that thou diddeft once dare to moue hir to loue, whom I alwayes feared to follicite in queftioning, afwel doubting to be grauelled by hir quicke and readye witte, as to bee confuted, by hir graue and wyfe aunfweres.

But thou wilt faye, fhe was of no great birth, of meaner parentage then thy felfe. I but *Philautus* they be moft noble who are commended more for their perfection, then their petegree, and let this fuffice thee that hir honour confifted in vertue, bewtie [beautie], witte, not bloode, aunceftors, antiquitie. But more of this at our next meeting ; where I thinke I fhal bee merry to heere the difcourfe of thy madneffe, for I imagine to my felfe that fhee handled thee verye hardely, confidering both the place fhee ferued in, and the perfon that ferued hir. And fure I am fhee did not hang for thy mowing.

A *Phœnix* is no foode for *Philautus*, that dayntie toothe of thine muft bee pulled out, elfe wilt thou furfecte [furfet] with defire, and that Eagles eye pecked out, els wilt [will it] bee dafeled with delyght. My counfaile muft rule thy conceipte, leaft thou confounde vs both.

I will this euening come to thy lodging, where wee will conferre. And till then, I commende mee to thee.

Thine euer to vfe, if
thou be thine owne.
Euphues.

THis letter was fo thankefully receiued of *Philautus*, that he almoft ranne beyonde himfelfe for ioye, preparing all thinges neceffary for the entertainement of his friende, who at the houre appointed fayled not.

Many embracings there were, much ftraunge curtefie, many pretie glaunces, being almoft for the time but ftraungers bicaufe of their long abfence.

But growing to queftioning one with another, they

fell to the whole difcourfe of *Philautus* loue, who left out nothing that before I put in, which I muft omitte, leaft I fet before you, Colewortes twife fodden, whiche will both offende your eares which I feeke to delight and trouble my hande which I couet to eafe.

But this I am fure that *Euphues* conclufion was this, betweene waking and winking, that our Englifh Ladies and Gentlewomen were fo cunning in loue, that the labour were more eafie in *Italie* to wed one and burie hir, then heere to wooe one and marrie hir. And thus they with long talking waxed wearie, wher I leaue them, not willing to talke any longer, but to fleepe their fills till morning.

Now Gentlewomen I appeale in this controuerfie to your confciences, whether there be in you an art to loue, as *Euphues* thinketh, or whether it breede in you as it doth in men : by fight, if one bee bewtifull [beautifull], by hearing, if one be wittie, by defertes if one be curteous, by defire, if one be vertuous, which I woulde not knowe, to this intent that I might bee inftructed howe to winne any of you, but to the ende I might wonder at you all : For if there be in loue an arte, then doe I not meruaile to fee men that euerie way are to bee beloued, fo oftentimes to be reiected. But fo fecreate is this matter, that* perteyning nothing to our fex, I will not farther enquire of it, leaft happily in geffing what art woemen vfe in loue, I fhould minifter an art they neuer before knewe : And fo in thinking to bewray the bayte that hath caught one, I giue them a nette to drawe many, putting a fworde into the hande, where there is but a fheath, teaching them to ftrike, that put vs to our tryings by warding, whiche woulde double our perrill, who without art cannot allure them, and encreafe their tyrany [tirannie], who with-out they torment, will come to no parley.

But this I admonifh you, that as your owne bewties [beauties] make you not couetous of your almes towardes true louers, fo other mens flatterie make you not prodigall of your honours towardes diffemblers. Let not them that fpeake faireſt be beleeued fooneſt, for true loue lacketh a tongue, and is tryed by the eyes, whiche in a hearte that meaneth well, are as farre from wanton glaunces, as the minde is from idle thoughts.

And this art I will giue you, which we men doe commonly praſctife, if you beholde any one that either your curtefie hath allured, or your beautie, or both, triumph not ouer him, but the more earneſt you fee him, the more re[a]die be to followe him, and when he thinketh himfelfe neereſt, let him be fartheſt off: Then if he take that with patience, aſſure your felfe he cannot be faithleſſe.

He that Angleth plucketh the bayte away when he is neere a byte, to the ende the fifh may be more eager to fwallowe the hooke, birds are trayned with a fweet call, but caught with a broade nette: and louers come with fayre lookes, but are entangled with difdainfull eyes.

The Spaniel that fawneth when he is beaten, will neuer forfake his maiſter, the man that do[a]teth when he is difdained, will neuer foregoe his miſtres.

But too much of this ſtring which fowndeth too much out of fquare, and returne we to *Euphues* and *Philautus.*

The next morning when they were ryfen they went into a gallerie, where *Euphues*, who perceiued *Philautus* grieuoufly perplexed for the loue of *Camilla*, beganne thus betweene ieſt and earneſt to talke with him.

P *Hilautus* I haue well nigh all this night beene difputing with my felfe of thy diſtreſſe, yet can I refolue my felfe in nothing that either may content mee, or quiet thee.

What mettall art thou made of *Philautus* that thinkeſt of nothing but loue, and art rewarded with nothing leſſe then loue: *Lucilla* was too badde, yet diddeſt thou court hir, thy fweete heart now in *Naples* is none of the beſt, yet diddeſt thou follow hir, *Camilla* exceeding all, where thou waſt to haue leaſt hope, thou haſt woed, not without great hazard to thy perſon, and griefe to mine.

I haue peruſed hir letters which in my ſimple iudgment are ſo far from al[l]owing thy ſuit, that they ſeeme to loath thy ſeruice. I wil not flatter thee in thy follies, ſhe is no match for thee, nor thou for hir, the one wanting liuing to mainteine a wife, the other birth to aduance an huſbande. *Surius* whome I remember thou diddeſt name in thy diſcourſe, I remember in the court, a man of great byrth and noble blood, ſinguler witte, and a* rare perſonage, if he go about to get credite, I muſe what hope thou couldeſt conceiue to haue a good countenaunce. Well *Philautus* to ſet downe precep[t]s againſt thy loue, will nothing preuaile, to perſwade thee to go forward, were very perillous, for I know in the one loue will regarde no lawes, and in the other perſwaſions can purchaſe no libertie. Thou art too heddie [headie] to enter in where no heed can helpe one out.

Theſeus woulde not goe into the Laborinth without a threede that might ſhew him the way out, neither any wife man enter into the crooked corners of loue, vnleſſe he knew by what meanes he might get out. Loue which ſhould continue for euer, ſhould not be begon [begun] in an houre, but ſlowly be taken in hande, and by length of time finiſhed: reſembling *Zeuxis*, that wife Painter, who in things that he would haue laſt long, tooke greateſt leaſure.

I haue not forgotten one Miſtres *Frauncis*, which the Ladye *Flauia* gaue thee for a Violet, and by thy diſcription, though ſhe be not equall with *Camilla*, yet is ſhe fitter for *Philautus*. If thy humour be ſuch that nothing can feede it but loue, caſt thy minde on hir, conferre the impoſſibilytie thou haſt to winne *Camilla*, with the lykelyhoode thou mayſt haue to enioy thy Violet: and in this I will endeauour both my wit and my good will, ſo that nothing ſhall want in mee, that may work eaſe in thee. Thy Violet if ſhe be honeſt, is worthy of thee, beautiful thou ſayſt ſhe is, and ther·

fore too worthy: Hoat fire is not onely quenched by
ye cleere Fountaine, nor loue onely fatiffied by the
faire face.   Therefore in this tell me thy minde, that
either we may proceede in that matter, or feeke a newe
medicine. *Philautus* thus replyed.

OH my good *Euphues*, I haue neither the power to
forfake mine owne *Camilla*, nor the heart to
deny thy counfaile, it is eafie to fall into a Nette, but
hard to get out.   Notwithftanding I will goe againft
the haire in all things, fo I may pleafe thee in anye
thing, O my *Camilla.*   With that *Euphues* ftayed him
faying.

HE that hath fore eyes muft not behold the candle,
nor he that would leaue his Loue, fall to the
remembring of his Lady, ye one caufeth the eye to
fmart, the other the heart to bleede, wel quoth *Philautus*, I am content to haue the wounde fearched,
yet vnwilling to haue it cured, but fithens that ficke
men are not to prefcribe diets but to keepe them, I
am redie to take potions, and if we[a]lth ferue to paye
thee for them, yet one thing maketh [mee] to feare, that
in running after two Hares, I catch neither.

And certeinelye quoth *Euphues*, I knowe manye
good Hunters, that take more delyght to haue the
Hare on foote, and neuer catch it, then to haue no
crye and yet kill in the Fourme: where-by I geffe, there
commeth greater delyght in the hunting, then in the
eating. It may be fayd *Philautus*, but I were then verye
vnfit for fuch paftimes, for what fporte foeuer I haue
all the day, I loue to haue the game in my difh at
night.

And trulye aunfwered *Euphues*, you are worfe made
for a hound then a hunter, for you marre your fent
with carren, before you ftart your game, which maketh
you hunt oftentimes counter, wher-as if you had kept
it pure, you might ere this time haue tour[ned] the
Hare you winded, and caught the game you courfed.

Why then I perceiue quoth *Philautus,* that to talke
with Gentlewomen, touching the difcourfes of loue,
to eate with them, to conferre with them, to laugh with
them, is as great pleafure as to enioye them, to the
which thou mayft by fome fallacie driue me, but neuer
perfwade me: For then were it as pleafaunt to behold
fruit, as to eate them, or to fee fayre bread, as to
taft it. Thou erreft *Philautus,* fayd *Euphues,* if thou
be not of that minde, for he that commeth into fine
gardens, is as much recreated to fmell the flower[s], as
to gather it. And many we fee more delyghted with
pictures, then defirous to be Painters: the effect of
loue is faith, not luft, delightfull conference, not deteftable concupifcence, which beginneth with folly and
endeth with repentaunce. For mine owne part I would
wifh nothing, if againe I fhould fall into that vaine,
then to haue the company of hir in common conference
that I beft loued, to heare hir fober talke, hir wife
aunfweres, to behold hir fharpe capacitie, and to bee
perfwaded of hir conftancie: and in thefe things do
we only differ from brute beafts, who haue no pleafure,
but in fenfuall appetite. You preach Herefie, quoth
Philautus, and befides fo repugnant to the text you
haue taken, that I am more ready to pull thee out of
thy Pulpit, than to beleeue thy glofes.

I loue the company of women well, yet to haue
them in lawfull Matrimony, I lyke much better, if thy
reafons fhould goe as currant, then were Loue no torment, for hardlye doeth it fall out with him, that is
denyed the fighte and talke of his Ladye.

Hungry ftomackes are not to be fed with fayings againft furfettings, nor thirft to be quenched with
fentences againft drunkenneffe. To loue women and
neuer enioy them, is as much as to loue wine, and
neuer taft it, or to be delighted with fair apparel, and
neuer weare it. An idle loue is that, and fit for him
that hath nothing but eares, that is fati[f]fied to heare
hir fpeak, not defirous to haue himfelfe fpeede. Why
then *Euphues,* to haue the picture of his Lady, is as

much, as to enioy hir prefence, and to reade hir letters
of as great force as to heare hir aunfweres : which if
it be, my fuite in loue fhould be as much to [as] the
painter to draw hir with an amyable face, as to my
Lady to write an amorous letter, both which, with
little fuite being obteined, I may lyue with loue, and
neuer wet my foot, nor breake my fleepes, nor waft
my money, nor torment my minde.

But this worketh as much delyght in the minde of
a louer, as the Apples that hang at *Tantalus* nofe, or
the Riuer that runneth clofe by his chinne. And in
one word, it would doe me no more good, to fee my
Lady and not[to] embrace hir, in the heate of my defire,
then to fee fire, and not warme me in the extremitie
of my colde. No, no *Euphues*, thou makeft Loue
nothing but a continual wooing, if thou barre it of
the effect, and then is it infinite, or if thou allow it,
and yet forbid it, a perpetuall warfare, and then is it
intollerable.

From this opinion no man fhall with-drawe mee,
that the ende of fifhing is catching, not anglyng : of
birding, taking, not whiftlyng : of loue, wedding, not
wooing. Other-wife it is no better then hanging.

Euphues fmilyng to fee *Philautus* fo earneft, vrged
him againe, in this manner.

WHy *Philautus*, what harme were it in loue, if
the heart fhould yeelde his right to the eye,
or the fancie his force to the eare. I haue read of
many, and fome I know, betweene whom there was
as feruent affection as might be, that neuer defired
any thing, but fweete talke, and continuall company
at bankets, at playes, and other affemblyes, as *Phrigius*
and *Pieria*, whofe conftant faith was fuch, that there
was neuer word nor thought of any vncleanneffe.
Pigmalion loued his Iuory Image, being enamoured
onely by the fight, and why fhould not the chaft loue
of others, be builded rather in agreeing in heauenly
meditations, then temporall actions. Beleeue me

Philautus, if thou kneweſt what it were to loue, thou wouldeſt bee as farre from the opinion thou holdeſt, as I am. *Philautus* thinking no greater abſurditie to be held in the world then this, replyed before the other coulde ende, as followeth.

IN deede *Euphues,* if the King would refigne his right to his Legate, then were it not amiſſe for the heart to yeelde to the eyes. Thou knoweſt *Euphues* that the eye is the meſſenger of loue, not the Maſter, that the eare is the caryer of newes, the hearte the difgeſter. Beſides this fuppofe one haue neither eares to heare his Ladie fpeake, nor eyes to fee hir beautie, ſhall he not therefore be fubiect to the impreſſion of loue. If thou aunſwere no, I can alledge diuers both deafe and blinde that haue beene wounded, if thou graunt it, then confeſſe the heart muſt haue his hope, which is neither feeing nor hearing, and what is the thirde?

Touching *Phrigius* and *Peria,* thinke them both fooles in this, for he that keepeth a Hen in his houfe to cackle and not lay, or a Cocke to crowe and not to treade, is not vnlike vnto him that hauing fowen his wheat neuer reapeth it, or reaping it neuer threaſheth it, taking more pleaſure to fee faire corne, then to eate fine bread : *Pigmalion* maketh againſt this, for Venus feeing him fo earneſtly to loue, and fo effectually to pray, graunted him his requeſt, which had he not by importunate fuit obtained, I doubt not but he would rather haue hewed hir in peeces then honoured hir with paſſions, and fet hir vp in fome Temple for an image, not kept hir in his houfe for a wife. He that defireth onely to talke and viewe without any farther fuit, is not farre different from him, that liketh to fee a paynted rofe better then to fmell a perfect Violet, or to heare a birde finge in a buſh, rather then to haue hir at home in his owne cage.

This will I followe, that to pleade for loue and re-queſt nothing but lookes, and to deferue workes, and

liue only by words, is as one fhould plowe his ground
and neuer fowe it, grinde his coulours and neuer paint,
faddle his horfe and neuer ryde.

As they were thus communing there came from the
Ladie *Flauia* a Gentleman who inuited them both that
night to fupper, which they with humble thankes giuen
promifed to doe fo, and till fupper time I leaue them
debating their queftion.

Nowe Gentlewomen in this matter I woulde I knewe
your mindes, and yet I can fomewhat geffe at your
meaninges, if any of you fhoulde loue a Gentleman of
fuch perfection as you can wifh, woulde it content you
onely to heare him, to fee him daunce, to marke his
perfonage, to delight in his witte, to wonder at all his
qualities, and defire no other folace?   If you like to
heare his pleafant voyce to fing, his fine fingers to play,
his proper perfonage to vndertake any exployt, woulde
you couet no more of your loue?   As good it were to
be filent and thinke no, as to blufhe and fay I.

I muft needes conclude with *Philautus*, though I
fhoulde cauill with *Euphues*, that the ende of loue is
the full fruition of the partie beloued, at all times and
in all places.   For it cannot followe in reafon, that
bicaufe the fauce is good which fhoulde prouoke myne
appetite, therefore I fhoulde for-fake the meate for which
it was made.   Beleeue me the qualities of the minde, the
bewtie [beautie] of the bodie, either in man or woman, are
but the fauce to whette our ftomakes, not the meate to
fill them.   For they that liue by the v[i]ew of beautie ftil
looke very leane, and they that feede onely vpon vertue at boorde will go with an hungry belly to bedde.

But I will not craue herein your refolute aunfwere,
bicaufe betweene them it was not determined, but
euery one as he lyketh and then.

Euphues and *Philautus* being nowe againe fent for
to the Lady *Flauia* hir houfe, they came prefently,
where they founde the worthy Gentleman *Surius*,
Camilla, Miftres *Frauncis*, with many other Gentlemen and Gentlewomen.

At their firſt entrance doing their duetie, they ſa-luted all the companie, and were welcommed.

The Lady *Flauia* entertayned them both very louingly, thanking *Philautus* for his laſt company, ſaying be merry Gentleman, at this time of the yeare, a Violette is better then a Roſe, and ſo ſhee aroſe and went hir way, leauing *Philautus* in a muſe at hir wordes, who before was in a maze at *Camillas* lookes. *Camilla* came to *Euphues* in this manner.

I am ſory *Euphues* that we haue no greene Ruſhes, conſidering you haue beene ſo great a ſtraunger, you make me almoſt to thinke that of you which com-monly I am not accuſtomed to iudge of any, that either you thought your ſelfe too good, or our cheere too badde, other cauſe of abſence I cannot imagine, vnleſſe ſe[e]ing vs very idle, you ſought meanes to be well imployed, but I pray you hereafter be bolde, and thoſe thinges which were amiſſe ſhall be redreſſed, for we will haue Quailes to amende your commons, and ſome queſtions to ſharpen your wittes, ſo that you ſhall neither finde faulte with your dyot [diet] for the groſe-neſſe, nor with your exerciſe for the eaſineſſe. As for your fellowe and friende *Philautus* we are bounde to him, for he would oftentimes ſee vs, but ſeldome eate with vs, which made vs thinke that he cared more for our company, then our meat.

Euphues as one that knewe his good, aunſwered hir in this wiſe.

Fayre Ladye, it were vnſeemely to ſtrewe greene ruſhes for his comming, whoſe companie is not worth a ſtrawe, or to accompt him a ſtraunger whoſe boldeneſſe hath bin ſtraunge to all thoſe that knew him to be a ſtraunger.

The ſmal[l] abilitie in me to requite, compared with the great cheere I receiued, might happlie make me refraine which is contrary to your conieᶜture : Whether [Neither] was I euer ſo buſied in any weightie affaircs, whiche I accompted not as loſt time in reſpeᶜt of the exerciſe I alwayes founde in your company, whiche maketh me thinke that your latter obieᶜtion proceeded rather to conuince mee for a treuant, then to manyfeſt a trueth.

As for the Quailes you promiſe me, I can be content with beefe, and for the queſtions they muſt be eaſie, els ſhall I not aunſwere them, for my wit will ſhew with what groſſe diot [diet] I haue beene brought vp, ſo that conferring my rude replyes with my baſe birth, you will thinke that meane cheare will ſerue me, and reſonable queſtions deceiue me, ſo that I ſhall neither finde fault for my repaſt, nor fauour for my reaſons. *Philautus* in deede taketh as much delight in good companie as in good cates, who ſhall anſwere for him-ſelfe, with that *Philautus* ſaide.

Truely *Camilla* where I thinke my ſelfe welcome, I loue to bee bolde, and when my ſtomake is filled I care for no meat, ſo that I hope you will not blame if I came often and eate little.

I doe not blame you by my faith quoth *Camilla*, you miſtake mee, for the oftener you come the better welcome, and the leſſe you eate, the more is ſaued.

Much talke paſſed which being onely as it were a repetition of former thinges, I omitte as ſuperfluous, but this I muſt note, that *Camilla* earneſtly deſired *Surius* to be acquainted with *Euphues*, who very willingly accompliſhed hir requeſt, deſiring *Euphues* for the good report he had harde [heard] of him, that he woulde be as bolde with him, as with any one in Englande, *Euphues* humbly ſhewing his duetie, promiſed alſo as occaſion ſhould ſerue, to trye him.

It now grew toward Supper time, when the table being couered, and the meate ſerued in, Ladye *Flauia* placed *Surius* ouer againſt *Camilla*, and *Philautus* next Miſtres *Frauncis*, ſhe tooke *Euphues* and the reſt, and placed them in ſuch order, as ſhe thought beſt. What cheere they had I know not, what talke they vſed, I heard not: but Supper being ended, they ſate ſtill, the Lady *Flauia* ſpeaking as followeth.

Entlemen and Gentlewomen theſe Lenten Euenings be long, and a ſhame it were to goe to bedde : colde they are, and therefore follye it were to walke abroad : to play at Cardes is common, at Cheftes tedious, at Dice vnfeemely, with Chriftmaffe games, vntimely. In my opinion therefore, to paffe awaye thefe long nights, I would haue fome paftime that might be pleafaunt, but not vnprofitable, rare, but not without reafoning : fo fhall we all accompt the Euening well fpent, be it neuer fo long, which otherwife would be tedious, were it neuer fo fhort.

Surius the beft in the companye, and therefore beft worthy to aunfwere, and the wifeft, and therefore beft able, replyed in this manner.

Ood Madame, you haue preuented my requeft with your owne, for as the cafe now ftandeth, there can be nothing either more agreeable to my humour, or thefe Gentlewomens defires, to vfe fome difcourfe, afwell to renue olde traditions, which haue bene heertofore vfed, as to encreafe friendfhip, which hath bene by the meanes of certeine odde perfons defaced. Euery one gaue his confent with *Surius,* yeelding the choyce of that nights paftime, to the dif- cretion of the Ladie *Flauia* who thus propofed hir minde.

Your tafke *Surius* fhall be to difpute wyth *Camilla,* and cho[o]fe your owne argumente, *Philautus* fhall argue with miftreffe *Frauncis, Martius* wyth my felfe. And all hauing finifhed their difcourfes, *Euphues* fhall be as iudge, who hath done beft, and whatfoeuer he fhal allot eyther for reward, to the worthieft, or for penance to the worft, fhal be prefently accomplifhed. This liked them all exceedingly. And thus *Surius* with a good grace, and pleafaunt fpeache, beganne to enter the liftes with *Camilla.*

Aire Ladie, you knowe I flatter not, I haue reade that the fting of an Afpe were incurable, had not nature giuen them dimme eyes, and the beautie of a woman no leffe infeftious, had not nature beftowed

C C

vpon them gentle hearts, which maketh me ground my
reafon vpon this common place, that beautiful women
are euer mercifull, if mercifull, vertuous, if vertuous
conftant, if conftant, though no more than goddeffes,
yet no leffe than Saintes, all thefe things graunted, I
vrge my queftion without condition.

If *Camilla*, one wounded with your beautie (for
vnder that name I comprehende all other vertues)
fhold fue to open his affeftion, ferue to trie it, and
driue you to fo narrow a point, that were you neuer fo
incredulous, he fhould proue it, yea fo farre to be from
fufpition of deceite, that you would confeffe he were
cleare from diftruft, what aunfweare woulde you make,
if you gaue your confent, or what excufe if you deny
hys curtefie.

Camilla who defired nothing more than to be queftioning with *Surius*, with a modeft countenaunce, yet
fomewhat bafhefull (which added more commendation
to hir fpeache then difgrace) replyed in thys manner.

THough ther be no caufe noble gentleman to fufpect an iniurie where a good turne hath bene receyued, yet is it wifdome to be carefull, what aunfwere
bee made, where the queftion is difficult.

I haue hearde that the Torteife in *India* when the
Sunne fhineth, fwimmeth aboue the water wyth hyr
back, and being delighted with the faire weather, forgetteth hir felfe vntill the heate of the Sunne fo harden hir fhell, that fhe cannot fincke when fhe woulde,
whereby fhe is caught. And fo maye it fare with me,
that in this good companye, difplaying my minde, hauing more regarde to my delight in talkyng, then to the
eares of the hearers, I forget what I fpeake and fo be
taken in fome thing, I fhoulde [would] not vtter, whiche
happilye the itchyng eares of young gentlemen woulde
fo canuas, that when I woulde call it in, I cannot,
and fo be caughte with the Torteife, when I would
not.

Therefore if anything be fpoken eyther vnwares or

vniuftly, I am to craue pardon for both : hauyng but
a weake memorie, and a worfe witte, which you can
not denye me, for that we faye, women are to be borne
withall if they offende againfte theyr wylles, and not
muche to be blamed, if they trip with theyr willes,
the one proceeding of forgetfulneffe, the other, of their
natural weakeneffe, but to the matter.

I F my beautie (whiche God knowes how fimple it
is) fhoulde entangle anye wyth defyre, then fhold
I thus thinke, yat either he were enflamed with luft
rather then loue (for yat he is moued by my countenance not enquiring of my conditions,) or els that I
gaue fome occafion of lightneffe, bicaufe he gathereth
a hope to fpeede, where he neuer had the heart to
fpeake. But if at the laft I fhould perceiue, that his
faith were tried lyke golde in the fire, that his affeftion
proceeded from a minde to pleafe, not from a mouth
to delude, then would I either aunfwer his loue with
lyking, or weane him from it by reafon. For I hope
fir you will not thinke this, but that there fhould be in
a woman afwell a tongue to deny, as in a man to defire,
that as men haue reafon to lyke for beautie, where
they loue, fo women haue wit to refufe for fundry
caufes, where they loue not.

Other-wife were we bounde to fuch an inconuenience, that whofoeuer ferued vs, we fhould aunfwere
his fuite, when in euery refpeft we miflyke his conditions, fo that Nature might be fayd to frame vs for
others humours not for our owne appetites. Wherein
to fome we fhould be thought very courteous, but to
the moft, fcarce honeft. For mine owne part if ther
be any thing in me to be lyked of any, I thinke it
reafon to beftow on fuch a one, as hath alfo fomewhat
to content me, fo that where I knowe my felfe loued,
and doe loue againe, I woulde vppon iuft tryall of his
conftancie, take him.

Surius with-out any ftoppe or long paufe, replyed
prefently.

L Ady if the Torteyfe you fpake off in *India,* wer as
   cunning in fwimming, as you are in fpeaking,
hee would neither feare the heate of the Sunne, nor
the ginne of the Fifher.   But that excufe was brought
in, rather to fhewe what you could fay, then to craue
pardon, for that you haue fayd.   But to your aunfwere.

What your beautie is, I will not heere difpute, leaft
either your modeft eares fhoulde glowe to heare your
owne prayfes, or my fmo[o]th tongue trippe in being
curious to your perfeftion, fo that what I cannot commende fufficiently, I will not ceafe continually to
meruaile at.   You wander in one thing out of the way,
where you fay that many are enflamed with the countenance, not enquiring of the conditions, when this
pofition was before grounded, that there was none beautifull, but fhe was alfo mercifull, and fo drawing by the
face of hir bewtie [beautie] all other morrall vertues, for
as one ring [thing] being touched with the Loadftone
draweth another, and that his fellow, til it come to a
chaine, fo a Lady endewed with bewtie [beautie],
pulleth on curtefie, curtefie mercy, and one vertue
linkes it felfe to another, vntill there be a rare perfeftion.

Befides touching your owne lightneffe, you muft
not imagine that loue breedeth in the heart of man by
your lookes, but by his owne eyes, neyther by your
wordes when you fpeake wittily, but by his owne eares,
which conceiue aptly.   So that were you dumbe and
coulde not fpeak, or blinde and coulde not fee, yet
fhoulde you be beloued, which argueth plainely, that the
eye of the man is the arrow, the bewtie [beautie] of the
woman the white, which fhooteth not, but receiueth,
being the patient, not the agent : vppon triall you confeffe you woulde truft, but what triall you require you
conceale, whiche maketh me fufpeft that either you
woulde haue a triall without meane, or without end,
either not to bee fuftained being impoffible, or not to
be fynifhed being infinite.   Wherein you would haue
one runne in a circle, where there is no way out, or
builde in the ayre, where there is no meanes howe.

Th:s triall *Camilla*, mufl be fifted to narrower pointes, leaft in feeking to trie your louer like a Ienet, you tyre him like a Iade.

Then you require this libertie (which truely I can not denie you) that you may haue the choyce as well to refufe, as the man hath to offer, requiring by that reafon fome quallities in the perfon you would beftow your loue on : yet craftily hyding what properties eyther pleafe you beft, or like woemen well : where-in againe you moue a doubt, whether perfonage, or we[a]lth, or witte, or all are to be required : fo that what with the clofe tryall of his fayth, and the fubtill wifhinge of his quallities, you make eyther your Louer fo holy, that for fayth hee muft be made all of trueth, or fo exquifite that for fhape hee muft be framed in wax : which if it be your opinion, the beautie you haue will be withered before you be wedded, and your wooers good old Gentlemen before they be fpeeders.

Camilla not permitting *Surius* to leape ouer the hedge, which fhe fet for to keepe him in, with a fmiling countenaunce fhaped him this aunfwer.

IF your pofition be graunted, that where beautie is, there is alfo vertue, then myght you adde that where a fayre flower is, there is alfo a fweete fauour, which how repugnant it is to our common experience, there is none but knoweth, and how contrary the other is to trueth, there is none but feeth. Why then do you not fet downe this for a rule which is as agreeable to reafon, that *Rhodope* beeing beautifull (if a good complection and fayre fauour be tearmed beautie) was alfo vertuous : that *Lais* excelling was alfo honeft ? that *Phrine* furpaffing them both in beautie, was alfo curteous ? But it is a reafon among your Philofophers, that the difpofition of the minde, followeth the compofition of the body, how true in arguing it maye bee, I knowe not, how falfe in tryall it is, who knoweth not ? Beautie, though it bee amiable, worketh many things contrarye to hir fayre fhewe, not vnlyke vnto Syluer,

which beeing white, draweth blacke lynes, or refembling the tall trees in *Ida* which allured many to reſt
in them vnder their ſhadow, and then infeċted them
with their ſent.

Nowe where-as you ſette downe, that loue commeth
not from the eyes of the woeman, but from the glaunces
of the man (vnder correċtion be it ſpoken) it is as
farre from the trueth, as the head from the toe.   For
were a Lady blinde, in what can ſhe be beautifull? if
dumbe, in what manifeſt hir witte? when as the eye
hath euer bene thought the Pearle of the face, and
the tongue the Ambaſſadour of the heart? If ther
were ſuch a Ladie in this company *Surius*, that ſhould
wincke with both eyes when you would haue hir ſee
your amorous lookes, or be no blabbe of hir tongue,
when you would haue aunſwere of your queſtions, I
can-not thinke, that eyther hir vertuous conditions,
or hir white and read [red] compleċtion coulde moue
you to loue.

Although this might ſom[e]what procure your liking,
that doing what you lyſt ſhee will not ſee it, and ſpeaking what you would, ſhe will not vtter it, two notable
vertues and rare in our ſex, patience and ſilence.

But why talke I about Ladyes that haue no eies,
when there is no manne that will loue them if hee
him-ſelfe haue eyes.   More reaſon there is to wooe one
that is doumbe [dumb], for that ſhe can-not deny your
ſuite, and yet hauing eares to heare, ſhe may as well giue
an anſwer with a ſigne, as a ſentence. But to the purpoſe.

Loue commeth not from him that loueth, but from
the partie loued, els muſt hee make his loue vppon no
cauſe, and then it is luſt, or think him-ſelfe the cauſe,
and then it is no loue.   Then muſt you conclude
thus, if there bee not in woemen the occaſion, they are
fooles to truſt men that praiſe them, if the cauſe bee in
them, then are not men wiſe to arrogate it to themſelues.

It is the eye of the woman that is made of Adamant,
the heart of the man that is framed of yron, and I cannot thinke you wil ſay that the vertue attraċtiue is in

the yron which is drawen by force, but in the Adamant that fercheth it perforce.

And this is the reafon that many men haue beene entangled againſt their wills with loue, and kept in it with their wills.

You knowe *Surius* that the fire is in the flinte that is ſtriken, not in the ſteele that ſtriketh, the light in the Sunne that lendeth, not in the Moone that bor-[r]oweth, the loue in the woman that is ſerued, not in the man that ſueth.

The fimilitude you brought in of the arrowe, flewe no-thing right to beautie, wherefore I muſt ſhute [ſhoot] that ſhafte at your owne breſt. For if the eye of man be the arrow, and beautie the white (a faire mark for him that draweth in cupids bow) then muſt it neceſſarily enſue, that the archer defireth with an ayme to hitte the white, not the white the arrowe, that the marke allu-reth the archer, not the ſhooter the marke, and therfore is *Venus* faide in one eye to haue two Apples, which is commonly applied to thoſe that witch with the eyes, not to thoſe that wooe with their eyes.

Touching tryall, I am neither ſo fooliſh to defire thinges impoſſible, nor ſo frowarde to requeſt yat which hath no ende. But wordes ſhall neuer make me beleeue without workes, leaſt in following a faire ſhadowe, I loofe the firme fubſtance, and in one worde ſet downe the onely triall that a Ladie requireth of hir louer, it is this, that he performe as much as he ſware, that euery o[a]the be a deede, euery gloaſe a goſpell, promiſing nothing in his talke, that he performe not in his triall.

The qualities that are required of the minde are good conditions, as temperance not to exceede in dyot [diet], chaſtitie not to ſinne in defire, conſtancie not to couet chaunge, witte to delight, wiſdome to inſtruct, myrth to pleaſe without offence, and modeſtie to go-uerne without prefifenes [precifeneſſe].

Concerning the body, as there is no Gentlewoman ſo curious to haue him in print, ſo is there no one ſo careles to haue him a wretch, onlye his right ſhape to fhew him a man, his Chriftendom[e] to proue his faith, indifferent wealth to maintaine his family, expecting al[l] things neceffary, nothing fuperfluous. And to conclude with you *Surius*, vnleffe I might haue fuch a one, I had as leaue be buried as maried, wifhing rather to haue no beautie and dye a chaft virgin, then no ioy and liue a curfed wife.

Surius as one daunted hauing little to aunfwere, yet delighted to heare hir fpeak, with a fhort fpeech vttered thefe words.

I Perceiue *Camilla*, that be your cloath neuer fo badde it will take fome colour, and your caufe neuer fo falfe, it will beare fome fhew of probabilytie, wherein you manifeft the right nature of a woman, who hauing no way to winne, thinketh to ouercome with words. This I gather by your aunfwere, that beautie may haue faire leaues, and foule fruite, that al that are amiable are not honeft, that loue proceedeth of the womans perfection, and the mans follies, that the triall lo[o]ked for, is to performe whatfoeuer they promife, that in minde he be vertuous, in bodye comelye, fuche a hufband in my opinion is to be wifhed for, but not looked for. Take heede *Camilla*, that feeking al the Woode for a ftreight fticke you chufe not at the laft a crooked ftaffe, or prefcribing [defcribing] a good counfaile to others, thou thy felfe follow the worft: much lyke to *Chius*, who felling the beft wine to others, drank him felfe of the lees.

Truly quoth *Camilla*, my Wooll was blacke, and therefore it could take no other colour, and my caufe good, and therefore admitteth no cauill: as for the rules I fet downe of loue, they were not coyned of me, but learned, and being fo true, beleeued. If my fortune bee fo yll that fe[a]rching for a wande, I gather a camocke, or felling wine to other, I drinke vineger my felfe, I muft be content, that of ye worft poore helpe patience, which by fo much the more is to be borne, by howe much the more it is perforce.

As *Surius* was fpeaking, the Ladie *Flauia* preuented him, faying, it is time that you breake off your
fpeach, leaft we haue nothing to fpeak, for fhould you
wade anye farther, you woulde both wafte the night
and leaue vs no time, and take our reafons, and leaue
vs no matter, that euery one therefore may fay fome
what, we commaunde you to ceafe, that you haue both
fayd fo well, we giue you thankes. Thus letting *Surius*
and *Camilla* to whifper by themfelues (whofe talke
we wil[l] not heare) the Lady began in this manner to
greet *Ma[r]tius.*

We fee *Martius* that where young folkes are they
treat of loue, when fouldiers meete they conferre of
warre, painters of their coulours. Mufitians of their
crochets, and euery one talketh of that moft he liketh
beft. Which feeing it is fo, it behoueth vs yat haue
more yeres, to haue more wifdome, not to meafure
our talk by the affections we haue had, but by thofe
we fhould haue.

In this therefore I woulde know thy minde whether
it be conuenient for women to haunt fuch places
where Gentlemen are, or for men to haue acceffe to
gentlewomen, which me thinketh in reafon cannot be
tollerable, knowing yat there is nothing more perni-
cious to either, then loue, and that loue breedeth by
nothing fooner then lookes. They that feare water
will come neere no wells, they that ftande in dreade
of burning flye from the fire : and ought not they that
woulde not be entangled with defire to refraine company? If loue haue ye panges which the paffionate
fet downe, why do they not abftaine from the caufe?
if it be pleafant why doe they difpraife it.

We fhunne the place of peftilence for feare of infection, the eyes of *Cathritiufs* [*Catherifmes*], bicaufe
of difeafes, the fight of the *Bafilifk*, for dreade of death,
and fhall wee not efchewe the companie of them that
may entrappe vs in loue, which is more bitter then any
diftruction ?

If we flye theeues that fteale our goods, fhall wee

followe murtherers yat cut our throates : If we be
heedie to come where Wafpes be, leaft we be ftong,
fhal wee hazarde to runne where *Cupid* is, where
we fhall bee ftifeled? Truely *Martius* in my opinion
there is nothing either more repugnant to reafon, or
abhorring from nature, then to feeke that we fhoulde
fhunne, leauing the cleare ftreame to drinke of the mud-
dye ditch, or in the extremitie of heate to lye in the
parching Sunne, when he may fleepe in the colde
fhadow or being free from fancy, to feeke after loue,
which is as much as to coole a hott[e] Liuer with ftrong
wine, or to cure a weake ftomake with raw flefh. In
this I would heare thy fentence, induced ye rather to
this difcourfe, for that *Surius* and *Camilla* haue begunne it, then that I like it : Loue in mee hath neither
power to commaunde, nor perfwafion to entreate.
Which how idle a thing it is, and how peftilent to
youth, I partly knowe, and you I am fure can geffe.

Martius not very young to difcourfe of thefe matters,
yet defirous to vtter his minde, whether it were to
flatter *Surius* in his will, or to make triall of the Ladies
witte : Began thus to frame his aunfwere.

M Adame, ther[e] is in *Chio* the Image of *Diana*,
which to thofe that enter feemeth [feeme] fharpe
and fower, but returning after their fuites made, lo[o]keth
with a merrie and pleafaunt countenaunce. And it
maye bee that at the ent[e]raunce of my difcourfe yee
will bende your browes as one difpleafed, but hearing
my proofe be delighted and fatiffied.

The queftion you mo[o]ue, is whether it be requifite,
that Gentlemen and Gentlewomen fhould meete. Truly
among Louers it is conuenient to augment defire,
among[e]ft thofe that are firme, neceffary to maintaine
focietie. For to take away all meeting for feare of
loue, were to kindle amongft all, the fire of hate.
There is greater daunger Madame, by abfence, which
breedeth melancholy, then by prefence, which engendreth affeftion.

If the fight be fo perillous, that the company fho[u]ld be barred, why then admit you thole to fee banquets, that may there-by furfet, or fuffer them to eate their meate by a candle that haue fore eyes? To be fepe-rated from one I loue, would make me more conftant, and to keepe company with hir I loue not, would not kindle defire. Loue commeth as well in at the eares, by the report of good conditions, as in at the eyes by the amiable countenaunce, which is the caufe, that diuers haue loued thofe they neuer faw, and feene thofe they neuer loued.

You alleadge that thofe who feare drowning, come neere no wells, nor they that dread burning, neere no fire. Why then let them ftand in doubt alfo to wafhe their handes in a fhallow brooke, for that *Serapus* fallyng into a channell was drowned : and let him that is colde neuer warme his hands, for that a fparke fell into the eyes of *Actina*, whereoff fhe dyed. Let none come into the companye of women, for that diuers haue bene allured to loue, and being refufed, haue vfed vyolence to them-felues.

Let this be fet downe for a law, that none walke abroad in the daye but men, leaft meeting a beautifull woman, he fall in loue, and loofe his lybertie.

I thinke Madam you will not be fo precife, to cut off al conferrence, bicaufe loue commeth by often communication, which if you do, let vs all now pre-fentlye departe, leaft in feeing the beautie which dafeleth our eies, and hearing the wifdom which tick-leth our ears, we be enflamed with loue.

But you fhall neuer beate the Flye from the Candell though he [fhe] burne, nor the Quaile from Hemlocke, though it be[e] poyfon, nor the Louer from the com-panye of his Lady though it be perillous.

It falleth out fundry tymes, that company is the caufe to fhake off loue, working the effects of the roote *Rubarbe*, which beeinge full of choler, purgeth choler, or of the Scorpions fting, which being full of poyfon, is a remedy for poyfon.

But this I conclude, that to barre one that is in loue of the companye of his lady, maketh him rather madde, then mortified, for him to refraine that neuer knewe loue, is eyther to fufpect him of folly with-out caufe, or the next way for him to fall into folly when he knoweth the caufe. A Louer is like [lyke] ye hearb *Helio-tropium*, which alwaies enclyneth to that place where the Sunne fhineth, and being depriued of the Sunne, dieth. For as *Lunaris* hearbe, as long as the Moone wax-eth, bringeth forth leaues, and in the waning fhaketh them of: fo a Louer whilft he is in the company of his Lady, wher al ioyes encreafe, vttereth manye plea-faunt conceites, but banyfhed from the fight of his Miftris, where all mirth decreafeth, eyther lyueth in Melancholie, or dieth with defperation.

The Lady *Flauia* fpeaking in his caft, proceeded in this manner.

TRuely *Martius* I had not thought that as yet your coltes tooth ftucke in your mouth, or that fo olde a trewant in loue, could hether-to remember his leffon. You feeme not to inferre that it is requi-fite they fhould meete, but being in loue that it is conuenient, leaft falling into a mad moode, they pine in their owne peuifhneffe. Why then let it follow, that the Drunckarde which furfeiteth with wine be alwayes quaffing, bicaufe hee liketh it, or the *Epicure* which glutteth him-felfe with meate be euer eating, for that it contenteth him, not feeking at any time the meanes to redreffe their vices, but to renue them. But it fareth with the Louer as it doth with him that powreth in much wine, who is euer more thirftie, then he that drinketh moderately, for hauing once tafted the delightes of loue, he defireth moft the thing that hurteth him moft, not laying a playfter to the wounde, but a corafiue.

I am of this minde, that if it bee daungerous, to laye Flaxe to the fyre, Salte to the eyes, *Sulphure* to the nofe, that then it can-not bee but perillous to let one

Louer come in prefence of the other. For† *Surius* ouer-hearing the Lady, and feeing hir fo earneft, although hee were more earneft in his fuite to *Camilla,* cut hir off with thefe wordes.

Ood Madame giue mee leaue eyther to departe, or to fpeake, for in trueth you gall me more with thefe tearmes, then you wift, in feeming to inueigh fo bitterly againft the meeting of Louers, which is the onely Marrow of loue, and though I doubt not but that *Martius* is fufficiently armed to aunfwere you, yet would I not haue thofe reafons refelled, which I loath to haue repeated. It maye be you vtter them not of malice you beare to loue, but only to moue controuerfie where ther is no queftion : For if thou enuie to haue Louers meete, why did you graunt vs, if allow it, why feeke you to feperate vs ?

The good Lady could not refraine from laughter, when fhe faw *Surius* fo angry, who in the middeft of his own tale, was troubled with hirs, whome fhe thus againe aunfwered.

I crye you mercie Gentleman, I had not thought to haue catched you, when I fifhed for an other, but I perceiue now that with one beane it is eafie to gette two Pigions [Pigeons], and with one baight to haue diuers bits. I fee that others maye geffe where the fhooe wringes, befides him that weares it. Madame quoth *Surius* you haue caught a Frog, if I be not deceiued, and therefore as good it were not to hurt him, as not to eate him, but if all this while you angled to haue a bytte at a Louer, you fhould haue vfed no bitter medicines, but pleafaunt baightes.

I can-not tell anfwered *Flauia,* whether my baight were bytter or not, but fure I am I haue the fifhe by the gill, that doth mee good. *Camilla* not thinking to be filent, put in hir fpoke as fhe thought into the beft wheele, faying.

Lady your cunning maye deceiue you in fifhing

† 'This 'For' is in both editions, but is evidently a slip of the pen.

with an Angle, therfore to catch him you would haue, you were beſt to vſe a net. A net quoth *Flauia*, I neede none, for my fiſhe playeth in a net already, with that *Surius* beganne to winche, replying immediately, ſo doth many a fiſhe good Ladye that ſlyppeth out, when the Fyſher thinketh him faſt in, and it may be, that eyther your nette is too weake to houlde him, or your hand too wette. A wette hande quoth *Flauia* will holde a dead Hearing [Herring]: I quoth *Surius*, but Eeles are no Hearinges [Herrings], but Louers are, ſayde *Flauia*

　Surius not willing to haue the graſſe mowne, whereof hee meant to make his haye, beganne thus to conclude.

Good Lady leaue off fiſhing for this time, and though it bee Lent, rather breake a ſtatute which is but penall, then ſew a pond that maye be perpetuall. I am content quoth *Flauia* rather to faſt for once, then to want a pleaſure for euer : yet *Surius* betwixte vs two, I will at large proue, that there is nothinge in loue more venemous then meeting, which filleth the minde with grief and the body with defeaſes : for hauing the one, he can-not fayle of the other. But now *Philautus* and Neece *Frauncis*, ſince I am cut off, beginne you : but be ſhorte, bicauſe the time is ſhort, and that I was more ſhort then I would.

　Frauncis who was euer of witte quicke, and of nature pleaſaunt, ſeeing *Philautus* all this while to be in his dumpes, beganne thus to playe with him.

Gentleman either you are muſing who ſhal be your ſeconde wife, or who ſhall father your firſt childe, els would you not all this while hang your head, nei- ther attending to the diſcourſes that you haue h[e]ard, nor regarding the company you are in : or it may be (which of both conieⅽtures is likelieſt) that hearing ſo much talke of loue, you are either driuen to the re- membrance of the Italian Ladyes which once you ſerued, or els to the ſeruice of thoſe in Englande which you haue ſince your comming ſeene, for **as**

Andromache when fo euer fhe faw the Tombe of *Hector* coulde not refraine from weeping, or as *Laodamia* could neuer beholde the picture of *Protefilaus* in wax, but fhe alwayes fainted, fo louers when-foeuer they viewe the image of their Ladies, though not the fame fubftance, yet the fimilitude in fhadow, they are fo benummed in their ioints, and fo bereft of their wittes, that they haue neither the power to moue their bodies to fhew life, nor their tongues to make aunfwere, fo yat I thinking that with your other fences, you had alfo loft your fmelling, thought rather to be a thorne whofe point might make you feele fomewhat, then a Violet whofe fauour could caufe you to fmell nothing.

Philautus fe[e]ing this Gentlewoman fo pleafantly difpofed, replyed in this manner.

Entlewoman, to ftudie for a feconde wife before I knowe my firft, were to refemble the good Hufwife in *Naples*, who tooke thought to bring fo[o]rth hir chi[c]kens before fhe had Hens to lay Eg[ge]s, and to mufe who fhould father my firft childe, wer to doubt when the cowe is mine, who fhould owe the calfe. But I will neither be fo haftie to beate my braines about two wiues, before I knowe where to get one, nor fo ie[a]lous to miftruft hir fidelitie when I haue one. Touching the view of Ladies or the remembrance of my loues [loue], me thinketh it fhould rather fharpe the poynt in me then abate the edge. My fences are not loft though my labour bee, and therefore my good Violet, pricke not him forwarde with fharpeneffe, whom thou fhouldeft rather comfort with fauours. But to put you out of doubt that my witts were not al[l] this while a wo[o]l-gathering, I was debating with my felfe, whether in loue it were better to be conftant, bewraying all the counfailes, or fecreat being ready euery hour to flinch : And fo many reafons came to confirme either, that I coulde not be refolued of any. To be conftant what thing more requifite in loue, when it fhall alwayes be greene like the Iuie, though the Sun parch it, that fhal euer be hard like ye true Diamond, though the hammer beate it, that ftill groweth with the good vine, though the knife cut it. Conftancy is like vnto the *Storke*, who wherefoeuer fhe flye commeth into no neaft but hir owne, or the Lapwinge, whom nothing can driue from hir young ones, but death : But to reueale the fecreats of loue, the counfailes, the conclufions, what greater difpite to his Ladie, or more fhamefull difcredite to himfelfe, can be immagined, when there fhall no letter paffe but it fhalbee difclofed, no talke vttered but it fhall bee againe repeated, nothing done but it fhall be reuealed : Which when I confidered, mee thought it better to haue one that fhoulde be fecreate though fickle, then a blab[be] though conftant.

For what is there in the worlde that more deli[gh]teth a louer then fecrecie, whiche is voyde of feare without fufpition, free from enuie: the onely hope a woeman hath to builde both hir honour and honeftie vppon.

The tongue of a louer fhould be like the poynt in the Diall, which though it go, none can fee it going, or a young tree which though it growe, none can perceiue it growing, hauing alwayes the ftone in their mouth which the Cranes vfe when they flye ouer mountaines, leaft they make a noyfe, but to be fylent, and lyghtly to efteeme of his Ladye, to fhake hir off though he be fecreat, to chaunge for euerything though he bewray nothing, is the onely thing that cutteth the heart in peeces of a true and conftant louer, which deepely waying with my felfe, I preferred him that woulde neuer remoue, though he reueiled [reueale] all before him that woulde conceale all, and euer be flyding, thus wafting† to[o] and fro, I appeale to you my good Violet, whether in loue be more required, fecrecie, or conftancy.

Frauncis with hir accuftomable boldnes, yet modeftly, replyed as followeth.

Entleman if I fhoulde afke you whether in the making of a good fworde, yron were more to bee required, or fteele, fure I am you woulde aunfwere that both were neceffarie : Or if I fhoulde be fo curious to demaunde whether in a tale tolde to your Ladyes difpofition, or mention moft conuenient, I cannot thinke but you woulde iudge them both expedient, for as one mettall is to be tempored [tempered] with another in fafhioning, a good blade leaft either, being all of fteele it quickly breake, or all of yron it neuer cutte, fo fareth it in fpeach, which if it be not feafoned as well with witte to mo[o]ue delight, as with art, to manifeft cunning, there is no eloquence, and in no other manner ftandeth it with loue, for to be fecreate [fecret] and not conftant, or conftant and not fecret, were to builde a houfe of morter without ftones, or a wall of ftones without morter.

There is no liuely picture drawen without [with one] colour, no curious Image wrought with one toole, no perfect Mufike played with one ftring, and wouldeft thou haue loue, the patterne of eternitie, couloured either with conftancie alone, or onely fecrecie ?

There muft in euery triangle be three lines, the firft beginneth, the feconde augmenteth, the third concludeth it a figure. So in loue three vertues, affection which draweth the heart, fecrecie which increafeth the hope, conftancie, which finifh[eth] the worke : without any of thefe lynes there can be no triangle, without any of thefe vertues, no loue.

There is no man that runneth with one legge, no birde that flyeth with one winge, no loue that lafteth with one lym [limme]. Loue is likened to the *Emerald* which cracketh rather then confenteth to any difloyaltie, and can there be any greater villany then being fecreat, not to be conftant, or being conftant not to be fecret. But it falleth out with thofe that being conftant and yet full of bab[b]le, as it doth with the ferpent Iaculus and the Viper, who burft with their owne brood, as [and] thefe are torne with their owne tongues.

D D

It is no queſtion *Philautus* to aſke which is beſt, when being not ioyned there is neuer a good. If thou make a queſtion where there is no doubt, thou muſt take an aunſwere where there is no reaſon. Why then alſo doeſt thou not enquire whether it were better for a horſe to want his forelegg[e]s or his hinder, when hauing not all he cannot trauell [trauaile]: why art thou not inquiſitiue, whether it were more conuenient for the wraſtlers in the games of *Olympia* to be without armes or without feete, or for trees to want rootes or lacke tops when either is impoſſible? Ther[e] is no true louer beleeue me *Philautus*, ſence telleth me ſo, not triall, that hath not faith, ſecrecie, and conſtancie. If thou want either it is luſt, no loue, and that thou haſt not them all, thy profound queſtion aſſureth me : which if thou diddeſt aſke to trie my wit, thou thoughteſt me very dull, if thou reſolue thy ſelfe of a doubt, I cannot thinke thee very ſharpe.

Philautus that perceiued hir to be ſo ſharp, thought once againe like a whetſton[e] to make hir ſharper, and in theſe wordes returned his aunſwere.

MY ſweete violet, you are not vnlike vnto thoſe, who hauing gotten the ſtarte in a race, thinke none to be neere their heeles, bicauſe they be formoſt : For hauing the tale in your mouth, you imagine it is all trueth, and that none can controll it.

Frauncis who was not willing to heare him goe forward in ſo fond an argument, cut him off before he ſhould come to his concluſion.

GEntle-man, the faſter you runne after me, the farther you are from me : therefore I would wiſh you to take heede, yat in ſeeking to ſtrik[e] at my heeles, you trippe not vp your owne. You would faine with your witte caſt a white vpon blacke, where-in you are not vnlike vnto thoſe, that ſe[e]ing their ſhadow very ſhort in the Sunne, thinke to touch their head with their heele, and putting forth their legge are farther from it, then when they ftoode ftill. In my opinion it were better to fit on the ground with [a] little eafe, then to ryfe and fall with great daunger.

Philautus beeing in a maze to what end this talke fhould tende, thought that eyther *Camilla* had made hir priuie to his loue, or that fhe meant by fufpition to entrappe him : Therfore meaning to leaue his former queftion, and to aunfwere hir fpeach proceeded thus.

MIftris *Frauncis*, you refemble in your fayings the Painter *Tamantes*, in whofe pictures there was euer more vnderftoode then painted : for with a glofe you feeme to fhadow yat, which in coulours you wil[l] not fhewe. It can-not be, my violet, that the fafter I run after you, the farther I fhoulde bee from you, vnleffe that eyther you haue wings tyed to your heeles, or I thornes thruft into mine. The laft dogge oftentimes catcheth the Hare, though the fleeteft turne him, the flow Snaile clymeth [climbeth] the tower at laft, though the fwift Swallowe mount it, the lafieft winneth the go[a]le, fomtimes, though the lighteft be neere it. In hunting I had as liefe ftand at the receite, as at the loofing, in running rather endure long with an eafie amble, then leaue off being out of winde, with a fwifte gallop : Efpecially when I runne as *Hippomanes* did with *Atlanta*, who was laft in the courfe, but firft at the crowne : So that I geffe that woemen are eyther eafie to be out ftripped [tripped], or willing.

I feeke not to trippe at you, bicaufe I might fo hynder you and hurt my felf : for in letting your courfe by ftriking at your fhorte heeles, you woulde when I fhould craue pardon, fhew me a high inftep. As for my fhadowe, I neuer go about to reach it, but when the Sunne is at the higheft, for then is my fhadowe at the fhorteft, fo that it is not difficult to touch my head with my heele, when it lyeth almofte vnder my heele.

You fay it is better to fit ftill then to aryfe and fall, and I faye hee that neuer clymbeth for feare of fall-

ing, is like vnto him that neuer drincketh for feare
of furfeting.

If you thinke eyther the ground fo flipperie, wherin
[whereon] I runne, that I muft needes fall, or my feete
fo chill that I muft needes founder, it maye be I will
chaunge my courfe here-after, but I meane to ende it
now: for I had rather fall out of a lowe window to
the ground, then hang in midde way by a bryer.

Frauncis who tooke no little pleafure to heare *Philautus* talke, began to come on roundly in thefe tearmes.

I T is a figne Gentleman that your footemanfhip is
    better then your ftomacke: for what-foeuer you
fay, me thinketh you had rather be held in a flippe,
then let flippe, where-in you refemble the graye-hounde,
that feeing his game, leapeth vpon him that holdeth
him, not running after that he [fhee] is held for: or the
Hawke which being caft off at a Partridge, taketh a
ftand to prune hir fe[a]thers, when fhe fhould take hir
flight.   For you [it] feeme[th] you beare good will to
the game you can-not play at, or will not, or dare not,
where-in you imitate the Cat that leaueth the Moufe,
to follow the milk-pan: for I perceiue that you let
the Hare go by, to hunt the Badger.

Philautus aftonied at this fpeache [fpeech], knew not
which way to frame his aunfwere, thinking now that
fhee perceiued his tale to be adreffed to hir, though his
loue were fixed on *Camilla* : But to rydde [rid] hir of fufpition, though loth that *Camilla* fhould conce[i]ue any
inckling, he played faft and loofe in this manner.

Gentle[wo]man you miftake me very much, for I haue
beene better taught then fedde, and therefore I knowe
how to follow my game, if it be for my gaine: For
wer[e] there two Hares to runne at, I would endeauor
not to catch the firft that I followed, but the laft that
I ftarted: yet fo as the firfte fhoulde not fcape, nor
the laft be caught.

You fpeake contraries, quoth *Frauncis*, and you wil[l]
worke wonders, but take heede your cunning in hunting, make you not to loofe both.

Both faid *Philautus*, why I feeke but for one, and
yet of two quoth *Frauncis*, you can-not tell which to
follow, one runneth fo faft you will neuer catch hir, the
other is fo at the fquat, you can neuer finde hir.

The Ladie *Flauia*, whether defirous to fleepe, or
lo[a]th[e] thefe iefts fhould be too broad as moderater
commaunded them both to filence, willing *Euphues* as
vmper in thefe matters, briefly to fpeake his minde.
Camilla and *Surius* are yet talking, *Frauncis* and *Philautus* are not idle, yet all attentiue to heare *Euphues*,
as well for the expectation they had of his wit, as to
knowe the drift of theyr difcourfes, who thus began
the conclufion of all their fpeaches.

I T was a lawe among the *Perfians*, that the Mufitian
fhould not iudge of the Painter, nor anye one
meddle in that handy craft, where-in hee was not expert, which maketh me meruaile good Madam yat
you fhould appoynt him to be an vmper in loue, who
neuer yet had fkill in his lawes. For although I feemed
to confent by my filence before I knewe the argument
where-of you would difpute, yet hearing nothing but
reafons for loue, I muft eyther call backe my promyfe,
or call in your difcourfes, and better it were in my
opinion not to haue your reafons concluded, then to
haue them confuted. But fure I am that neyther a
good excufe will ferue, where authority is rigorous,
nor a bad one be h[e]ard, where neceffitie compelleth.
But leaft I be longer in breaking a web then the Spider
is in weauing it, Your pardons obteyned, if I offend
in fharpneffe, and your patience graunted, if moleft in
length, I thus beginne to conclude againft you all,
not as one finguler in his owne conceite, but to be
tryed by your gentle conftructions.

S *Vrius* beginneth with loue, which proce[e]deth by
beautie, (vnder the whiche hee comprehendeth
all other vertues) Ladye *Flauia* moueth a queftion,
whether the meeting of Louers be tollerable. *Philau-*

tus commeth in with two braunches in his hande, as though there were no more leaues on that tree, afking whether conftancie or fecrecie be moft to be required, great holde there hath beene who fhoulde proue his loue beft, when in my opinion there is none good. But fuch is the vanitie of youth, that it thinketh nothing worthie either of commendation, or conference but onely loue, whereof they fowe much and reape little, wherein they fpende all and gaine nothing, where-by they runne into daungers before they wift, and repent their defires before they woulde. I doe not difcommende honeft affeĉtion, which is grounded vppon vertue as the meane, but difordinate fancie whiche is builded vppon luft as an extremitie: and luft I muft tearme that which is begunne in an houre and ended in a minuit [minute], the common loue in this our age, where Ladyes are courted for beautye, not for vertue, men loued for proportion in bodie, not perfeĉtion in minde.

It fareth with louers as with thofe that drinke of the ryuer *Iellus* in *Phrigia*, whereof fipping moderately is a medecine, but fwilling with exceffe it breedeth madneffe.

Lycurgus fet it downe for a lawe, that where men were commonly dronken, the vynes fhoulde bee deftroyed, and I am of that minde, that where youth is [are] giuen to loue, the meanes fhoulde be remoued. For as the earth wherein the Mynes of Siluer and golde are hidden is profitable for no other thing but mettalles, fo the heart wherein loue is harboured, receiueth no other feede but affeĉtion. Louers feeke not thofe thinges which are moft profitable, but moft pleafant, refembling thofe that make garlands, who choofe the fayreft flowers, not the [w]hol[e]fomeft, and beeing once entangled with defire, they alwayes haue ye difeafe, not vnlike vnto the Goat, who is neuer without an aigue [Ague], then beeing once in, they followe the note of the Nightingale, which is faide with continual ftrayning to finge, to perifhe in hir fweete layes, as they doe in their fugred liues: where is it poffible either to eate

or drinke, or walke but he ſhal[l] heare ſome queſtion of
loue ? in ſomuch that loue is become ſo common, that
there is no artificer of ſo baſe a crafte, no clowne ſo
ſimple, no begger ſo poore, but either talketh of loue,
or liueth in loue, when they neither know the meanes
to come by it, nor the wiſedome to encreaſe it : And
what can be the cauſe of theſe louing wormes, but.
onely idleneſſe ?

But to ſet downe as a moderator the true perfection of loue, not like an enemie to talke of the infeċtion,
(whiche is neither the part of my office, nor pleaſaunt
to your eares,) this is my iudgement.

True and vertuous loue is to be grounded vppon
Time, Reaſon, Fauour and Vertue. Time to make
trial, not at the firſt glaunce ſo to ſettle his minde, as
though he were willing to be caught, when he might
eſcape, but ſo by obſeruation and experience, to builde
and augment his deſires, that he be not deceaued
with beautie, but perſwaded with conſtancie. Reaſon,
that all his doings and proceedings ſeeme not to flowe
from a minde enflamed with luſt, but a true* h[e]art
kindled with loue. Fauour, to delight his eyes, which
are the firſt meſſengers of affeċtion, Vertue to allure
the ſoule, for the which all thinges are to be deſired.

The arguments of faith in a man, are conſtancie
not to be remo[o]ued, ſecrecie not to vtter, ſecuritie not
to miſtruſt, credulitie to beleeue : in a woman patience
to endure, ie[a]louſie to ſuſpeċt, liberalitie to beſtowe,
feruency, faithfulnes, one of the which braunches if
either the man want, or the woman, it may be a lyking
betweene them for the time, but no loue to continue
for euer. Touching *Surius* his queſtion whether loue
come from the man or the woman, it is manifeſt that
it beginneth in both, els can it not ende in both.

To the Lady *Flauias* demaunde concerning companie, it is requiſite they ſhoulde meete, and though
they be hindered by diuers meanes, yet is it impoſſible
but that they will meete.

Philautus muſt this [thus] thinke, that conſtancie

without fecrecie auaileth little, and fecrecie without conftancie profiteth leffe.

Thus haue I good maddame according to my fimple fkill in loue fet downe my iudgement, which you may at your Ladifhippes pleafure correcte, for hee that neuer tooke the* oare in hand muft not think fcorne to be taught. Well quoth the Lady, you can fay more if you lift, but either you feare to offende our eares, or to bewray your own follies, one may eafily perceiue yat you haue bene of late in the painters fhop, by ye colours that fticke in your coate, but at this time I will vrge nothing though I fufpect fomewhat.

Surius gaue *Euphues* thanks, allowing his iudgment in the defcription of loue, efpecially in this, yat he would haue a woman if fhe were faithful to be alfo ielious [iealous], which is [was] as neceffary to be required in them as conftancie.

Camilla fmiling faide that *Euphues* was deceiued, for he would haue faide that men fhould haue bene ielious [iealous], and yet that had bene but fuperfluous, for they are neuer otherwife.

Philautus thinking *Camilla* to vfe that fpeach to girde him for that all that night he v[i]ewed hir with a. fufpitious eye, anfwered that ie[a]loufie in a man was to be pardoned, bicaufe there is no difference in the looke of a louer, that can diftinguifh a ielious [iealous] eye, from a louing.

Frauncis who thought hir part not to be the leaft, faide that in all thinges *Euphues* fpake gofpel fauing in that he bounde a woman to patience, which is [was] to make them fooles.

Thus euery one gaue his verdit, and fo with thanks to the Lady *Flauia*, they all tooke their leaue for that night. *Surius* went to his lodging, *Euphues* and *Philautus* to theirs, *Camilla* accompan[i]ed with hir women [woman] and hir wayting maide, departed to hir home, whome I meane to bring to hir chamber, leauing all the reft to their reft.

Camilla no fooner had entred in* hir chamber, but

fhe began in ftraunge tearmes to vtter this ftraunge tale, hir doore being cloofe fhutte, and hir chamber voyded.

AH *Camilla*, ah wretched wench *Camilla*, I per-ceiue nowe, that when the Hoppe groweth high it muft haue a pole, when ye Iuie fpreadeth, it cleau-eth to ye flint, when the Vine rifeth it wre[a]theth about ye Elme, when virgins wax[e] in yeares, they follow that which belongeth to their appeti[t]es, loue, loue? Yea loue *Camilla*, the force whereof thou knoweft not, and yet muft endure the furie. Where is that precious herbe *Panace* which cureth all difeafes? Or that herbe *Nepenthes* that procureth all delights? No no *Camilla*: loue is not to bee cured by herbes which commeth by fancy, neither can plaifters take away the griefe, which is growen fo great by perfwafions. For as the ftone *Draconites* can by no meanes be polifhed vnleffe the Lapidarie burne it, fo the mind [of] *Camilla* can by no meanes be cured, except *Surius* eafe it.

I fee that loue is not vnlike vnto the ftone *Pantura*, which draweth all other ftones, be they neuer fo h[e]auie, hauing in it the three rootes which they attri-but[e] to Muficke, Mirth, Melancholie, Madneffe.

I but *Camilla* diffemble thy loue, though it fhorten thy lyfe, for better it were to dye with griefe, then lyue with fhame. The Spunge is full of water, yet is it not feene, the hearbe *Adyaton* though it be wet, looketh alwayes drye, and a wife Louer be fhe neuer so much tormented, behaueth hir felfe as though fhee were not touched. I but fire can-not be hydden in the flaxe with-out fmoake, nor Mufke in the bofome with-out fmell, nor loue in the breaft with-out fuf-pition: Why then confeffe thy loue to *Surius, Camilla*, who is ready to afk before thou graunt. But it fareth in loue, as it doth with the roote of ye Reede, which being put vnto the ferne taketh away all his ftrength, and likewife the Roote of the Ferne put to the Reede, depriueth it of all his force: fo the lookes of *Surius* hauing taken all freedome from the eyes of *Camilla*, it may be the glaunces of *Camilla* haue bereaued *Surius* of all libertie, which if it wer[e] fo, how happy fhouldeft thou be, and that it is fo, why fhouldeft not thou hope. I but *Surius* is noble, I but loue regardeth no byrth, I but his friendes will not confent, I but loue knoweth no kindred, I but he is not willing to loue, nor thou worthy to bee wooed, I but loue maketh the proudeft to ftoupe, and to court the pooreft.

Whylft fhe was thus debating, one of hir Maidens chaunced to knocke, which fhe hearing left off that, which al[l] you Gentlewomen would gladly heare, for no doubt fhe determined to make a long fermon, had not fhe beene interrupted : But by the preamble you may geffe to what purpofe the drift tended. This I note, that they that are moft wife, moft vertuous, moft beau- tiful, are not free from the impreffions of Fancy : For who would haue thought that *Camilla*, who feemed to difdaine loue, fhould fo foone be entangled. But as ye ftraighteft wands are to be bent when they be fmall, fo the prefifeft [precifeft] Virgins are to be won when they be young. But I will leaue *Camilla*, with whofe loue I haue nothing to meddle, for that it maketh nothing to my matter. And returne we to *Euphues*, who muft play the laft parte.

E*Vphues* beftowing his time in the Courte, began to marke diligentlye the men, and their manners, not as one curious to mifconfter, but defirous to be inftructed. Manye dayes hee vfed fpeach with the Ladyes, fundrye tymes with the Gentle-women, with all became fo familyar, that he was of all earneftly beloued.

Philautus had taken fuch a fmacke in the good entertainment of the Ladie *Flauia*, that he beganne to look afkew vppon *Camilla*, driuing out the remem- brance of his olde loue, with the recording of the new. Who now but his violet, who but Miftris *Frauncis*, whom if once euery day he had not feene, he wo[u]ld haue beene fo folen, that no man fhould haue feene him.

Euphues who watched his friend, demaunded how his loue proce[e]ded with *Camilla*, vnto whom *Philautus* gaue no aunfwere but a fmile, by the which *Euphues* thought his affeĉtion but fmall. At the laft thinking it both contrary to his o[a]th and his honeftie to conceale anye thinge from *Euphues*, he confeffed, that his minde was chaunged from *Camilla* to *Frauncis.* Loue quoth *Euphues* will neuer make thee mad, for it commeth by fits, not like a quotidian, but a tertian.

In deede quoth *Philautus*, if euer I kill my felfe for loue, it fhall be with a figh, not with a fworde.

Thus they paffed the time many dayes in *England,* *Euphues* commonlye in the court to learne fafhions, *Philautus* euer in the countrey to loue *Frauncis* : fo fweete a violet to his nofe, that he could hardly fuffer it to be an houre from his nofe.

But nowe came the tyme, that *Euphues* was to trye *Philautus* trueth, for it happened that letters were di-reĉted from *Athens* to *London*, concerning ferious and waightie affayres of his owne, which incited him to haften his departure, the contentes of the which when he had imparted to *Philautus*, and requefted his com-pany, his friende was fo faft tyed by the eyes, that he found thornes in his heele, which *Euphues* knewe to be though[t]es in his heart, and by no meanes he could perfwade him to goe into *Italy*, fo fweete was the very fmoke of *England.*

Euphues knowing the tyde would tarrye for no man, and feeing his bufineffe to require fuch fpeede, beeing for his great preferment, determined fodeinly to de parte, yet not with-out taking of his leaue curteouflye, and giuing thankes to all thofe which fince his com-ming had vfed him friendlye : Which that it myght be done with one breath, hee defired the Merchaunt with whome all this while he foiournied to inuite a great number to dynner, fome of great calling, manye of good credit, amonge the which *Surius* as chiefe, the

Ladie *Flauia*, *Camilla* and Miſtris *Frauncis* were not forgotten.

The time being come of meeting, he ſaluted them all in this manner.

I was neuer more deſirous to come into *England* then I am loth to departe, ſuch curteſie haue I found, which I looked not for, and ſuch qualities as I could not looke for, which I ſpeake not to flatter any, when in trueth it is knowne to you all. But now the time is come that *Enphues* muſt packe from thoſe, whome he beſt loueth, and go to the Seas, which he hardlye brooketh. But I would Fortune had de[a]lt ſo fauourably with a poore *Grecian*, that he might haue eyther beene borne heere, or able to liue heere : which ſeeing the one is paſt and can-not be, the other vnlik[e]ly, and therfore not eaſie to be, I muſt endure the crueltie of the one, and with patience beare the neceſſitie of the other.

Yet this I earneſtly craue of you all, that you wil[l] in ſteede of a recompence accept thankes, and of him that is able to giue nothing, take prayer for payment. What my good minde is to you all, my tongue can-not vtter, what my true meaning is, your heartes can-not conceiue : yet as occaſion ſhall ſerue, I will ſhewe that I haue not forgotten any, though I may not requit[e] on[e]. *Philautus* not wiſer then I in this, though bolder, is determined to tarry behinde : for hee ſayth that he had as liefe be buried [burned] in *England*, as married in *Italy* : ſo holy doth he thinke the ground heere, or ſo homely the women ther[e], whome although I would gladly haue with me, yet ſeeing I can-not, I am moſt earneſtlye to requeſt you all, not for my ſake, who ought to deſire nothing, nor for his ſake who is able to deſerue little, but for the curteſies ſake of *England*, that you vſe him not ſo well as you haue done, which wold make him proud, but no worſe then I wiſh him, which wil[l] make him pure : for tho[u]gh I ſpeak before his face, you ſhall finde true behinde his backe, that **he is yet but wax**, which muſt be wrought whileſt the

water is warme, and yron which being hot, is apt
either to make a key or a locke.

It may be Ladies and Gentlewoemen all, that
though *England* be not for *Euphues* to dwell in, yet it
is for *Euphues* to fend to.

When he had thus fayd, he could fcarfe fpeake for
weeping, all the companye were forye to forgoe him,
fome proffered him mony, fome lands, fome houfes,
but he refufed them all, telling them that not the
neceffitie of lacke caufed him not* to departe, but of
importance.

This done they fate downe all to dinner, but *Euphues* could not be merry, for yat he fhould fo foone
depart, ye feaft being ended, which was very fumptuous, as Merchaunts neuer fpare for coft, when they
haue ful[l] coffers, they al heartely tooke their leaues of
Euphues, *Camilla* who liked verie well of his company, taking him by the hande, defired him that being
in *Athens*, he woulde not forget his friends in Englande, and the rather for your fake quoth fhe, your
friende fhalbe better welcome, yea, and to me for his
owne fake quoth *Flauia*, where at *Philautus* reioyced
and *Frauncis* was not forie, who began a little to
liften to the lure of loue.

Euphues hauing all thinges in a re[a]dineffe went immediately toward Douer, whether *Philautus* alfo accompanied him, yet not forgetting by the way to
vifite the good olde father *Fidus*, whofe curtefie they
receaued [receiued] at their comming. *Fidus* glade to fee
them made them great cheare according to his abilitie,
which had it beene leffe, woulde haue bene aunfwerable to either [their] defires. Much communication
they had of the court, but *Euphues* cryed quittance, for
he faide thinges that are commonly knowne it were
folly to repeat, and fecretes, it were againft mine honeftie to vtter.

The next morning they went to Douer where *Eu
phues* being readie to take fhip, he firft tooke his farewell of *Philautus* in thefe wordes.

PHilautus the care that I haue had of thee, from time to time, hath beene tried by the counfaile I haue alwayes giuen thee, which if thou haue for-gotten, I meane no more to write in water, if thou remember imprint it ftill. But feeing my departure from thee is as it were my death, for that I knowe not whether euer I fhall fee thee, take this as my laft teftament of good will.

Bee humble to thy fuperiours, gentle to thy equalls, to thy inferiours fauourable, enuie not thy betters, iuftle not thy fellowes, oppreffe not the poore.

The ftipende that is allowed to maintaine thee vfe wifely, be neither prodigall to fpende all, nor couetous to keepe all, cut thy coat according to thy cloth, and thinke it better to bee accompted thriftie among the wife, then a good companion among the riotous.

For thy ftudie or trade of life, vfe thy booke in the morning, thy bowe after dinner or what other exer-cife fhall pleafe thee beft, but alwayes haue an eye to the mayne, what foeuer thou art chaunced at the buy.

Let thy practife be lawe, for the practife of Phifike is too bafe for fo fyne a ftomacke as thine, and diuini-tie too curious for fo fickle a heade as thou haft.

Touching thy proceedings in loue, be conftant to one, and trie but one, otherwife thou fhalt bring thy credite into queftion, and thy loue into derifion.

Weane thy felfe from *Camilla*, deale wifely with *Frauncis*, for in Englande thou fhalt finde thofe that will decypher thy dealings be they neuer fo politique, be fecret to thy felfe, and truft none in matters of loue as thou loueft thy life.

Certifie me of the [thy] proceedings by thy letters, and thinke that *Euphues* cannot forget *Philautus*, who is as deare to mee as my felfe. Commende me to all my friendes: And fo farewell good *Philautus*, and well fhalt thou fare if thou followe the counfell of *Euphues*.

PHilautus the water ftanding in his eyes, not able to aunfwere one worde vntill he had well wepte, replyed at the laſt as it were in one worde, ſaying, that his counſaile ſhoulde bee engrauen in his heart, and hee woulde followe euerie thing that was pre ſcribed him, certifying him of his ſucceſſe as either occaſion, or opportunitie ſhould ſerue.

But when friendes at departing [parting] woulde vtter moſt, then teares hinder moſt, whiche breake off both his aunſwere, and ſtayde *Euphues* replye, ſo after many millions of embracinges, at the laſt they departed. *Philautus* to London where I leaue him, *Euphues* to *Athens* where I meane to followe him, for hee it is that I am to goe with, not *Philautus.*

THere was nothing that happened on the Seas worthie the writing, but within fewe dayes *Euphues* hauing a merrye winde arryued at *Athens*, where after hee had viſited his friendes, and ſet an order in his affayres, he began to addreſſe his letters to *Liuia* touching the ſtate of Englande in this manner.

L*Iuia* I ſalute thee in the Lorde, &c. I am at length returned out of Englande, a place in my opinion (if any ſuch may be in the earth) not inferiour to a Paradiſe.

I haue here incloſed ſent thee the diſcription, the manners, the conditions, the gouernement and enter- tainement of that countrie.

I haue thought it good to dedicate it to the Ladies of *Italy*, if thou thinke it worthy, as thou canneſt not otherwiſe, cauſe it to be imprinted, that the praiſe of ſuch an Iſle, may cauſe thoſe yat dwell els where, both to commende it, and maruell at it.

Philautus I haue left behinde me, who like an olde dogge followeth his olde ſent, loue, wiſer he is then he was woont, but as yet nothing more fortunate. I am in he[a]lth, and that thou art ſo, I heare nothing to the contrarie, but I knowe not howe it fareth with me, for I cannot as yet brooke mine owne countrie, I am ſo delighted with another.

Aduertiſe me by letters what eſtate thou art in, alſo

howe thou likeſt the ſtate of Englande, which I haue
ſent thee.   And ſo farewell.

Thine to vſe Euphues.

*To the Ladyes and Gentlewomen of
Italy: Euphues wiſheth he[a]lth
and honour.*

I F I had brought (Ladyes) little dogges from *Malta*,
   or ſtraunge ſtones from *India*, or fine carpets from
Turkie, I am ſure that either you woulde haue wo[o]ed
me to haue them, or wiſhed to ſee them.

But I am come out of Englande with a Glaſſe,
wherein you ſhall behold the things which you neuer
ſawe, and maruel at the ſightes when [which] you haue
ſeene. Not a Glaſſe to make you beautiful, but to make
you bluſh, yet not at your vices, but others vertues, not
a Glaſſe to dreſſe your haires but to redreſſe your
harmes, by the which if you euery morning correcte
your manners, being as carefull to amend faultes in
your hearts, as you are curious to finde faults in your
heads, you ſhall in ſhort time be as much commended
for vertue of the wiſe, as for beautie of the wanton.

Yet at the firſt ſight [if] you ſeeme deformed by looking in this glaſſe, you muſt not thinke that the fault is
in the glaſſe, but in your manners, not reſembling
Lauia, who ſeeing hir beautie in a true glaſſe to be
but deformitie, waſhed hir face, and broke the glaſſe.

Heere ſhall you ſee beautie accompanyed with virginitie, temperaunce, mercie, iuſtice, magnanimitie,
and all other vertues whatſo[e]uer, rare in your ſex, and
but one, and rarer then the *Phœnix* where I thinke
there is not one.

In this glaſſe ſhall you ſee that the glaſſes which
you carrye in your fannes of fethers, ſhewe you to be
lyghter then fethers, that the Glaſſes wher-in you
carouſe your wine, make you to be more wanton then

Bacchus, that the new found glaffe Cheynes that you
weare about your neckes, argue you to be more brittle
then glaffe. But your eyes being too olde to iudge of
fo rare a fpectacle, my counfell is that you looke with
fpectacles: for ill can you abyde the beames of the
cleere Sunne, being fkant [fcant] able to view the blafe
of a dymme candell. The fpectacles I would haue you
vfe, are for the one eie iudg[e]ment with-out flattering
your felues, for the other eye, beliefe with-out miftruft-
ing of mee.

And then I doubte not but you fhall both thanke
mee for this Glaffe (which I fende alfo into all places
of *Europe*) and thinke worfe of your garyfhe Glaffes,
which maketh you of no more price then broken
Glaffes.

Thus fayre Ladyes, hoping you will be as willing
to prye in this Glaffe for amendement of manners, as
you are to prancke your felues in a lookinge Glaffe,
for commendation of menne, I wifhe you as much
beautie as you would haue, fo as you woulde endeuo[u]r
to haue as much vertue as you fhould haue. And fo
farewell.

Euphues.

¶ *Euphues Glaffe for
Europe.*

THere is an Ifle lying in the *Ocean* Sea, directly
againft that part of *Fraunce,* which containeth
Picardie and *Normandie,* called now *England,* heeretofore named *Britaine,* it hath *Ireland* vpon the Weft fide,
on the North the maine Sea, on the Eaft fide, the *Germanie Germaine*] *Occan.* This Ifla[n]de is in circuit
1720. myles, in forme like vnto a Triangle, beeing broad-
eft in the South part, and gathering narrower and narrower till it come to the fartheft poynt of Cathneffe,
Northward, wher it is narroweft, and ther[e] endeth in

E E

manner of a Promonterie. To repeate the auncient manner of this Ifland, or what fundry nations haue inhabited there, to fet downe the Giauntes, which in bygneffe [highneffe] of bone haue paffed the common fife, and almoft common creditte, to rehearfe what di- uerfities of Languages haue beene vfed, into how many kyngdomes it hath beene deuided, what Religions haue beene followed before the comming of Chrift, although it would breede great delight to your eares [eyes], yet might it happily feeme tedious : For that honnie taken exceffiuelye cloyeth the ftomacke though it be honnie.

But my minde is briefly to touch fuch things as at my being there I gathered by myne owne ftudie and enquirie, not meaning to write a Chronocle [Cronicle], but to fet downe in a word what I heard by conference.

It hath in it twentie and fixe Cities, of the which the chiefeft is named *London*, a place both for the beautie of buyldinge, infinite riches, varietie of all things, that excelleth all the Cities in the world : infomuch that it maye be called the Store-houfe and Marte of all *Europe*. Clofe by this Citie runneth the famous Ryuer called the Theames [*Thames*], which from the head wher[e] it ryfeth named *Ifis*, vnto the fall [full] middway it is thought to be an hundred and forefcore myles. What can there be in anye place vnder the heauens, that is not in this noble Citie eyther to be bought or borrowed?

It hath diuers Hofpitals for the relieuing of the poore, fix-fcore fayre Churches for diuine feruice, a gloryous Burfe which they call the Ryoll Exchaung [*Royall Ex- change*], for the meeting of Merchants of all countries where anye traffique is to be had. And among al[l] the ftraung[e] and beautifull fhowes, mee thinketh there is none fo notable, as the Bridge which croffeth the Theames [*Thames*], which is in manner of a continuall ftreete, well replenyfhed with large and ftately houfes on both fides, and fituate vpon twentie Arches, where-of each one is made of excellent free ftone fquared, euerye one of them being three-fcore foote in h[e]ight, and full twentie in diftaunce one from an other.

To this place the whole Realme hath his recourfe, wher-by it feemeth fo populous, that one would fcarfe think fo many people to be in the whole Ifland, as he fhall fee fomtymes in *London.*

This maketh Gentlemen braue, and Merchaunts rich, Citifens to purchafe, and foiourn[er]s to morgage, fo that it is to be thought, that the greateft wealth and fubftaunce of the whole Realme is couched with-in the walles of *London*, where they that be rich keepe it from thofe that be ryotous, not deteining it from the luftie youthes of *England* by rigor, but encreafing it vntill young men fhall fauor of reafon, wherein they fhew them-felues Trefurers [treaforers] for others, not horders for them-felues, yet although it be fure enough, woulde they had it, in my opinion, it were better to be in the Gentle-mans purfe, then in the Merchants handes.

There are in this Ifle two and twentie Byfhops, which are as it wer[e] fuperentendaunts ouer the church, men of great zeale, and deepe knowledge, diligent Preachers of the worde, earneft followers of theyr doctrine, carefull watchmenne that the Woulfe deuoure not the Sheepe, in ciuil gouernment politique, in ruling the fpirituall fworde (as farre as to [in] them vnder their Prince apperteineth) iuft, cutting of thofe members from the Church by rigor, that are obftinate in in their herifies, and inftructing thofe that are ignoraunt, appoynting godlye and learned Minifters in euery of their Seas, that in their abfence maye bee lightes to fuch as are in darkeneffe, falt to thofe that are vnfauorie, leauen to fuch as are not feafoned.

Vifitations are holden oftentymes, where-by abufes and diforders, eyther in the laitie for negligence, or in the clergie for fuperftition, or in al, for wicked liuing three [there] are punyfhements, by due execution wherof the diuine feruice of God is honoured with more puritie, and followed with greater finceritie.

There are alfo in this Iflande two famous Vniuerfities, the one *Oxforde*, the other *Cambridge*, both for the profeffion of al[l] fciences, for Diuinitie, phificke, Lawe, and all kinde of learning, excelling all the Vniuerfities in Chriftendome.

I was my felfe in either of them, and like them both fo well, that I meane not in the way of controuerfie to preferre any for the better in Englande, but both for the beft in the world, fauing this, that Colledges in *Oxenford* are much more ftately for the building, and *Cambridge* much more fumptuous for the houfes in the towne, but the learning neither lyeth in the free ftones of the one, nor the fine ftreates of the other, for out of them both do dayly proceede men of great wifedome, to rule in the common we[a]lth, of learning to inftruct the common people, of all finguler kinde of profeffions to do good to all. And let this fuffice, not to enquire which of them is the fuperiour, but that neither of them haue their equall, neither to afke which of them is the moft auncient, but whether any other bee fo famous.

But to proceede in Englande, their buildings are not very ftatelye vnleffe it be the houfes of noble men and here and there, the place of a Gentleman, but much amended, as they report yat haue told me. For their munition they haue not onely great ftore, but alfo great cunning to vfe them, and courage to practife them, there armour is not vnlike vnto that which in other countries they vfe, as Corfelets, Almaine Riuetts, fhirts of male, iack[e]s quilted and couered ouer with Leather, Fuftion or Canuas, ouer thicke plates of yron that are fowed in [to] the fame.

The ordinaunce they haue is great, and thereof great ftore.

Their nauie is deuided as it were into three forts, of the which the one ferueth for warres, the other for burthen, the thirde for fifhermen. And fome veffels there be (I knowe not by experience, and yet I beleeue by circumftance) that will faile nyne hundred myles in a weeke, when I fhould fcarce thinke that a birde could flye foure hundred.

Touching other commodities, they haue foure bathes, the firſt called Saint *Vincents*: the ſeconde, *Hallie well*, the third *Buxton*, the fourth (as in olde time they reade) *Cair Bledud*, but nowe taking his name of a town neere adioyning it, is called the *Bath*.

Befides this many wonders there are to be found in this Ifland, which I will not repeat bicaufe I my felfe neuer fawe them, and you haue hearde of greater.

Concerning their dyot [diet], in number of diſhes and chaung[e] of meate, ye nobilitie of England do exceed moſt, hauing all things yat either may be bought for money, or gotten for the feafon : Gentlemen and merchaunts feede very finely, and a poore man it is that dineth with one diſh, and yet ſo content with a little, that hauing halfe dyned, they fay as it were in a prouerbe, yat they are as well fatiffied as the Lorde Maior of London whom they think to fare beſt, though he eate not moſt.

In their meales there is great filence and grauitie, vfing wine rather to eafe the ſtomacke, then to load it, not like vnto other nations, who neuer thinke that they haue dyned till they be dronken.

The attire they vfe is rather ledde by the imitation of others, then their owne inuention, ſo that there is nothing in Englande more conſtant, then the inconſtancie of attire, nowe vfing the French faſhion, nowe the Spaniſh, then the Morifco gownes, then one thing, then another, infomuch that in drawing of an Engliſh man ye paynter fetteth him downe naked, hauing in ye one hande a payre of ſheares, in the other a piece of cloath, who hauing cut his choler [cholar] after the french guife is readie to make his fleeue after the Barbarian manner. And although this were the greateſt enormitie that I coulde fee in Englande, yet is it to be excufed, for they that cannot maintaine this pride muſt leaue of neceſſitie, and they that be able, will leaue when they fee the vanitie.

The lawes they vfe are different from ours for although the Common and Ciuil lawe be not aboliſhed, yet are they not had in fo greate reputation as their owne common lawes which they tearme the lawes of the Crowne.

The regiment that they haue dependeth vppon ftatute lawe, and that is by Parl[i]ament which is the higheft court, confifting of three feueral fortes of people, the Nobilitie, Clergie, and Commons of the Realme, fo as whatfoeuer be among them enacted, the Queene ftriketh the ftroke, allowing fuch things as to hir maiefty feemeth beft. Then vpon common law, which ftandeth vpon Maximes and principles, yeares and tearmes, the cafes in this lawe are called plees, or actions, and they are either criminall or ciuil, ye meane to determine are writts, fome originall, fome iudiciall : Their trials and recoueries are either by verdect [verdit], or demur, confeffion or default, wherin if any fault haue beene committed, either in proceffe or forme, matter or iudgement, the partie greeued may haue a write [writ] of errour.

Then vpon cuftomable law, which confifteth vppon laudable cuftomes, vfed in fome priuate countrie.

Laft of all vppon prefcription, whiche is a certeine cuftome continued time out of minde, but it is more particular then their cuftomary lawe.

Murtherers and theeues are hanged, witches burnt, al other villanies that deferue death punifhed with death, infomuch that there are very fewe haynous offences practifed in refpecte of thofe that in other countries are commonly vfed.

Of fauage beaftes and vermyn they haue no great ftore, nor any that are noyfome, the cattell they keepe for profite, are Oxen, Horfes, Sheepe, Goats, and Swine, and fuch like, whereof they haue abundance, wildfo[u]le and fifh they want none, nor any thing that either may ferue for pleafure or profite.

They haue more ftore of pafture then tillage, their meddowes better then their corne field[s], which maketh more grafiors then Cornemungers, yet fufficient ftore of both.

They excel for one thing, there [their] dogges of al forts, fpan[i]els, hounds, maiſtiffes, and diuers fuch, the one they keepe for hunting and hawking, the other for neceſſarie vfes about their houfes, as to drawe water, to watch theeues, &c. and there-of they deriue the worde maſtiffe of Mafe and thiefe.

There is in that Iſle Salt made, and Saffron, there are great quarries of ſtone[s] for building, fundrie minerals of Quickſiluer, Antimony, Sulphur, blacke Lead and Orpiment redde and yellowe. Alſo there groweth ye fineſt Alum yat is, Vermilion, Bittament, Chrifocolla, Coporus [Coperus], the mineral ſtone whereof Petreolum is made, and that which is moſt ſtraunge, the minerall pearle, which as they are for greatneſſe and coulour moſt excellent, ſo are they digged out of the maine lande, in places farre diſtant from the ſhoare.

Befides thefe, though not ſtraunge, yet neceſſarie, they haue Cole mines, falt Peter for ordinance, Salt Sode for Glaſſe.

They want no Tinne nor Leade, there groweth Yron, Steele and Copper, and what not, ſo hath God bleſſed that countrie, as it ſhoulde feeme not onely to haue fufficient to ferue their owne turnes, but alfo others neceſſities, whereof there was an olde faying, all countries ſtande in neede of *Britaine*, and *Britaine* of none.

Their Aire is very wholfome and pleafant, their ciuilitie not inferiour to thofe that deferue beſt, their wittes very ſharpe and quicke, although I haue heard that the *Italian* and the *French-man* haue accompted them but grofe and dull pated, which I think came not to paſſe by the proofe they made of their wits, but by the Englifhmans reporte.

For this is ſtraunge (and yet how true it is, there is none that euer trauailed thether but can reporte) that it is alwayes incident to an Englifh-man, to thinke worſt of his owne nation, eyther in learning, experience, common reafon, or wit, preferring alwaies a ſtraunger rather for the name, then the wifdome. I for mine owne parte thinke, that in all *Europe* there are not

Lawyers more learned, Diuines more profound, Phifitions more expert, then are in *England.*

But that which moſt allureth a ſtraunger is their curteſie, their ciuilitie, and good entertainment. I ſpeake this by experience, that I found more curteſie in *England* among thoſe I neuer knewe, in one yeare, then I haue done in *Athens* or *Italy* among thoſe I euer loued, in twentie.

But hauing entreated ſufficiently of the countrey and their conditions, let me come to the Glaſſe I promiſed being the court, where although I ſhould as order requireth beginne with the chiefeſt, yet I am enforced with the Painter, to reſerue my beſt colours to end *Venus*, and to laie the ground with the baſeſt.

Firſt then I muſt tell you of the graue and wiſe Counſailors, whoſeforeſight in peace warranteth ſaf[e]tie in warre, whoſe prouiſion in plentie, maketh ſufficient in dearth, whoſe care in health is as it were a preparatiue againſt ſickneſſe, how great their wiſdom hath beene in all things, the twentie two yeares peace doth both ſhew and proue. For what ſubtilty hath ther[e] bin wrought ſo cloſly, what priuy attempts ſo craftily, what rebellions ſtirred vp ſo diſorderly, but they haue by policie bewrayed, preuented by wiſdome, repreſſed by iuſtice? What conſpiracies abroad, what confederacies at home, what iniuries in anye place hath there beene contriued, the which they haue not eyther foreſeene before they could kindle, or quenched before they could flame?

If anye wilye *Vlyſſes* ſhould faine madneſſe, there was amonge them alwayes ſome *Palamedes* to reueale him, if any *Thetis* went about to keepe hir ſonne from the doing of his countrey ſeruice, there was alſo a wiſe *Vlyſſes* in the courte to bewraye it: If *Sinon* came with a ſmoothe tale to bringe in the horſe into *Troye*, there hath beene alwayes ſome couragious *Lacaon* to throwe his ſpeare agaynſt the bowelles, whiche beeing not bewitched with *Lacaon*, hath vnfoulded that, which *Lacaon* ſuſpeſted.

If *Argus* with his hundred eyes went prying to
vndermine *Iupiter*, yet met he with *Mercurie*, who
whif[t]elled all his eyes out : in-fomuch as ther[e] coulde
neuer yet any craft preuaile againſt their policie, or
any chalenge againſt their courage. There hath alwayes beene *Achilles* at home, to buckle with *Hedor*
abroad, *Neſtors* grauitie to counteruaile *Priams* coun-
fail, *Vliſſes* fubtilties to ma[t]ch with *Antenors* policies.
England hath al[l] thofe, yat can and haue wreſtled with
al others, wher-of we can require no greater proofe
then experience.

Befides they haue al[l] a ze[a]lous care for the encreafing of true religion, whofe faiths for the moſt part hath
bin [beene] tried through the fire, which they had felt,
had not they fledde ouer the water. More-ouer the great
ſtudie they bend towards fchooles of learning, both
fufficiently declare, that they are not onely furtherers
of learning, but fathers of the learned. O thrife [thrice]
happy *England* where fuch Counfaylours are, where
fuch people liue, where fuch vertue fpringeth.

Amonge thefe fhall you finde *Zopirus* that will
mangle him-felfe to do his country good, *Achates* that
will neuer ſtart an ynch from his Prince *Aeneas*,
Naufícla that neuer wanted a fhift in extremitie, *Cato*
that euer counfayled to the beſt, *Ptolomeus Phila-
delphus* that alwaies maintained learning. Among
the number of all which noble and wife counfailors,
(I can-not but for his honors fake remember) the moſt
prudent and right honourable ye Lorde *Burgleigh*,
high Treafurer of that Realme, no leffe reuerenced for
his wifdome, than renowmed for his office, more loued
at home then feared abroade, and yet more feared for
his counfayle amonge other nations, then fworde or
fyre, in whome the faying of *Agamemnon* may be verified, who rather wifhed for one fuch as *Neſtor*, then
many fuch as *Aiax*.

This noble man I found fo ready being but a
ſtraunger, to do me good, that neyther I ought to for-.
get him, neyther ceafe to pray for him, that as he hath

the wifdome of *Neſtor*, fo he may haue the age, that
hauing the policies of *Vlyſſes*, he may haue his honor,
worthye to lyue long, by whome fo manye lyue in
quiet, and not vnworthy to be aduaunced, by whofe
care fo many haue beene preferred.

Is not this a Glaſſe fayre Ladyes for all other
countrie[s] to beholde, wher[e] there is not only an agreement in fayth, religion, and counfayle, but in friend-
fhyppe, brother-hoode and lyuing? By whofe good
endeuours vice is punyſhed, vertue rewarded, peace
eſtablyſhed, forren broyles repreſſed, domeſticall cares
appeafed? what nation can of Counſailors defire
more? what Dominion, yat excepted, hath fo much?
when neither courage can preuaile againſt their chiual-
rie, nor craft take place agaynſt their counfayle, nor
both ioyned in one be of force to vndermine their
country, when you haue dafeled your eies with this
Glaſſe, behold here an other. It was my fortune to be
acquainted with certaine Engliſh Gentlemen, which
brought mee to the court, wher[e] when I came, I was
driuen into a maze to behold the luſty and braue gallants, the be[a]utiful and chaſt Ladies, ye rare and
godly orders, fo as I could not tel whether I ſhould
moſt commend vertue or brauery. At the laſt comming oft[e]ner thether, then it befeemed one of my
degree, yet not fo often as they defired my company,
I began to prye after theyr manners, natures, and
lyues, and that which followeth I faw, where-of who fo
doubteth, I will fweare.

The Ladyes fpend the morning in deuout prayer,
not refembling the Gentlewoemen in *Greece* and *Italy*,
who begin their morning at midnoone, and make their
euening at midnight, vfing fonets for pfalmes, and
paſtymes for prayers, reading ye Epiſtle of a Louer,
when they ſhould perufe the Gofpell of our Lorde,
drawing wanton lynes when death is before their face,
as *Archimedes* did triangles and circles when the
enimy was at his backe. Behold Ladies in this glaſſe,
that the feruice of God is to be preferred before all

things, imitat[e] the Englyfh Damofelles, who haue theyr bookes tyed to theyr gyrdles, not fe[a]thers, who are as cunning in ye fcriptures, as you are in *Ariofto* or *Petrack* or anye booke that lyketh you beft, and be- commeth you moft.

For brauery I cannot fay that you exceede them, for certainly it is ye moft gorgeouft [gorgious] court that euer I haue feene, read, or heard of, but yet do they not vfe theyr apperell fo nicelye as you in *Italy*, who thinke fcorn to kneele at feruice, for feare of wrinckles in your filks, who dare not lift vp your head to heauen, for feare of rumpling ye rufs in your neck, yet your hands I confeffe are holden vp, rather I thinke to fhewe your ringes, then to manifeft your righteouf- neffe. The brauerie they vfe is for the honour of their Prince, the attyre you weare for the alluring of your pray, the ritch apparell maketh their beautie more feene, your difguifing caufeth your faces to be more fufpected, they refemble in their rayment the *Eftrich* who being gafed on, clofeth hir winges and hideth hir fethers, you in your robes are not vnlike the pecocke, who being prayfed fpreadeth hir tayle, and bewrayeth hir pride. Veluetts and Silkes in them are like golde about a pure Diamond, in you like a greene hedge, about a filthy dunghill. Thinke not Ladies that bi- caufe you are decked with golde, you are endued with grace, imagine not that fhining like the Sunne in earth, yea fhall climbe the Sunne in heauen, looke diligently into this Englifh glaffe, and then fhall you fee that the more coftly your apparell is, the greater your curtefie fhould be, that you ought to be as farre from pride, as you are from pouertie, and as neere to princes in beautie, as you are in brightnes. Bi- caufe you are braue, difdaine not thofe that are bafe, thinke with your felues that ruffet coates haue their Chriftendome, that the Sunne when he is at his h[e]ight fhineth afwel vpon courfe carfie, as cloth of tiffue, though you haue pearles in your eares, Iewels in your breaftes, preacious ftones on your fingers, yet difdaine not the ftones in the ftreat, which although they are nothing fo noble, yet are they much more neceffarie. Let not your robes hinder your deuotion, learne of the Englifh Ladies, yat God is worthy to be worfhipped with the moft price, to whom you ought to giue all praife, then fhall you be like ftars to ye wife, who now are but ftaring ftockes to the foolifh, then fhall you be prayfed of moft, who are now pointed at of all, then fhall God beare with your folly, who nowe abhorreth your pride.

As the Ladies in this bleffed Iflande are deuout and braue, fo are they chaft and beautifull, infomuch that when I firft behelde them, I could not tell whether fome mift had bleared myne eyes, or fome ftra[u]ng[e] enchauntmentaltered my minde, for it may bee, thought I, that in this Ifland, either fome *Artimedorus* or *Lifi-mandro*, or fome odd *Nigromancer* did inhabit, who would fhewe me Fayries, or the bodie of *Helen*, or the new fhape of *Venus*, but comming to my felfe, and feeing that my fences were not chaunged, but hindered, that the place where I ftoode was no enchaunted caftell, but a gallant court, I could fcarce reftraine my voyce from crying, *There is no beautie but in England.* There did I behold them of pure complexion, exceeding the lillie, and the rofe, of fauour (wherein ye chiefeft beautie confifteth) furpaffing the pictures that were feyned [fained], or the Magition that would faine, their eyes pe[a]rcing like the Sun beames, yet chaft, their fpeach pleafant and fweete, yet modeft and curteous, their gate comly, their bodies ftraight, their hands white, al[l] things that man could wifh, or women woulde haue, which howe much it is, none can fet downe, when as ye one defireth as much as may be, the other more. And to thefe beautifull mouldes, chaft mindes: to thefe comely bodies temperance, modeftie, mildeneffe, fo-brietie, whom I often beheld merrie yet wife, conferring with courtiers yet warily : drinking of wine yet moderately, eating of delicat[e]s yet but their eare ful, lift[en]ing to difcourfes of loue but not without reafoning of learning : for there it more delighteth them to talke of Robin hood, then to fhoot in his bowe, and greater pleafure they take, to heare of loue, then to be in loue. Heere Ladies is a Glaffe that will make you blufh for fhame, and looke wan for anger, their beautie commeth by nature, yours by art, they encreafe their fauours with faire water, you maintaine yours with painters colours, the haire they lay out groweth vpon their owne heads, your feemelines hangeth vpon others, theirs is alwayes in their owne keeping, yours often in the Dyars, their bewtie [beautie] is not loft with a fharpe blaft, yours fadeth with a foft breath: Not vnlike vnto Paper Floures [flowers], which breake as foone as they are touched, refembling the birds in *Aegypt* called *Ibes*, who being handled, loofe their feathers, or the ferpent *Serapie*, which beeing but toucht with a brake, burfteth. They vfe their beautie, bicaufe it is commendable, you bicaufe you woulde be common, they if they haue little, doe not feeke to make it more, you that haue none endeauour to befpeake moft, if theirs wither by age they nothing efteeme it, if yours waft by yeares, you goe about to keepe it, they knowe that beautie muft faile if life continue, you fweare that it fhall not fade if coulours laft.

But to what ende (Ladies) doe you alter the giftes of nature, by the fhiftes of arte ? Is there no colour good but white, no Planet bright but *Venus*, no Linnen faire but Lawne ? Why goe yee about to make the face fayre by thofe meanes, that are moft foule, a thing loathfome to man, and therefore not louely, horrible before God, and therefore not lawefull.

Haue you not hearde that the beautie of the Cradell is moft brighteft, that paintings are for pictures with out fence, not for perfons with true reafon. Follow at the laft Ladies the Gentlewomen of *England*, who being beautifull doe thofe thinges as fhall beecome fo amyable faces, if of an indifferent h[i]ew[e], thofe things as they fhall make them louely, not adding an ounce to beautie, that may detract a dram from vertue. Befides this their chaftitie and temparance [temperaunce] is as

rare, as their beautie, not going in your footesteppes, that
drinke wine before you rise to encrease your coulour,
and swill it when you are vp, to prouoke your lust:
They vse their needle to banish idlenes, not the pen
to nourish it, not spending their times in answering ye
letters of those that woe them, but forswearing the companie of those that write them, giuing no occasion
either by wanton lookes, vnseemely gestures, vnaduised
speach, or any vncomly behauiour, of lightnesse, or
liking. Contrarie to the custome of many countries,
where filthie wordes are accompted to sauour of a
fine witte, broade speach, of a bolde courage, wanton
glaunces, of a sharpe eye sight, wicked deedes, of a
comely gesture, all vaine delights, of a right curteous
curtesie.

And yet are they not in England presise [precise],
but wary, not disdainefull to conferre, but careful [feare-
full] to offende, not without remorse where they perceiue trueth, but without replying where they suspect
tre[a]cherie, when as among other nations, there is
no tale so lothsome to chast eares but it is heard with
great sport, and aunswered with great speade [speede].

Is it not then a shame (Ladyes) that that little
Island shoulde be a myrrour to you, to Europe, to the
whole worlde?

Where is the temperance you professe when wine
is more common then water? where the chastity when
lust is thought lawful, where the modestie when your
mirth turneth to vncleanes, vncleanes to shamelesnes,
shamelesnesse to al sinfulnesse? Learne Ladies though
late, yet at length, that the chiefest title of honour in
earth, is to giue all honour to him that is in heauen,
that the greatest brauerie in this worlde, is to be burning lampes in the worlde to come, that the clearest
beautie in this life, is to be amiable to him that shall
giue life eternall: Looke in the Glasse of England, too
bright I feare me for your eyes, what is there in your
sex that they haue not, and what that you should not
haue?

They are in prayer deuoute, in brauery humble, in beautie chaft, in feafting temperate, in affeċtion wife, in mirth modeft, in al[l] their aċtions though courtlye, bicaufe woemen, yet Aungels, [Angels] bicaufe virtuous.

Ah (good Ladies) good, I fay, for that I loue you, I would yee [you] could a little abate that pride of your ftomackes, that loofeneffe of minde, that lycentious behauiour which I haue feene in you, with no fmal[l] forrowe, and can-not remedy with continuall fighes.

They in *England* pray when you play, fowe when you fleep, faft when you feaft, and weepe for their fins, when you laugh at your fenfualitie.

They frequent the Church to ferue God, you to fee gallants, they deck them-felues for cle[a]nlineffe, you for pride, they maintaine their beautie for their owne lyking, you for others luft, they refraine wine, bicaufe they fear to take too much, you bicaufe you can take no more. Come Ladies, with teares I call you, looke in this Glaffe, repent your fins paft, refrain your pre-fent vices, abhor vanities to come, fay thus with one voice, *we can fee our faults only in the Englifh Glaffe*: a Glas of grace to them, of grief to you, to them in the* fteed of righteoufnes, to you in place of repentance. The Lords and Gentlemen in ye [that] court are alfo an example for all others to fol[l]ow, true tipes [types] of nobility, the only ftay and ftaf[fe] to [of] honor, braue courtiers, ftout foldiers, apt to reuell in peace, and ryde in warre. In fight fearce [fierce], not dreading death, in friendfhip firme, not breaking promife, curteous to all that deferue well, cruell to none that deferue ill. Their aduerfaries they truft not, that fheweth their wifdome, their enimies they feare not, that argueth their courage. They are not apt to proffer iniuries, nor fit to take any : loth to pick quarrels, but longing to reuenge them.

Aċtiue they are in all things, whether it be to wreftle in the games of *Olympia,* or to fight at Barriers in *Paleftra,* able to carry as great burthens as *Milo,* of ftrength to throwe as byg ftones as *Turnus,* and what not that eyther man hath done or may do, worthye of

fuch Ladies, and none but they, and Ladies willing
to haue fuch Lordes, and none but fuch.

This is a Glaffe for our youth in *Greece*, for your
young ones in *Italy*, the Englifh Glaffe, behold it
Ladies and* Lordes, and all, that eyther meane to haue
pietie, vfe brauerie, encreafe beautie, or that defire
temperancie, chaftitie, witte, wifdome, valure, or any
thing that may delight your felues, or deferue praife of
others.

But an other fight there is in my Glaffe, which
maketh me figh for griefe I can-not fhewe it, and yet
had I rather offend in derogating from my Glaffe, then
my good will.

Bleffed is that Land, that hath all commodities to
encreafe the common wealth, happye is that Iflande
that hath wife counfailours to maintaine it, vertuous
courtiers to beautifie it, noble Gentle-menne to aduance it, but to haue fuche a Prince to gouerne it, as
is their Soueraigne queene, I know not whether I
fhould thinke the people to be more fortunate, or the
Prince famous, whether their felicitie be more to be
had in admiration, that haue fuch a ruler, or hir vertues to be honoured, that hath fuch royaltie : for fuch
is their eftat[e] ther[e], that I am enforced to think
that euery day is as lucky to the Englifhmen, as the
fixt daye of Februarie hath beene to the *Grecians*.

But I fee you gafe vntill I fhew this Glaffe, which
you hauing once feene, wil make you giddy : Oh Ladies
I know not when to begin, nor where to ende :
for the more I go about to expreffe the brightnes, the
more I finde mine eyes bleared, the neerer I defire to
come to it, the farther I fe[e]me from it, not vnlike vnto
Simonides, who being curious to fet downe what God
was, the more leyfure he tooke, the more loth hee was
to meddle, faying that in thinges aboue reach, it was
eafie to catch a ftraine, but impoffible to touch a Star :
and ther[e]fore fcarfe tollerable to poynt at that, which
one can neuer pull at. When *Alexander* had commaunded that none fhoulde paint him but *Appelles*,

none carue him but *Lyfippus*, none engraue him but
Pirgotales [*Pergotales*], *Parrhafius* framed a Table
fquared, euerye way twoo hundred foote, which in the
borders he trimmed with frefh coulours, and limmed
with fine golde, leauing all the other roume [roome]
with-out knotte or lyne, which table he prefented to
Alexander, who no leffe meruailing at the bignes, then
at the barenes, demaunded to what ende he gaue him
a frame with-out face, being fo naked, and with-out
fafhion being fo great. *Parrhafius* aunfwered him, let
it be lawful for *Parrhafius*, O *Alexander*, to fhew a
Table wherin he would paint *Alexander*, if it were not
vnlawfull, and for others to fquare Timber, though
Lyfippus carue it, and for all to caft braffe though
Pirgoteles [*Pergoteles*] ingraue it. *Alexander* perceiuing
the good minde of *Parrhafius*, pardoned his boldneffe,
and preferred his arte: yet enquyring why hee framed
the table fo bygge, hee aunfwered, that hee thought
that frame to bee but little enough for his Picture,
when the whole worlde was to little for his perfonne,
faying that *Alexander* muft as well bee prayfed, as
paynted, and that all hys victoryes and vertues, were
not for to bee drawne in the Compaffe of a Sygnette,
[Signet] but in a fielde.

This aunfwer *Alexander* both lyked and rewarded,
infomuch that it was lawful euer after for *Parrhafius*
both to praife that noble king and to paint him.

In the like manner I hope, that though it be not
requifite that any fhould paynt their Prince in *England*, that can-not fufficiently perfect hir, yet it fhall
not be thought rafhneffe or rudeneffe for *Euphues*, to
frame a table for *Elizabeth*, though he prefume not to
paynt hir. Let *Appelles* fhewe his fine arte, *Euphues*
will manifeft his faythfull heart, the one can but proue
his conceite to blafe his cunning, the other his good
will to grinde his coulours : hee that whetteth the
tooles is not to bee mifliked, though hee can-not carue
the Image, the worme that fpinneth the filke, is to be
efteemed, though fhe cannot worke the fampler, they

F F

that fell tymber for fhippes, are not to be blamed, bi-caufe they can-not builde fhippes.

He that caryeth morter furthereth the building, though hee be no expert Mafon, hee that diggeth the garden, is to be confidered, though he cannot treade the knottes, the Golde-fmythes boye muſt haue his wages for blowing the fire, though he can-not fafhion the Iewell.

Then Ladyes I hope poore *Euphues* fhalt not bee reuiled, though hee deferue not to bee rewarded. I will fet downe this *Elizabeth*, as neere as I can : And it may be, that as the *Venus* of *Appelles*, not finifhed, the *Tindarides* of *Nichomachus* not ended, the *Medea* ot *Timomachus* not perfeǒted, the table of *Parrhafius* not couloured, brought greater defire to them, to con-fumate them, and to others to fee them : fo the *Eliza-beth* of *Euphues*, being but fhadowed for others to vernifh, but begun for others to ende, but drawen with a blacke coale, for others to blafe with a bright cou-lour, may worke either a defire in *Euphues* heereafter if he liue, to ende it, or a minde in thofe that are better able to amende it, or in all (if none can worke it) a wil[1] to wifh it. In the meane feafon I fay as *Zeuxis* did when he had drawen the piǒture of *Ata-lanta*, more wil enuie me then imitate me, and not commende it though they cannot amende it. But I come to my *England*.

There were for a long time ciuill wars in this [the] countrey, by reafon of feueral claymes to the Crowne, betweene the two famous and noble houfes of *Lancaſter* and *Yorke*, either of them pretending to be of the royall bloude, which caufed them both to fpende their vitall bloode, thefe iarres continued long, not without great loffe, both to the Nobilitie and Communaltie, who ioyning not in one, but diuers parts, turned the realme to great ruine, hauing almoſt deſtroyed their countrey before they coulde annoynt a king.

But the lyuing God who was loath to oppreffe *England*, at laſt began to repreffe iniuries, and to giue

an ende by mercie, to thofe that could finde no ende of
malice, nor looke for any ende of mifchiefe. So tender
a care hath he alwaies had of that *England*, as of a
new *Ifrael*, his chofen and peculier [peculiar] people.

This peace began by a marriage folemnized by
Gods fpeciall prouidence, betweene *Henrie* Earle of
Ritchmond heire of the houfe of *Lancafter*, and *Eliza-*
beth daughter to *Edward* the fourth, the vndoubted
iffue and heire of the houfe of *Yorke*, where by (as they
tearme it) the redde Rofe and the white, were vnited
and ioyned together. Out of thefe Rofes fprang two
noble buddes, Prince *Arthur* and *Henrie*, the eldeft
dying without iffue, the other of moft famous memorie, leauing behinde him three children, Prince *Ed-*
warde, the Ladie *Marie*, the Ladie *Elizabeth*. King
Edwarde liued not long, which coulde neuer for that
Realme haue liued too long, but fharpe froftes bite
forwarde fpringes, Eafterly windes blafteth towardly
bloffoms, cruell death fpareth not thofe, which we
our felues liuing cannot fpare.

The elder fifter the Princes *Marie*, fucceeded as
next heire to the crowne, and as it chaunced nexte
heire to the graue, touching whofe life, I can fay little
bicaufe I was fcarce borne, and what others fay, of me
fhalbe forborne.

This Queene being defeafed [deceafed], *Elizabeth*
being of the age of xxij. yeares, of more beautie then
honour, and yet of more honour then any earthly creature, was called from a prifoner to be a Prince, from
the caftell [Caftle] to the crowne, from the feare of
loofing hir heade, to be fupreame heade. And here
Ladies it may be you wil[l] moue a queftion, why this
noble Ladie was either in daunger of death, or caufe of
diftreffe, which had you thought to haue paffed in filence,
I would notwithftanding haue reueiled [reuealed].

This Ladie all the time of hir fifters reigne was
kept clofe, as one that tendered not thofe proceedings,
which were contrarie to hir confcience, who hauing
diuers enemies, endured many croffes, but fo patiently

as in hir deepeſt ſorrow, ſhe would rather ſigh for the libertie of the goſpel, then hir own freedome. Suffering hir inferiours to triumph ouer hir, hir foes to threaten hir, hir diſſembling friends to vndermine hir, learning in all this miſerie onely the patience that *Zeno* taught *Eretricus* to beare and forbeare, neuer ſeeking reuenge but with good *Lycurgus*, to looſe hir owne eye, rather then to hurt an others eye.

But being nowe placed in the ſeate royall, ſhe firſt of al[l] eſtabliſhed religion, baniſhed poperie, aduaunced the worde, that before was ſo much defaced, who hauing in hir hande the ſworde to reuenge, vſed rather bountifully to reward : Being as farre from rigour when ſhee might haue killed, as hir enemies were from honeſtie when they coulde not, giuing a general pardon, when ſhe had cauſe to vſe perticuler puniſhments, preferring the name of pittie before the remembrance of perils, thinking no reuenge more princely, then to ſpare when ſhe might ſpill, to ſtaye when ſhe might ſtrike, to profer to ſaue with mercie, when ſhe might haue deſtroyed with iuſtice.   Heere is the clemencie worthie commendation and admiration, nothing inferiour to the gentle diſpoſition of *Ariſtides*, who after his exile did not ſo much as note them that baniſhed him, ſaying with *Alexander* that there can be nothing more noble then to doe well to thoſe, that deſerue yll.

This mightie and merciful Queene, hauing many bils [billes] of priuate perſons, yat ſought before time to betray hir, burnt them all, reſembling *Iulius Cæſar*, who being preſented with ye like complaints of his commons, threw them into ye fire, ſaying that he had rather, not knowe the names of rebels, then haue occaſion to reueng[e], thinking it better to be ignorant of thoſe that hated him, then to be angrie with them.

This clemencie did hir maieſtie not onely ſhew at hir comming to the crowne, but alſo throughout hir whole gouernement, when ſhe hath ſpared to ſhedde their bloods, that ſought to ſpill hirs, not racking the lawes to extremitie, but mittigating the rigour with

mercy infomuch as it may be faid of yat royal Monarch as it was of *Antonius*, furnamed ye godly Emperour, who raigned many yeares with-out the effufion of
blood. What greater vertue can there be in a Prince
then mercy, what greater praife then to abate the
edge which fhe fhould wette, to pardon where fhe
fhoulde punifh, to rewarde where fhe fhould reuenge.
I my felfe being in *England* when hir maieftie was
for hir recreation in hir Barge vpon ye Thames, hard
of a Gun that was fhotte off though of the partie vnwittingly, yet to hir noble perfon daungeroufly, which
faét fhe moft gracioufly pardoned, accepting a iuft
excufe before a great amends, taking more griefe for
hir poore Bargeman, that was a little hurt, then care
for hir felfe that ftoode in greateft hafarde : O rare example of pittie, O finguler fpeétacle of pietie.
Diuers befides haue there beene which by priuate
confpiracies, open rebellions, clofe wiles, cruel witch-
craftes, haue fought to ende hir life, which faueth all
their liues, whofe praétifes by the diuine prouidence
of the almightie, haue euer beene difclofed, infomuch
that he hath kept hir fafe in the whales belly when hir
fubieéts went about to throwe hir into the fea, preferued hir in the [hotte] hoat Ouen, when hir enimies
encreafed the fire, not fuffering a haire to fal[l] from hir,
much leffe any harme to faften vppon hir. Thefe iniuries and treafons of hir fubieéts, thefe policies and
vndermining of forreine nations fo littled moued hir,
yat.fhe woulde often fay, Let them knowe that though
it bee not lawfull for them to fpeake what they lift,
yet it is [is it] lawfull for vs to doe with them what we lift,
being alwayes of that mercifull minde, which was in
Theodofius, who wifhed rather that he might call the
deade to life, then put the liuing to death, faying with
Auguftus when fhe fhoulde fet hir hande to any con-
dempnation, I woulde to God we could not writ[e].
Infinite were the enfamples that might be alledged,
and almoft incredible, whereby fhee hath fhewed hir
felfe a Lambe in meekeneffe, when fhe had caufe to

be a Lion in might, proued a Doue in fauour, when fhe
was prouoked to be an Eagle in fierceneffe, requiting
iniuries with benefits, reuenging grudges with gifts,
in higheft maieftie bearing the loweft minde, forgiuing
all that fued for mercie, and forgetting all that deferued Iuftice.

O diuine nature, O heauenly nobilitie, what thing
can there more be required in a Prince, then in greateft power, to fhewe greateft patience, in chiefeft glorye,
to bring forth chiefeft grace, in abundaunce of all earthlye
pom[p]e, to manifeft aboundaunce of all heauenlye
pietie : O fortunate *England* that hath fuch a Queene,
vngratefull, if thou praye not for hir, wicked, if thou do
not loue hir, miferable, if thou loofe hir.

Heere Ladies is a Glaffe for all Princes to behold,
that being called to dignitie, they vfe moderation, not
might, tempering the feueritie of the lawes, with the
mildnes of loue, not executing al[l] they wil, but fhewing
what they may. Happy are they, and onely they that
are vnder this glorious and gracious Souereigntie : infomuch that I accompt all thofe abiects, that be not
hir fubiectes.

But why doe I treade ftill in one path, when I haue
fo large a fielde to walke, or lynger about one flower,
when I haue manye to gather : where-in I refemble
thofe that beeinge delighted with the little brooke,
neglect the fountaines head, or that painter, that
being curious to coulour *Cupids* Bow, forgot to paint
the ftring.

As this noble Prince is endued with mercie, pacience
and moderation, fo is fhe adourned with finguler beautie
and chaftitie, excelling in the one *Venus,* in the other
Vefta. Who knoweth not how rare a thing it is (Ladies)
to match virginitie with beautie, a chaft[e] minde with
an amiable face, diuine cogitations with a comelye
countenaunce ? But fuche is the grace beftowed vppon
this earthlye Goddeffe, that hauing the beautie that
myght allure all Princes, fhe hath the chaftitie alfo
to refufe all, accounting [accompting] it no leffe praife

to be called a Virgin, then to be efteemed a *Venus*, thinking it as great honour to bee found chaft[e], as thought amiable : Where is now *Electra* the chaft[e] Daughter of *Agamemnon*? Where is *Lala* that renoumed Virgin? Wher is *Aemilia*, that through hir chaftitie wrought wonders, in maintayning continuall fire at the Altar of *Vefla*? Where is *Claudia*, that to manifeft hir virginitie fet the Shippe on float with hir finger, that multitudes could not remoue by force? Where is *Tufcia* one of the fame order, that brought to paffe no leffe meruailes, by carrying water in a fiue, not fhedding one drop from *Tiber* to the Temple of *Vefla*? If Virginitie haue fuch force, then what hath this chaft Virgin *Elizabeth* don[e], who by the fpace of twenty and odde yeares with continuall peace againft all policies, with fundry myracles, contrary to all hope, hath gouerned that noble Ifland. Againft whome neyther forre[i]n force, nor ciuill fraude, neyther difcorde at home, nor confpiracies abroad, could preuaile. What greater meruaile hath happened fince the beginning of the world, then for a young and tender Maiden, to gouern ftrong and valiaunt menne, then for a Virgin to make the whole worlde, if not to ftand in awe of hir, yet to honour hir, yea and to liue in fpight of all thofe that fpight hir, with hir fword in the fhe[a]th, with hir armour in the Tower, with hir fouldiers in their gownes, infomuch as hir peace may be called more bleffed then the quiet raigne of *Numa Pompilius*, in whofe gouernment the Bees haue made their hiues in the foldiers helmettes. Now is the Temple of *Ianus* remoued from *Rome* to *England*, whofe dore hath not bene opened this twentie yeares, more to be meruayled at, then the regiment of *Debora*, who ruled twentie yeares with religion, or *Semeriamis* [*Semyramis*] that gouerned long with power, or *Zenobia* that reigned fix yeares in profperitie.

This is the onelye myracle that virginitie euer wrought, for a little Ifland enuironed round about with warres, to ftande in peace, for the walles of

Fraunce to burne, and the houfes of *England* to freefe, for all other nations eyther with ciuile [cruell] fworde to bee deuided, or with forren foes to be inuaded, and that countrey neyther to be molefted with broyles in their owne bofomes, nor threatned with blafts of other borderers : But alwayes though not laughing, yet looking through an Emeraud at others iarres.

Their fields haue beene fowne with corne, ftraungers theirs pytched with Camps, they haue their men reaping their harueft, when others are muftring in their harneis, they vfe their peeces to fowle for pleafure, others their Caliuers for feare of perrill. O bleffed peace, oh happy Prince, O fortunate people : The lyuing God is onely the Englyfh God, wher[e] he hath placed peace, which bryngeth all plentie, annoynted a Virgin Queene, which with a wand ruleth hir owne fubiects, and with hir worthineffe, winneth the good willes of ftraungers, fo that fhe is no leffe gratious among hir own, then glorious to others, no leffe loued of hir people, then merua[i]led at of other nations.

This is the bleffing that Chrift alwayes gaue to his people, peace : This is the curfe that hee giueth to the wicked, there fhall bee no peace to the vngodlye : This was the onelye falutation hee vfed to his Difciples, *peace be vnto you* : And therefore is hee called the G O D of loue, and peace in hollye [holy] writte.

In peace was the Temple of the Lorde buylt by *Salomon*, Chrift would not be borne, vntill there were peace through-out the whole worlde, this was the only thing that *Efechias* prayed for, let there be trueth and peace, O Lorde in my dayes. All which examples doe manifeftly proue, that ther[e] can be nothing giuen of God to man more notable then peace.

This peace hath the Lorde continued with great and vnfpeakeable goodneffe amonge his chofen peopie of *England*. How much is that nation bounde to fuch a Prince, by whome they enioye all benefits of peace, hauing their barnes full, when others famifh, their cof[f]ers ftuffed with gold, when others haue no filuer, their wiues without daunger, when others are defamed, their daughters chaft, when others are defloured, theyr houfes furnifhed, when others are fired, where they haue all thinges for fuperfluitie, others nothing to fuftaine their neede. This peace hath God giuen for hir vertues, pittie, moderation, virginitie, which peace, the fame God of peace continue for his names fake.

TOuching the beautie of this Prince, hir countenaunce, hir perfonage, hir maieftie, I can-not thinke that it may be fufficiently commended, when it can-not be too much meruailed at : So that I am conftrained to faye as *Praxitiles* did, when hee beganne to paynt *Venus* and hir Sonne, who doubted, whether the worlde could affoorde coulours good enough for two fuch fayre faces, and I whether our tongue canne yeelde wordes to blafe that beautie, the perfection where-of none canne imagine, which feeing it is fo, I muft doe like thofe that want a cleere fight, who being not able to difcerne the Sunne in the Skie are inforced to beholde it in the water. *Zeuxis* hauing before him fiftie faire virgins of *Sparta* where by to draw one amiable *Venus*, faid, that fiftie more fayrer than thofe coulde not minifter fufficient beautie to fhewe the Godeffe of beautie, therefore being in difpaire either by art to fhadow hir, or by imagination to comprehend hir, he drew in a table a faire temple, the gates open, and *Venus* going in, fo as nothing coulde be perceiued but hir backe, wherein he vfed fuch cunning, that *Appelles* himfelfe feeing this worke, wifhed yat *Venus* woulde turne hir face, faying yat if it were in all partes agreeable to the backe, he woulde become apprentice to *Zeuxis*, and flaue to *Venus*. In the like manner fareth it with me, for hauing all the Ladyes in *Italy* more then fiftie hundered, whereby to coulour *Elizabeth*, I muft fay with *Zeuxis*, that as many more will not fuffife, and therefore in as great an agonie paint hir court with hir back towards you,'for yat I cannot by art portraie hir beautie, wherein though I want the fkill to doe it as *Zeuxis* did, yet v[i]ewing it narrowly, and comparing it wifely, you all will fay yat if hir face be aunfwerable to hir backe, you wil[l] like my handi-crafte, and become hir hand-maides. In the meane feafon I leaue you gafing vntill fhe turne hir face, imagining hir to be fuch a one as nature framed to yat end, that no art fhould imitate, wherein fhee hath proued hir felfe to bee exquifite, and painters to be Apes.

This Beautifull moulde when I behelde to be en-dued, with chaflitie, temperance, mildneffe, and all other good giftes of nature (as hereafter fhall appeare) when I faw hir to furpaffe all in beautie, and yet a virgin, to excell all in pietie, and yet a prince, to be inferiour to none in all the liniaments of the bodie, and yet fuperiour to euery one in all giftes of the minde, I beegan thus to pray, that as fhe hath liued fortie yeares a virgin in great maieflie, fo fhe may lyue fourefcore yeares a mother, with great ioye, that as with hir we haue long time hadde peace and plentie, fo by hir we may euer haue quietneffe and aboun-daunce, wifhing this euen from the bottome of a heart that wifheth well to *England*, though feareth ill, that either the world may ende before fhe dye, or fhe lyue to fee hir childrens children in the world : otherwife, how tickle their flate is yat now triumph, vpon what a twift they hang that now are in honour, they yat lyue fhal fee which I to thinke on, figh. But God for his mercies fake, Chrifl for his merits fake, ye holy Ghofl for his names fake, graunt to that realme, comfort with-out anye ill chaunce, and the Prince they haue without any other chaunge, that ye longer fhe liueth the fweeter fhe may fmell, lyke the bird *Ibis*, that fhe maye be tri-umphant in victories lyke the Palme tree, fruitfull in hir age lyke the Vyne, in all ages profperous, to all men gratious, in all places glorious : fo that there be no ende of hir praife, vntill the ende of all flefh.

Thus did I often talke with my felfe, and wifhe with mine whole foule [heart].

What fhould I talke of hir fharpe wit, excellent wifedome, exquifite learning, and all other qualities of the minde, where-in fhe feemeth as farre to excell thofe that haue bene accompted fingular, as the learned haue furpaffed thofe, that haue bene thought fimple.

In queftioning not inferiour to *Nicaulia* the Queene of *Saba*, that did put fo many hard doubts to *Salomon*, equall to *Nicoftrata* in the *Greeke* tongue, who was thought to giue precepts for the better perfection: more learned in the *Latine*, then *Amalafunta* : paffing *Afpafia* in Philofophie, who taught *Pericles* : exceeding in iudgement *Themiftoclea*, who inftructed *Pithagoras*, adde to thefe qualyties, thofe, that none of thefe had, the *French* tongue, the *Spanifh*, the *Italian*, not meane in euery one, but excellent in all, readyer to correct efcapes in thofe languages, then to be controlled, fitter to teach others, then learne of anye, more able to adde new rules, then to erre in ye olde : Infomuch as there is no Embaffadour, that commeth into hir court, but fhe is willing and able both to vnderftand his meffage, and vtter hir minde, not lyke vnto ye Kings of *Afsiria*, who aunfwere[d] Embaffades by meffengers, while they themfelues either dally in finne, or fnort in fleepe. Hir godly zeale to learning, with hir great fkil, hath bene fo manifeftly approoued, yat I cannot tell whether fhe deferue more honour for hir knowledge, or admiration for hir curtefie, who in great pompe, hath twice directed hir Progreffe vnto the Vniuerfities, with no leffe ioye to the Students, then glory to hir State. Where, after long and folempne difputations in Law, Phificke, and Diuinitie, not as one we[a]ried with Schollers arguments, but wedded to their orations, when euery one feared to offend in length, fhe in hir own perfon, with no leffe praife to hir Maieftie, then delight to hir fubiects, with a wife and learned conclufion, both gaue them thankes, and put felfe to paines. O noble patterne of a princelye minde, not like to ye kings of *Perfia*, who in their progreffes did nothing els but cut ftickes to driue away the time, nor like ye delicate liues of the *Sybarites*, who would not admit any Art to be exercifed within their citie, yat might make ye leaft noyfe. Hir wit fo fharp, that if I fhould repeat the apt aunfweres, ye fubtil quef tions, ye fine fpeaches, ye pithie fentences, which on ye fodain fhe hath vttered, they wold rather breed admiration then credit. But fuch are ye gifts yat ye liuing God hath indued hir with-all, that looke in what Arte or Language, wit or learning, vertue or beautie, any one hath particularly excelled moft, fhe onely hath generally exceeded euery one in al, infomuch, that there is nothing to bee added, that either man would wifh in a woman, or God doth giue to a creature.

I let paffe hir fkill in Muficke, hir knowledg[e] in al[l] ye other fciences, when as I feare leaft by my fimplicity I fhoulde make them leffe then they are, in feeking to fhewe howe great they are, vnleffe I were praifing hir in the gallerie of *Olympia*, where gyuing forth one worde, I might heare feuen.

But all thefe graces although they be to be wondered at, yet hir politique gouernement, hir prudent counfaile, hir zeale to religion, hir clemencie to thofe that fubmit, hir ftoutneffe to thofe that threaten, fo farre exceede all other vertues, that they are more eafie to be meruailed at, then imitated.

Two and twentie yeares hath fhe borne the fword with fuch iuftice, that neither offenders coulde complaine of rigour, nor the innocent of wrong, yet fo tempered with mercie, as malefaÄours haue beene fometimes pardoned vpon hope of grace, and the iniuried requited to eafe their griefe, infomuch that in ye whole courfe of hir glorious raigne, it coulde neuer be faide, that either the poore were oppreffed without remedie, or the guiltie repreffed without caufe, bearing this engrauen in hir noble heart, that iuftice without mercie were extreame iniurie, and pittie without equitie plaine partialitie, and that it is as great tyranny not to mitigate Laws, as iniquitie to breake them.

Hir care for the flouriſhing of the Goſpell hath wel appeared, when as neither the curſes of the Pope, (which are bleſſings to good people) nor the threatenings of kings, (which are perillous to a Prince) nor the perſwaſions of Papiſts, (which are Honny to the mouth) could either feare hir, or allure hir, to violate the holy league contraᵍcted with Chriſt, or to maculate the blood of the aunciente Lambe, whiche is Chriſt. But alwayes conſtaunt in the true fayth, ſhe hath to the exceeding ioye of hir ſubieᵍtes, to the vnſpeakeable comforte of hir foule, to the great glorye of God, eſtablyſhed that religion, the mayntenance where-of, ſhee rather ſeeketh to confirme by fortitude, then leaue off for feare, knowing that there is nothing that ſmelleth ſweeter to the Lorde, then a ſounde ſpirite, which neyther the hoſtes of the vngodlye, nor the horror of death, can eyther remo[o]ue or moue.

This Goſpell with inuincible courage, with rare conſtancie, with hotte zeale ſhee hath maintained in hir owne countries with-out chaunge, and defended againſt all kingdomes that ſought chaunge, in-ſomuch that all nations rounde about hir, threatninge alteration, ſhaking ſwordes, throwing fyre, menacing famyne, murther, deſtruᵍtion, deſolation, ſhee onely hath ſtoode like a Lampe [Lambe] on the toppe of a hill, not fearing the blaſtes of the ſharpe winds, but truſting in his prouidence that rydeth vppon the winges of the foure windes. Next followeth the loue ſhee beareth to hir ſubieᵍtes, who no leſſe tendereth them, then the apple of hir owne eye, ſhewing hir ſelfe a mother to the a[f]fliᵍted, a Phiſition to the ſicke, a Souereigne and mylde Gouerneſſe to all.

Touchinge hir Magnanimitie, hir Maieſtie, hir Eſtate royall, there was neyther *Alexander*, nor *Galba* the Emperour, nor any that might be compared with hir.

This is fhe that refembling the noble Queene of *Nauarr*[e], vfeth the Marigolde for hir flower, which at the rifing of the Sunne openeth hir leaues, and at the fetting fhutteth them, referring all hir actions and endeuours to him that ruleth the Sunne. This is that *Cæfar* that firft bound the Crocodile to the Palme tree, bridling thofe, that fought to raine [rayne] hir : This is that good Pelican that to feede hir people fpareth not to rend hir owne perfonne : This is that mightie Eagle, that hath throwne duft into the eyes of the Hart, that went about to worke deftruction to hir fubiectes, into whofe winges although the blinde Beetle would haue crept, and fo being carryed into hir neft, deftroyed hir young ones, yet hath fhe with the vertue of hir fethers, confumed that flye in his owne fraud.

She hath exiled the Swallowe that fought to fpoyle the Grafhopper, and giuen bytter Almondes to the rauenous Wolues, that ende[a]uored to deuoure the filly Lambes, burning euen with the breath of hir mouth like ye princ[e]ly Stag, the ferpents yat wer[e] engendred by the breath of the huge Elephant, fo that now all hir enimies, are as whift as the bird *Attagen*, who neuer fingeth any tune after fhe is taken, nor they beeing fo ouertaken.

But whether do I wade Ladyes as one forgetting him-felfe, thinking to found the dep[t]h of hir vertues with a few fadomes, when there is no bottome : For I knowe not how it commeth to paffe, that being in this Laborinth, I may fooner loofe my felfe, then finde the ende.

Beholde Ladyes in this Glaffe a Queene, a woeman, a Virgin in all giftes of the bodye, in all graces of the minde, in all perfection of eyther, fo farre to excell all men, that I know not whether I may thinke the place too badde for hir to dwell amonge men.

To talke of other thinges in that Court, wer[e] to bring Egges after apples, or after the fetting out of the Sunne, to tell a tale of a Shaddow.

But this I faye, that all offyces are looked to with

great care, that vertue is embraced of all, vice hated, religion daily encreafed, mànners reformed, that who fo feeth the place there, will thinke it rather a Church for diuine feruice, then a Court for Princes delight.

This is the Glaffe Ladies wher-in I woulde haue you gafe, wher-in I tooke my whole delight, imitate the Ladyes in *England,* amende your manners, rubbe out the wrinckles of the minde, and be not curious about the weams in the face. As for their *Elizabeth,* fith you can neyther fufficiently meruaile at hir, nor I prayfe hir, let vs all pray for hir, which is the onely duetie we can performe, and the greateſt that we can proffer.

Yours to commaund
Euphues.

¶ *Iouis Elizabeth.*

Allas, *Iuno, Venus, cum Nympham numine plenam*
     Speɛtarunt, noſtra hæc, quæque triumphat, erit.
Contendunt auidè, ſic tandem regia Iuno,
     Eſt mea, de magnis ſtemma petiuit auis.
Hoc leue, (nec fperno tantorum inſignia patrum)
     Ingenio pollet, dos mea, Pallas ait.
Dulce Venus riſit, vultusque; in lumina fixit,
     Hæc mea dixit erit, nam quod ametur habet.
Iudicio Paridis, cum ſit prælata venuſtas:
     Ingenium Pallas ? Iuno quid vrget auos ?
Hæc Venus : impatiens veteris Saturnia damni,
     Arbiter in cœlis, non Paris, inquit erit.
Intumuit Pallas nunquam paſſura priorem,
     Priamides Helenem, dixit adulter amet.
Riſit, et erubuit, mixto Cytherea colore,
     Iudicium dixit Iuppiter ipfe ferat.
Aſſenfere, Iouem, compellant vocibus vltrò,
     Incipit affari regia Iuno Iouem.
Iuppiter, Elizabeth veſtras ſi venit ad aures

(Quam certe omnino cælica turba stupent)
Hanc propriam, et merito semper vult esse Monarcham,
   Quæque sanam, namque est pulchra, deserta, potens.
Quod pulchra, est Veneris, quod polleat arte, Mineruæ,
   Quod Princeps; Nympham quis neget esse meam ?
Arbiter istius, modo vis, certaminis esto,
   Sin minus, est nullum lis habitura modum.
Obstupet Omnipotens, durum est quod poscitis, inquit,
   Est tamen arbitrio res peragenda meo.
Tu soror et coniux Iuno, tu filia Pallas,
   Es quoque quid simulem ? ter mihi chara Venus.
Non tua da veniam Iuno, nec Palladis illa est,
   Nec Veneris, credas hoc licet alma Venus.
Hæc Iuno, hæc Pallas, Venus hæc, et quæque Dearum.
   Diuisum Elizabeth cum Ioue numen habet.
Ergo quid obstrepitis ? frustra contenditis inquit,
   Vltima vox hæc est, Elizabetha mea est.

Euphues

Es Iouis Elizabeth, nec quid Ioue maius habendum,
   Et Ioue teste Ioui es, Iuno, Minerua, Venus.

THese Verses *Euphues* sent also vnder his Glasse
which hauing once finished, he gaue him-self
to his booke, determininge to ende his lyfe in *Athens*
although he hadde a moneths minde to *England*, wh
at all tymes, and in all companies, was no niggard
of his good speach to that Nation, as one willyng
liue in that Court, and wedded to the manners of th
countrey.

It chaunced that being in *Athens* not passing on
quarter of a yeare, he receiued letters out of *England*
from *Philautus*, which I thought necessarye also t
insert, that I mightgiue some ende to the matters [o
in *England*, which at *Euphues* departure were bu
rawly left. And thus they follow.

Philautus to his owne
Euphues.

'ʃ Haue oftentimes (*Euphues*) ſince thy departure
complained, of the diſtance of place that I am ſo
&rre from thee, of the length of time that I coulde
ⁿt heare of thee, of the ſpite of Fortune, that I
tⁱght not ſende to thee, but time at length, and not
tₒ late, bicauſe at laſt, hath recompenſed the iniuries
y all, offering me both a conuenient meſſenger by
Pₒm to ſend, and ſtraung[e] newes whereof to write.
dₜThou knoweſt howe frowarde matters went, when
plu tookeſt ſhippe, and thou wouldeſt meruaile to
:re [heare] howe forwarde they were before thou
)keſt ſaile, for I had not beene long in London, ſure I
thou waſt not then at *Athens*, when as the corne
iche was greene in the blade, began to wax ripe in
: eare, when the ſeede which I ſcarce thought to
ıe taken roote, began to ſpring, when the loue of
_rius whiche hardly I would haue geſſed to haue a
ɔſſome, ſhewed a budde. But ſo vnkinde a yeare it
ath beene in *England*, that we felt the heate of the
ımmer, before we could diſcerne the temperature of
ₜ Spring, inſomuch that we were ready to make
ℐₜᵧe, before we coulde mowe graſſe, hauing in effeĉte
: Ides of May before the Calends of March, which
ℐₜing it is ſo forward in theſe things, I meruailed the
ᵉe to ſee it ſo re[a]dy in matters of loue, wher[e]
ℐᵗentimes they clap hands before they know the
:gaine, and ſeale the Oblygation, before they read
ℐₜ₃ condition.
ₜAt my being [at] in the houſe of *Camilla*, it happened
ℐₜound *Surius* accompanied with two knights, and the
ₐdy *Flauia* with three other Ladyes, I drew back as
⁴ıe ſomewhat ſhameſaſt, when I was willed to draw
:ere, as one that was wiſhed for. Who thinking of
⁴ɔthing leſſe then to heare a contraĉt for mar[r]iage,
ʰer[e] I only expeĉted a conceipt for mirth, I ſo-
⁴ᵢnly, yet ſolempnly, h[e]ard thoſe wordes of aſſurance

betweene *Surius* and *Camilla*, in the which I had
rather haue bene a partie, then a witnes, I was not a
lyttle amazed to fee them ftrike the yron which I
thought colde, and to make an ende before I could
heere [heare] a beginning. When they faw me as it
were in a traunce, *Surius* taking mee by the hand,
began thus to ieft.

You mufe *Philautus* to fee *Camilla* and me to bee
affured, not that you doubted it vnlikely to come to
paffe, but that you were ignorant of the practifes,
thinking the diall to ftand ftil[l], bicaufe you cannot perceiue it to moue. But had you bene priuie to all
proofes, both of hir good meaning towards me, and of my
good wil[l] towards hir, you wo[u]ld rather haue thought
great haft to be made, then long deliberation. For
this vnderftande, that my friends are vnwilling yat I
fho[u]ld match fo low, not knowing yat loue thinketh ye
Iuniper fhrub, to be as high as ye tal O[a]ke[s], or ye
Nightingales layes, to be more precious then ye
Oftriches feathers, or ye Lark yat breedeth in ye
ground, to be better then ye Hobby yat mounteth to
the cloudes. I haue alwaies hetherto preferred beautie
before riches, and honeftie before bloud, knowing
that birth is ye praife we receiue of our aunceftours,
honeftie the renowne we leaue to our fucceffours, and
of t[w]o brit[t]le goods, riches and beautie, I had rather
chufe that which might delyght me, then deftroy me.
Made mar[r]iages by friends, how daungerous they haue
bene I know, *Philautus*, and fome prefent haue proued,
which can be likened to nothing els fo well, then as if
a man fhould be conftrayned to pull on a fhoe by an
others laft, not by the length of his owne foote, which
beeing too little, wrings him that weares it, not him
yat made it, if too bigge, fhameth him that hath it, not
him that gaue it. In meates, I loue to carue wher[e] I
like, and in mar[r]iage fhall I be carued where I lyke
not? I had as liefe an other fho[u]ld take me[a]fure by
his back, of my apparel[l], as appoint what wife I fhall
[fhould] haue, by his minde.

In the choyce of a wife, fundry men are of fundry
mindes, one looketh high as one yat feareth no chips,
faying yat the oyle that fwimmeth in ye top is ye
wholfomeft, an other poreth in ye ground, as dreading
al daungers that happen in great ftock[e]s, alledging
that ye honny yat lieth in ye bottome is ye fweeteft, I
affent to neither, as one willing to follow the meane,
thinking yat the wine which is in the middeft to be
the fineft. That I might therefore match to mine owne
minde, I haue chofen *Camilla*, a virgin of no noble
race, nor yet the childe of a bafe father, but betweene
both, a Gentle-woman of an auncient and worfhipfull
houfe, in beautie inferio[u]r to none, in vertue fuperio[u]r
to a number. Long time we loued, but 1 either durft
fhe manifeft hir affeċtion, bicaufe I was noble, nor I
vtter myne, for feare of offence, feeing in hir alwayes a
minde more willing to car[r]y torches before *Vefta*, then
tapers before *Iuno*. But as fire when it burfteth out
catcheth hold fooneft of the dryeft wood, fo loue when
it is reueyled ⌐reuealed⌐, fafteneth vppon the eafieft
affeċtionate will, which came to paffe in both [of] vs, for
talking of Loue, of his lawes, of his delyghts, torments,
and all other braunches, I coulde neither fo diffemble
my liking, but that fhe efpied it, where at I [fhe] began
to figh, nor fhe fo cloake hir loue, but that I perceiued it,
where at fhee began to blufh : at the laft, though long
time ftrayning curtefie who fhould goe ouer the ftile,
when we had both haft, I (for that I knew women would
rather die, then feeme to defire) began firft to vnfolde
the extremities of my paffions, the caufes of my loue,
the conftancie of my faith, the which fhe knowing to
bee true, eafely beleeued, and replyed in the like manner, which I thought not certeine, not that I mif-
doubted hir faith, but that I coulde not perfwade my
felfe of fo good fortune. Hauing thus made e[a]ch other
priuie to our wifhed defires, I frequented more often
to *Camilla*, which caufed my friendes to fufpeċt that,
which nowe they fhall finde true, and this was the
caufe that we al[l] meete heere, that before this good

company. we might knit that knot with our tongues, that we ſhall neuer vndoe with our teeth.

This was *Surius* ſpeach vnto me, which *Camilla* with the reſt affirmed. But I *Euphues*, in whoſe h[e]art the ſtumpes of Loue were yet ſticking, beganne to chaunge colour, feelyng as it were newe ſtormes to ariſe after a pleaſaunt calme, but thinking with my ſelfe, that the time was paſt to wo[o]e hir, that an other was to wedde, I digeſted the Pill which had almoſt chockt [choakt] me. But time cauſed me to ſing a new Tune as after thou ſhalt heare.

After much talke and great cheere, I taking my leaue departed, being willed to viſite the Ladie *Flauia* at my leaſure, which worde was to me in ſteede of a welcome.

Within a while after it was noyſed that *Surius* was aſſured to *Camilla*, which bread [bred] great quarrells, but hee like a noble Gentle-man reioycing more in his Loue, then eſteeminge the loſſe of his friendes, maugre them all was mar[r]ied, not in a chamber priuatelye as one fearing tumultes, but openlye in the Church, as one ready to aunſwer any obiections.

This mar[r]iage ſolemniſed, could not be recalled, which cauſed his Allies to conſent, and ſo all parties pleaſed, I thinke them the happyeſt couple in the worlde.

NOw *Euphues* thou ſhalt vnderſtand, that all hope being cut off, from obtaining *Camilla*, I began to vſe the aduauntage of the word, that Lady *Flauia* caſt out, whome I viſited more lyke to a ſoiourner, then a ſtranger, being abſent at no time from breack-faſt, till euening.

Draffe was mine arrand [errand], but drinke I would, my great curteſie was to excuſe my greeuous tormentes : for I ceaſed not continuallye to courte my violette, whome I neuer found ſo coye as I thought, nor ſo curte-ous as I wiſhed .At the laſt thinking not to ſpend all my wooinge in ſignes, I fell to flatte ſayinges [flat ſaying],

reuealing the bytter fweetes that I fuftained, the ioy
at hir prefence, the griefe at hir abfence, with al[l]
fpeeches that a Louer myght frame : She not degene-
rating from the wyles of a woeman, feemed to accufe
men of inconftancie, that the painted wordes were but
winde, that feygned [fained] fighes, were but fleyghtes,
that all their loue, was but to laugh, laying baites to
catch the fifh, that they meant agayne to throw into
the ryuer, practifinge onelye cunninge to deceyue,
not curtefie, to tell trueth, where-in fhe compared all
Louers, to *Mizaldus* the Poet, which was fo lyght that
euery winde would blowe him awaye, vnleffe hee had
lead tyed to his heeles, and to the fugitiue ftone in
Cyzico [*Cicico*], which runneth away if it be not faftened
to fome poft.

Thus would fhe dally, a wench euer-more giuen to
fuch difporte : I aunfwered for my felfe as I could,
and for all men as I thought.

Thus oftentimes had we conference, but no conclufion, many meetinges, but few paftimes, vntill at the
laft *Surius* one that could quickly perceiue, on which
fide my bread was buttered, beganne to breake with
me touching *Frauncis*, not as though he had heard
any thing, but as one that would vnderftand fomething. I durft not feeme ftraunge when I founde him
fo curteous, knowing that in this matter he might
almofte worke all to my lyking.

I vnfolded to him from time to time, the whole
difcourfes I had with my Violet, my earneft defire to
obtaine hir, my landes, goodes, and reuenues, who
hearing my tale, promifed to further my fuite, where-in
he fo befturred his ftudie, that with-in one moneth, I
I was in poffibilitie to haue hir, I moft wifhed, and
leaft looked for.

It were too too long to write an hiftorie, being but
determined to fend a Letter : therefore I will diferre
all the actions and accidentes that happened, vntill
occafion fhall ferue eyther to meete thee, or minifter
leafure to me.

To this ende it grewe, that conditions drawen for the performaunce of a certaine ioynter (for the which I had manye *Italians* bounde) we were both made as fure as *Surius* and *Camilla.*

Hir dowrie was in re[a]dy money a thoufand pounds, and a fayre houfe, where-in I meane fhortelye to dwell. The ioynter I muft make is foure hundred poundes yearelye, the which I muft heere purchafe in *England*, and fell my landes in *Italy.*

Now *Euphues* imagine with thy felf that *Philautus* beginneth to chaunge, although in one yeare to mar[r]ie and to thr[i]ue it be hard.

But would I might once againe fee thee heere, vnto whome thou fhalt be no leffe welcome, then to thy beft friende.

Surius that noble Gentleman commendeth him vnto thee, *Camilla* forgetteth thee not, both earneftly wifh thy returne, with great promifes to do thee good, whether thou wifh it in the court or in the countrey, and this I durft fweare, that if thou come againe into *England*, thou wilt be fo friendly entreated, that either thou wilt altogether dwell here, or tarry here longer.

The Lady *Flauia* faluteth thee, and alfo my Violet, euery one wifheth thee fo well, as thou canft wifh thy felfe no better.

Other newes here is none, but that which lyttle apperteyneth to mee, and nothing to thee.

Two requeftes I haue to make, afwel from *Surius* as my felfe, the one to come into England, the other to heare thyne aunfwere. And thus in haft I byd the[e] farewell. From *London* the firft of *Februarie.* 1579.

Thyne or not his owne:
PHILAVTVS.

THis Letter being deliuered to *Euphues*, and well perufed, caufed him both to meruaile, and to ioy, feeing all thinges fo ftraung[e]ly concluded, and his friende fo happilye contracted : hauing therefore by the fame meanes opportunitie to fend aunfwere, by the whiche he had pleafure to receiue newes, he difpatched his letter in this forme.

¶ *Euphues to Philautus.*

THer[e] co[u]ld nothing haue come out of *England*, to *Euphues* more welcome then thy letters, vnleffe it had bin thy perfon, which when I had throughly perufed, I could not at ye firft, either beleeue them for ye ftraungnes, or at the laft for the happineffe : for vpon the fodaine to heare fuch alterations of *Surius*, paffed all credit, and to vnderftand fo fortunate fucceffe to *Philautus*, all expectation : yet confidering that manye thinges fall betweene the cup and the lippe, that in one lucky houre more rare things come to paffe, then fom[e]times in feuen yeare[s], that mar[r]iages are made in heauen, though confum[m]ated in yearth [earth], I was brought both to beleeue the euents, and to allow them. Touching *Surius* and *Camilla*, there is no doubt but that they both will lyue well in mar[r]iage, who loued fo well before theyr matching, and in my mind he de[a]lt both wif[e]ly and hono[u]rably, to prefer vertue before vainglory, and the go[o]dly ornaments of nature [vertue], before the rich armour of nobilitie ; for this muft we all think, (how well foeuer we think of our felues) that vertue is moft noble, by the which men became firft noble. As for thine own eftat[e], I will be bold to counfell thee, knowing it neuer to be more neceffary to vfe aduife then in mar[r]iag[e]. *Solon* gaue counfel[l] that before one affured him-felf he fhould be fo warie, that in tying him-felfe faft, he did not vndo him-felfe, wifhing them firft to eat a Quince peare, yat is, to haue [a] fweete conference with-out brawles ; then falt, to be wife with-out boafting.

In *Boetia* they couered the bride with *Afparagonia* the nature of the* which plant is, to bring fweete fruit out of a fharpe thorne, wher-by they noted, that al-though the virgin were fom[e]what fhrewifhe at the firft,

yet in time fhe myght become a fheepe. Therefore
Philautus, if thy Vyolet feeme in the firft moneth
either to chide or chafe, thou muft heare with out reply, and endure it with patience, for they that can-not
fuffer the wranglyngs of young mar[r]yed women, are
not vnlyke vnto thofe, that tafting the grape to be
fower before it be ripe, leaue to gather it when it is
ripe, refemblyng them, that being ftong [ftung] with
the Bee, forfake the Honny.

Thou muft vfe fweete words, not bitter checkes,
and though happely thou wilt fay that wandes are [bee]
to be wrought when they are greene, leaft they rather
break then bende when they be drye, yet know alfo,
that he that bendeth a twigge, bicaufe he would fee
if it wo[u]ld bow by ftrength, maye chaunce to haue a
crooked tree, when he would haue a ftreight.

It is pretelye noted of a contention betweene the
Winde, and the Sunne, who fhould haue the victorye.
A Gentleman walking abroad, the Winde thought to
blowe of[f] his cloake, which with great blaftes and
blufterings ftriuing to vnloofe it, made it to ftick fafter
to his backe, for the more the winde encreafed the
clofer his cloake clapt to his body, then the Sunne,
fhining with his hoat beames began to warme this
gentleman, who waxing fom[e]what faint in this faire
weather, did not onely put of[f] his cloake but his
coate, which the Wynde perceiuing, yeelded the conqueft to the Sunne.

In the* very* like manner fareth it with young wiues,
for if their hufbands with great threatnings, with iarres,
with braules, feeke to make them tractable, or bend
their knees, the more ftiffe they make them in the
ioyntes, the oftener they goe about by force to rule
them, the more froward they finde them, but vfing
milde words, gentle perfwafions, familyar counfaile,
entreatie, fubmiffion, they fhall not onely make them
to bow their knees, but to hold vp their hands, not
onely caufe them to honour them, but to ftand in awe
of them: for their ftomackes are al framed of Diamond,

which is not to be brufed with a hammer but bloode, not by force, but flatterie, refemblyng the Cocke, who is not to be feared by a Serpent, but a glead. They that feare theyr Vines will make too fharpe wine, muft not cutte the armes, but graft next to them Mandrage [Mendrage], which caufeth the grape to be more pleafaunt. They that feare to haue curft wiues, muft not with rigo[u]rfeeke [feeme] to calme [reclaime] them, but faying gentle words in euery place by them, which maketh them more quyet.

Inftruments found fweeteft when they be touched fofteft, women waxe wifeft, when they be [are] vfed mildeft. The horfe ftriueth when he is hardly rayned, but hauing ye bridle neuer ftirreth, women are ftarke mad if they be ruled by might, but with a gentle rayne they will beare a white mouth. Gal[l] was caft out from ye facrifice of *Iuno*, which betokened that the mar[r]iage bed fhould be without bitternes. Thou muft be a glaffe to thy wife for in thy face muft fhe fee hir owne, for if when thou laugheft fhe weepe, when thou mourneft fhe gig[g]le, the one is a manifeft figne fhe delighteth in others, the other a token fhe defpifeth thee. Be in thy behauiour modeft, temperate, fober, for as thou frameft thy manners, fo wil thy wife fit hirs. Kings that be wraftlers caufe their fubiects to exercife that feate. Princes that are Mufitians incite their people to vfe Inftruments, hufbands that are chaft and godly, caufe alfo their wiues to imitate their goodneffe.

For thy great dowry that ought to be in thine own handes, for as we call that wine, where-in there is more then halfe water, fo doe we tearme that, the goods of the hufband which his wife bringeth, though it be all.

Helen gaped for [his] goods, *Paris* for pleafure. *Vlyffes* was content with chaft *Penelope*, fo let it be with thee, that whatfoeuer others mar[r]ie for, be thou alwayes fatiffied with vertue, otherwife may I vfe that fpeach to thee that *Olympias* did to a young Gentleman who only tooke a wife for beautie, faying : this Gentleman hath onely mar[r]yed his eyes, but by that time he haue [hath] alfo wedded his eare, he wil[l] confeffe that a faire fhooe wrings, though it be fmoothe in the wearing.

Lycurgus made a law that there fhould be no dowry giuen with Maidens, to the ende that the vertuous might be mar[r]yed, who commonly haue lyttle, not the amorous, who oftentimes haue to much.

Behaue thy felf modeftly with thy wife before company, remembring the feueritie of *Cato*, who remoued *Manlius* from the Senate, for that he was feene to kiffe his wife in prefence of his daughter : olde men are feldome merry before children, leaft their laughter might breedě in them loofeneffe, hufbands fho[u]ld fcarce ieft before their wiues, leaft want of modeftie on their parts, be caufe of wantonnes on their wiues part. Imitate the Kings of *Perfia*, who when they were giuen to ryot, kept no company with their wiues, but when they vfed good order, had their Queenes euer at their [the] table. Giue no example of lyghtneffe, for looke what thou practifeft moft, yat will thy wife follow moft, though it becommeth hir leaft. And yet woulde I not haue thy wife fo curious to pleafe thee, yat fearing leaft hir hufband fho[u]ld thinke fhe painted hir face, fhe fho[u]ld not therefore wafh it, onely let hir refraine from fuch things as fhe knoweth cannot wel like thee, he yat commeth before an Elephant will not weare bright colo[u]rs, nor he that commeth to a Bul[l], red, nor he yat ftandeth by a Tiger, play on a Taber : for that by the fight or noyfe of thefe things, they are commonly much incenfed. In the lyke manner, there is no wife if fhe be honeft, that will practife thofe things, that to hir mate fhall feeme difpleafaunt, or moue him to cholar.

Be thriftie and warie in thy expences, for in olde time, they were as foone condemned by law that fpent their wiues dowry prodigally, as they that diuorced them wrongfully.

Flye that vyce which is peculiar to al thofe of thy countrey, *Ieloufie* [Iealoufie] : for if thou fufpect without

caufe, it is the next way to haue caufe, women are to
bee ruled by their owne wits, for be they chaft, no
golde canne winne them, if immodeft no griefe can
amende them, fo that all miftruft is either needeleffe
or bootleffe.

Be not too imperious ouer hir, that will make hir
to hate thee, nor too fubmiffe [demiffe], that will caufe
hir to difdaine thee, let hir neither be thy flaue, nor thy
fouereigne, for if fhe lye vnder thy foote fhe will neuer
loue thee, if clyme aboue thy head neuer care for thee:
the one will breed thy fhame to loue hir to [fo] little,
the other thy griefe to fuffer too much.

In gouerning thy houfeholde, vfe thine owne eye,
and hir hande, for hufwifery confifteth as much in feeing things as fetlyng things, and yet in that goe not
aboue thy latchet, for Cookes are not to be taught in
the Kitchen, nor Painters in their fhoppes, nor Hufwiues in their houfes. Let al[l] the keyes hang at hir
girdel, but the purffe at thine, fo fhalt thou knowe
what thou doft fpend, and how fhe can fpare.

Breake nothing of thy ftocke, for as the Stone
Thyrrenus [*Thirrennius*] beeing whole, fwimmeth,
but neuer fo lyttle diminifhed, finketh to the
bottome: fo a man hauing his ftocke full, is euer
afloat, but wafting of his ftore, becommeth bankerout
[bankrupt].

Enterteine fuch men as fhall be truftie, for if thou keepe
a Wolfe within thy doores to doe mifchiefe, or a Foxe

[Completed from the Bodleian copy, 1580.]

to worke craft and fubtiltie, thou fhalt finde it as perrilous, as if in thy barnes thou fhouldeft mainteyne Myce,
or in thy groundes Moles.

Let thy Maydens be fuch, as fhal[l] feeme readier to
take paynes, then follow pleafure, willinger to dreffe
vp theyr houfe, then their heades, not fo fine fingered,
to call for a Lute, when they fhoulde vfe the [a] diftaffe,

nor fo dainetie mouthed, that their filken thro[a]tes
fhould fwallow no packthre[e]d.

For thy dyet be not fumptuous, nor yet fimple : For
thy attyre not coftly, nor yet clownifh, but cutting thy
coat by thy cloth, go no farther then fhal become thy
eftate, leaft thou be thought proude, and fo enuied, nor
debafe not thy byrth, leaft thou be deemed poore, and
fo pittied.

Now thou art come to that honourable eftate, forget all thy former follyes, and debate with thy felfe,
that here-to-fore thou diddeft but goe about the world,
and that nowe, thou art come into it, that Loue
did once make thee to follow ryot, that it mufte
now enforce thee to purfue thrifte, that then there
was no pleafure to bee compared to the courting of
Ladyes, that now there can be no delight greater than
to haue a wife.

Commend me humbly to that noble man *Surius,*
and to his good Lady *Camilla.*

Let my duetie to the Ladie *Flauia* be remembred,
and to thy Violyt, let nothing that may be added, be
forgotten.

Thou wouldeft haue me come againe into *England,*
I woulde but I can-not: But if thou defire to fee
Euphues, when thou art willing to viffite thine vncle,
I will meete thee, in the meane feafon, know, that
it is as farre from *Athens* to *England,* as from *England*
to *Athens.*

Thou fayeft I am much wifhed for, that many
fayre promifes are made to mee : Truely *Philautus*
I know that a friende in the court is better then
a penney in the purfe, but yet I haue heard that
fuche a friend cannot be gotten in the court without
pence.

Fayre words fatte few, great promifes without
performance, delight for the tyme, but ye[a]rke euer
after.

I cannot but thank *Surius,* who wifheth me well,
and all thofe that at my beeing in *England* lyked me

wel[l]. And fo with my h[e]artie commendations vntili I heare from thee, I bid thee farewell.

Thine to vfe, if mari-
age chaunge not man-
ners Euphues.

This letter difpatched, *Euphues* gaue himfelfe to folitarineffe, determining to foiourne in fome vncauth [vncouth] place, vntil time might turne white falt into fine fugar : for furely he was both tormented in body and grieued in minde.

And fo I leaue him, neither in *Athens* nor els where that I know : But this order he left with his friends, that if any newes came or letters, that they fhould direct them to the Mount of *Silixfedra*, where I leaue him, eyther to his mufing or Mufes.

Gentlemen, *Euphues* is mufing in the bottome of the Mountaine *Silixfedra* : *Philautus*[is] marryed in the Ifle of *England* : two friendes parted, the one liuing in the delightes of his newe wife, the other in contemplation of his olde griefes.

What *Philautus* doeth, they can imagine that are newly married, how *Euphues* liueth, they may geffe that are cruelly martyred : I commit them both to ftande to their owne bargaines, for if I fhould meddle any farther with the marriage of *Philautus*, it might happely make him iealous, if with the melancholy of *Euphues*, it might caufe him to be cholaricke : fo the one would take occafion to rub his head, fit his hat neuer fo clofe, and the other offence, to gall his heart, be his cafe neuer fo quiet. I Gentlewomen, am indifferent, for it may be, that *Philautus* would not haue his life knowen which he leadeth in mar[r]iage, nor *Euphues*, his loue defcryed, which he beginneth in folitarineffe, leaft either

the one being too kinde, might be thought to doat,
or the other too conftant, might be iudged to be
madde. But were the trueth knowen, I am fure Gentlewomen it would be a hard queftion among Ladies, whether *Philautus* were a better wooer, or a hufband, whether *Euphues* were a better louer, or a fcholler. But
let the one marke the other, I leaue them both,
to conferre at theyr nexte meeting, and
committe you, to the Al-
mightie.

FINIS.

¶ Imprinted at London, by Thomas Eaft, for Gabriel
Cawood dwelling in Paules Churchyard. 1580.

1. EUPHUES AND HIS EPHŒBUS.—Profeffor Edward Dowden, of Trinity College, Dublin, informs me, under date 16 October, 1868, that his friend, Profeffor Rufhton of Queen's College, Cork, had pointed out to him, that 'Euphues and his Ephœbus' is almoft entirely a tranflation from Plutarch on 'Education.'

Mr. Dowden adds, "I did not compare Lyly with the Greek, but with Philemon Holland's *The Philofophie, commonly called The Morals written by the learned Philofopher Plutarch of Chæronea* [London. 1603. fol.], pp. 2 and onwards. Lyly and Holland read as different tranflations of the fame original, Lyly omitting paffages here and there, and making a few additions."

THE END.

o

www.ingramcontent.com/pod-product-compliance
Lightning Source LLC
Chambersburg PA
CBHW052339110726
47901CB00005B/1289